Quality Management

Divided into four main chapters, this book covers the inception through to the handover of a project and details the three main stages (study stage, design stage, and construction stage) involved with managing any type of project. The book discusses the sustainability framework and provides an overview of quality management with construction projects along with the most common quality tools used to manage quality and achieve sustainability in projects.

Quality Management: How to Achieve Sustainability in Projects takes the reader from start to finish with a focus on the sustainability elements needed to manage quality in projects and details the application of sustainability principles at different stages. The book discusses the quality tools used in managing sustainability and provides concise and complete information on how to easily achieve it through to the project handover stage.

The book is written for Project Management professionals such as Project Managers, Quality Managers, Industrial Engineers, and Construction Managers, as well as Design Management professionals, academics, trainers, and graduate students.

Quality Management and Risk Series

Series Editor: Abdul Razzak Rumane

Senior Engineering Consultant, Kuwait

This new series will include the latest and innovative books related to quality management and risk related topics. The definition of quality relating to manufacturing, processes, projects, and the service industries, is to meet the customer's need, satisfaction, fitness for use, conforms to requirements, and a degree of excellence at an acceptable price. With globalization and competitive markets, the emphasis on quality management has increased. Quality has become the most important single factor for the survival and success of any company or organization. The demand for better products and services at the lowest costs have put tremendous pressure on organizations to improve the quality of products, services, projects, and processes, in order to compete in the marketplace. Because of these changes, the ISO 9001 now lists that risk-based thinking must be incorporated into the management system by considering the context of the organization. Quality management and risk management now play an important role in the over all quality management system. This means that books which cover quality, need to also cover risk to update practices/processes, tools, and techniques, per ISO 9001. The goal of this new series is to include the books that will meet this need and demand.

Quality Management in Oil and Gas Projects
Abdul Razzak Rumane

Risk Management Applications to Sustain Quality in Projects
A Practical Guide
Abdul Razzak Rumane

Quality Management
How to Achieve Sustainability in Projects
Abdul Razzak Rumane

For more information on this series, please visit: https://www.routledge.com/Quality-Management-and-Risk-Series/book-series/CRCQMR

Quality Management
How to Achieve Sustainability in Projects

Abdul Razzak Rumane

CRC Press
Taylor & Francis Group
Boca Raton London New York

CRC Press is an imprint of the
Taylor & Francis Group, an **informa** business

First edition published 2024
by CRC Press
6000 Broken Sound Parkway NW, Suite 300, Boca Raton, FL 33487-2742

and by CRC Press
4 Park Square, Milton Park, Abingdon, Oxon, OX14 4RN

CRC Press is an imprint of Taylor & Francis Group, LLC

© 2024 Abdul Razzak Rumane

Reasonable efforts have been made to publish reliable data and information, but the author and publisher cannot assume responsibility for the validity of all materials or the consequences of their use. The authors and publishers have attempted to trace the copyright holders of all material reproduced in this publication and apologize to copyright holders if permission to publish in this form has not been obtained. If any copyright material has not been acknowledged please write and let us know so we may rectify in any future reprint.

Except as permitted under U.S. Copyright Law, no part of this book may be reprinted, reproduced, transmitted, or utilized in any form by any electronic, mechanical, or other means, now known or hereafter invented, including photocopying, microfilming, and recording, or in any information storage or retrieval system, without written permission from the publishers.

For permission to photocopy or use material electronically from this work, access www.copyright.com or contact the Copyright Clearance Center, Inc. (CCC), 222 Rosewood Drive, Danvers, MA 01923, 978-750-8400. For works that are not available on CCC please contact mpkbookspermissions@tandf.co.uk

Trademark notice: Product or corporate names may be trademarks or registered trademarks and are used only for identification and explanation without intent to infringe.

ISBN: 978-1-032-45438-2 (hbk)
ISBN: 978-1-032-45522-8 (pbk)
ISBN: 978-1-003-37737-5 (ebk)

DOI: 10.1201/9781003377375

Typeset in Times
by Deanta Global Publishing Services, Chennai, India

Dedication

To
My Parents

For their prayers and love

My prayers are always for my father and my mother who encouraged and inspired me at all the times.

I wish they would have been here to see this book and give me their blessings.

To
My Wife

I miss my wife, who stood with me all the times during the writing of my earlier books.

Contents

Figures xi
Tables xvii
Foreword xxi
Preface xxiii
Acknowledgments xxvii
About the Author xxix
Abbreviations xxxi
Synonyms xxxiii

Chapter 1   Sustainability Framework 1
            1.1   Sustainability 1
            1.2   Development of Sustainable Projects 3
                   1.2.1   Sustainability Framework 4
                   1.2.2   Sustainable Project Development Process 8
            1.3   Elements of Sustainable Project 8
                   1.3.1   Economical 9
                   1.3.2   Environmental 10
                   1.3.3   Social 11
            1.4   Sustainability Development Goals and Sustainable Project 11

Chapter 2   Overview of Quality in Construction Projects 13
            2.1   Construction Projects 13
            2.2   Quality Definition for Construction Projects 17
                   2.2.1   Quality Principles of Construction Projects ... 20
            2.3   Quality Management System (QMS) 23
                   2.3.1   Integrated Quality Management System 26
            2.4   Quality Management in Construction Projects 27
                   2.4.1   Quality Plan 27
                   2.4.2   Quality Assurance 29
                   2.4.3   Quality Control 29
                   2.4.4   Quality During the Study Stage 30
                   2.4.5   Quality During the Design Stage 31
                   2.4.6   Quality During the Bidding and Tendering Stage 31
                   2.4.7   Quality During the Construction Stage 35

Chapter 3   Quality Tools for Sustainability 37
            3.1   Introduction 37
            3.2   Categorization of Tools 37

		3.2.1	Classic Quality Tools	37
		3.2.2	Management and Planning Tools:	45
		3.2.3	Process Analysis Tools	59
		3.2.4	Process Improvement Tools	64
		3.2.5	Innovation and Creative Tools	69
		3.2.6	Lean Tools	75
		3.2.7	Cost of Quality	82
		3.2.8	Quality Function Deployment (QFD)	85
		3.2.9	Six Sigma	86
		3.2.10	TRIZ	99
	3.3	Tools for Project Development		102
		3.3.1	Tools for the Study Stage	105
		3.3.2	Tools for the Design Stage	105
		3.3.3	Tools for the Bidding and Tendering Stage	105
		3.3.4	Tools for the Construction Stage	115

Chapter 4 Management of Quality for Sustainability 122

	4.1	Introduction		122
	4.2	Process Group		125
		4.2.1	Initiating Process Group	126
		4.2.2	Planning Process Group	126
		4.2.3	Executing Process Group	131
		4.2.4	Monitoring and Controlling Process Group	131
		4.2.5	Closing Process Group	131
	4.3	Sustainability in the Study Stage		138
		4.3.1	Identification of Need	139
		4.3.2	Feasibility Study	144
		4.3.3	Project Goals and Objectives	146
		4.3.4	Regulatory Clearance	147
		4.3.5	Identification of Alternatives/Options	147
		4.3.6	Development of the Design Basis	148
		4.3.7	Establish Project Delivery and Contracting System	148
		4.3.8	Project Charter	150
	4.4	Sustainability in the Design Stage		153
		4.4.1	The Concept Design	155
		4.4.2	Preliminary Design	184
		4.4.3	Detail Design	203
		4.4.4	Construction Documents	237
	4.5	Sustainability at the Bidding and Tendering Stage		260
		4.5.1	Identify the Stakeholders	260
		4.5.2	Organize the Tendering Documents	261
		4.5.3	Identify the Tendering Procedure	263
		4.5.4	Identify the Bidders	263
		4.5.5	Manage the Tendering Process	263

Contents   ix

	4.5.6	Submit/Receive Bids 268
	4.5.7	Manage the Bidding and Tendering Quality 268
	4.5.8	Manage the Tendering Risk 269
	4.5.9	Review the Bid Documents 269
	4.5.10	Award the Contract 270
4.6	Sustainability in the Construction Stage 270
	4.6.1	Identify the Stakeholders 276
	4.6.2	Mobilization 281
	4.6.3	Development of the Project Site Facilities 286
	4.6.4	Identify the Project Execution/Installation Requirements 287
	4.6.5	Identify the Sustainability Requirements 289
	4.6.6	Develop the Project Execution Scope 289
	4.6.7	Project Planning and Scheduling 291
	4.6.8	Develop Management Plans 296
	4.6.9	Construction/Execution of the Works 314
	4.6.10	Monitoring and Controlling 329
	4.6.11	Inspection of the Executed Works/Systems 390
	4.6.12	Validate the Executed Works 392
4.7	Testing, Commissioning, and Handover 393
	4.7.1	Identify Stakeholders 396
	4.7.2	Identify Testing, Commissioning, and Handover Requirements 396
	4.7.3	Identify the Sustainability Requirements 399
	4.7.4	Develop Testing, Commissioning Scope 399
	4.7.5	Develop the Inspection and Testing Plan 400
	4.7.6	Execute the Commission Works/Systems 400
	4.7.7	Manage Testing, Commissioning Quality 403
	4.7.8	Develop Documents 405
	4.7.9	Monitor the Work Progress 408
	4.7.10	Train the Owner's/End-User's Personnel 409
	4.7.11	Handover of Project 409
	4.7.12	Issue the Substantial Completion Certificate 411
	4.7.13	Lessons Learned 411
	4.7.14	Settle Payments 413
	4.7.15	Settle Claims 413
	4.7.16	Close the Contract 413

Bibliography 415
Index 417

Figures

Figure 1.1 Sustainability 2
Figure 1.2 Sustainability elements 4
Figure 1.3 PDCA cycle for project development 8
Figure 2.1 Concept of traditional construction project organization ... 16
Figure 2.2 Construction project quality trilogy 19
Figure 2.3 QMS documentation pyramid 23
Figure 2.4 QMS documentation model 25
Figure 2.5 Flowchart diagram for development of IQMS 28
Figure 3.1 Cause and effect diagram for false ceiling rejection 39
Figure 3.2 Check sheet. Source: Abdul Razzak Rumane 40
Figure 3.3 Control chart for air handling unit air distribution (cfm) ... 40
Figure 3.4 Flowchart for contractor's staff approval 41
Figure 3.5 Histogram of employee reporting 42
Figure 3.6 Pareto analysis for construction cost 43
Figure 3.7 Pie chart for site staff 43
Figure 3.8 Run chart for manpower 44
Figure 3.9 Scatter diagram 45
Figure 3.10 Stratification chart 46
Figure 3.11 Activity network diagram 47
Figure 3.12 Arrow diagramming method for concrete foundation ... 47
Figure 3.13 Dependency relationship diagram 48
Figure 3.14 PDM diagramming method 48
Figure 3.15 Critical path method 51
Figure 3.16 Gantt chart for substation 52
Figure 3.17 Affinity diagram for concrete slab 54
Figure 3.18 Interelationship digraph 55
Figure 3.19 T-shaped matrix 56

Figure 3.20	Roof shaped matrix	57
Figure 3.21	Prioritization matrix	57
Figure 3.22	Process decision diagram chart	58
Figure 3.23	Tree diagram for no water in storage tank	58
Figure 3.24	Benchmarking process	60
Figure 3.25	Cause and effect diagram for masonary work	61
Figure 3.26	Failure mode and effects analysis (FMEA) process	62
Figure 3.27	FMEA recording form	63
Figure 3.28	Process mapping/flowcharting for approval of variation order	65
Figure 3.29	Root cause analysis for rejection of executed marble work	66
Figure 3.30	PDCA cycle for preparation of shop drawing	67
Figure 3.31	Statistical process control chart for generator frequency	69
Figure 3.32	Brainstorming process	71
Figure 3.33	Delphi technique process	72
Figure 3.34	Mind mapping	74
Figure 3.35	Cellular main switch board	76
Figure 3.36	Concurrent engineering for construction life cycle	77
Figure 3.37	Value stream mapping for emergency power system	81
Figure 3.38	House of quality for hospital building project	87
Figure 3.39	Six sigma roadmap	89
Figure 3.40	Logic flow diagram of activities in the study stage	108
Figure 3.41	Preliminary schedule for construction project	114
Figure 3.42	Major activities in the detailed design phase	116
Figure 3.43	Cost of quality during the design stage	117
Figure 3.44	PDCA cycle for construction projects(design phases)	117
Figure 3.45	Project monitoring and controlling process cycle	119
Figure 3.46	Root cause analysis for bad concrete	120
Figure 3.47	PDCA cycle (Deming Wheel) for execution of works	120
Figure 3.48	Flowchart for concrete casting	121
Figure 4.1	Flowchart for study stage	140

Figures

Figure 4.2	Major activities relating to study stage process groups	141
Figure 4.3	Evaluation and analysis of alternatives method to select preferred alternative	149
Figure 4.4	Flowchart for development of terms of reference	154
Figure 4.5	Development of project scope documents	155
Figure 4.6	Logic flowchart for concept design phase	156
Figure 4.7	Major activities relating to concept design process groups	157
Figure 4.8	Project design team organization chart	160
Figure 4.9	Concept of QFD "house of quality."	164
Figure 4.10	House of quality for college building project	168
Figure 4.11	Schedule: Classifications versus levels	176
Figure 4.12	Quality management procedure for development of concept design	181
Figure 4.13	Logic flowchart for preliminary design phase	185
Figure 4.14	Major activities relating to preliminary design process groups	186
Figure 4.15	Structural/civil design team organization chart	189
Figure 4.16	Value engineering study process activities	204
Figure 4.17	Logic flowchart for detail design phase	208
Figure 4.18	Major activities relating to detail design process	209
Figure 4.19	Design management team	211
Figure 4.20	Design review steps	238
Figure 4.21	Logic flowchart for construction documents phase	247
Figure 4.22	Major activities relating to construction documents process groups	248
Figure 4.23	Logic flowchart for bidding and tendering phase	261
Figure 4.24	Major activities relating to bidding and tendering process groups	262
Figure 4.25	Project team procurement strategy for short listing/registration	264
Figure 4.26	Bid clarification	268
Figure 4.27	Contract award procedure	271
Figure 4.28	Logic flowchart for construction phase	272
Figure 4.29	Major activities relating to construction process groups	273
Figure 4.30	Notice to proceed	274

Figure 4.31	Kickoff meeting agenda	275
Figure 4.32	Site staff selection procedure	279
Figure 4.33	Project staffing process	280
Figure 4.34	Schedule development process	294
Figure 4.35	Logic flowchart for development of contractor's quality control plan	297
Figure 4.36	Material management process for construction project	303
Figure 4.37	Transmittal form	310
Figure 4.38	Job site instruction	311
Figure 4.39	Logic flowchart for vendor selection procedure	319
Figure 4.40	Logic flowchart for material approval and procurement procedure	320
Figure 4.41	Site transmittal for material approval	321
Figure 4.42	Specification comparison statement	322
Figure 4.43	Shop drawing preparation and approval procedure	323
Figure 4.44	Site transmittal for work shop drawings	324
Figure 4.45	Builder's workshop drawing preparation and approval procedure	325
Figure 4.46	Composite/coordination shop drawing preparation and approval procedure	326
Figure 4.47	Sequence of execution of works	329
Figure 4.48	Flowchart for concrete casting	330
Figure 4.49	Process for structural concrete work	331
Figure 4.50	Logic flowchart for monitoring and controlling process	335
Figure 4.51	Scope validation process for construction project (design and bidding stage)	340
Figure 4.52	Scope control process	341
Figure 4.53	Request for information (RFI)	343
Figure 4.54	Process to resolve request for variation	344
Figure 4.55	Process to resolve scope change (owner initiated)	345
Figure 4.56	Schedule monitoring and controlling process	347
Figure 4.57	Traditional monitoring system	348
Figure 4.58	Digitized progress monitoring	349

Figures

Figure 4.59	S-Curve (work progress).	356
Figure 4.60	Contractor's procurement log.	365
Figure 4.61	Supply chain process in construction project.	366
Figure 4.62A	Submittal process (paper copy).	370
Figure 4.62B	Submittal process (electronic).	371
Figure 4.63	Contractor's submittal status log.	372
Figure 4.64	Contractor's shop drawing submittal log.	373
Figure 4.65	Typical flowchart for risk management procedure.	377
Figure 4.66	Typical flowchart for risk monitoring process.	378
Figure 4.67	Procurement management process stages for construction projects.	380
Figure 4.68	Contract management process.	381
Figure 4.69	Safety violation report.	384
Figure 4.70	Safety disciplinary notice.	385
Figure 4.71	Concept of safety disciplinary action.	386
Figure 4.72	Progress payment submission format.	387
Figure 4.73	Progress payment approval process.	388
Figure 4.74	Claim resolution process.	389
Figure 4.75	Check list.	391
Figure 4.76	Remedial note.	392
Figure 4.77	Non-conformance report.	393
Figure 4.78	Material inspection report.	394
Figure 4.79	Logic flowchart for testing, commissioning, and handover phase.	395
Figure 4.80	Major activities relating to testing, commissioning, and handover.	397
Figure 4.81	Logic flowchart for development of inspection and test plan.	404
Figure 4.82	Handing over of spare parts.	410
Figure 4.83	Handing over certificate.	412

Tables

Table 1.1	UN Sustainability Development Goals Agenda 2030	6
Table 1.2	UN Sustainability Development Goals Agenda 2030 and Sustainable Projects	11
Table 2.1	Types of Construction Projects	15
Table 2.2	Quality Principles of Construction Projects	21
Table 2.3	Major Quality Planning Activities During the Study Stage	30
Table 2.4	Major Quality Assurance Activities During the Study Stage	30
Table 2.5	Major Quality Control Activities During the Study Stage	31
Table 2.6	Major Quality Planning Activities During the Design Stage	32
Table 2.7	Major Quality Assurance Activities During the Design Stage	33
Table 2.8	Major Quality Control Activities During the Design Stage	34
Table 2.9	Contents of the Contractor's Quality Control Plan	35
Table 3.1	Classic Quality Tools	38
Table 3.2	Management and Planning Tools	46
Table 3.3	Activities to Construct a Substation Building	50
Table 3.4	L-Shaped Matrix	56
Table 3.5	Process Analysis Tools	59
Table 3.6	5Why Analysis for Cable Burning	64
Table 3.7	5W2H Analysis for Slab Collapse	64
Table 3.8	Process Improvement Tools	66
Table 3.9	SIPOC Analysis for an Electrical Panel	68
Table 3.10	Innovative and Creative Tools	69
Table 3.11	5W2H Analysis of a New Product	73
Table 3.12	Lean Tools	75
Table 3.13	5S for Construction Projects	78
Table 3.14	Mistake Proofing for Eliminating Design Errors	80
Table 3.15	Elements of the Cost of Quality	83

Table 3.16	Fundamental Objectives of Six Sigma-DMADV Tool	94
Table 3.17	Fundamental Objectives of Six Sigma DMAIC Tool	96
Table 3.18	Fundamental Objectives of the Six Sigma DMADDD Tool	100
Table 3.19	Level of Inventiveness	101
Table 3.20	Project Development Stages	103
Table 3.21	Major Elements of the Study Stage	106
Table 3.22	Need Assessment for Development of Project	109
Table 3.23	Major Considerations for Feasibility Study for Development of a Project	109
Table 3.24	Typical Contents of Terms of Reference (TOR) Documents	110
Table 3.25	5W2H Analysis for Project Need	111
Table 3.26	Major Elements of the Design Stage	112
Table 3.27	Major Items for Data Collection During the Concept Design Phase	112
Table 3.28	Development of the Concept Design	113
Table 3.29	Development of the Schematic Design for a Construction Project	115
Table 3.30	Major Elements of the Bidding and Tendering Stage	118
Table 3.31	Major Elements of the Construction Stage	118
Table 3.32	Major Activities by Contractor During the Construction Phase	119
Table 4.1	Benefits of Sustainability in the Design and Construction of the Project	123
Table 4.2	Major Project Activities Relating to the Initiating Process Group	126
Table 4.3	Major Project Activities Relating to the Planning Process Group	127
Table 4.4	Major Project Activities Relating to the Project Executing Group	132
Table 4.5	Major Project Activities Relating to Monitoring and Controlling Processes Group	134
Table 4.6	Major Project Activities Relating to the Closing Process Group	137
Table 4.7	5W2H Quality Tool for the Identification of Need	142
Table 4.8	Major Points to be Considered for Need Analysis	143
Table 4.9	Need Statement	144
Table 4.10	Categories of Project Delivery Systems	151
Table 4.11	Responsibilities of Various Participants During the Concept Design Phase	158

Tables

Table 4.12	Checklist for Owner's Requirements	167
Table 4.13	Elements to be Included in Concept Design Drawings	169
Table 4.14	Contents of Concept Design Report (Trade Name)	174
Table 4.15	Generic Schedule Classification Matrix	177
Table 4.16	Cost Estimation Levels for Construction Projects	178
Table 4.17	Quality Check for Cost Estimate During Concept Design	178
Table 4.18	Contents of Designer's Quality Control Plan	179
Table 4.19	Major Points for Review of Concept Design	183
Table 4.20	Responsibilities of Various Participants (Design-Bid-Build Type of Contracts) During the Schematic Design Phase	188
Table 4.21	Schematic Design Deliverables	194
Table 4.22	Major Points to Review for the Preliminary Design	205
Table 4.23	Responsibilities of Various Participants (Design--Bid-Build Type of Contracts) during the Detail Design Phase	210
Table 4.24	Detail Design Deliverables	216
Table 4.25	Mistake-Proofing for Eliminating Design Errors	235
Table 4.26	Interdisciplinary Coordination	239
Table 4.27	Checklist for Detail Design Review	246
Table 4.28	Responsibilities of Various Participants (Design-Bid-Build Type of Contracts) During the Construction Documents Phase	249
Table 4.29	Construction Document Deliverables	252
Table 4.30	Quality Check for Working Drawings	257
Table 4.31	Items to be Reviewed for Construction Documents	259
Table 4.32	Responsibilities of Various Participants (Design-Bid-Build Type of Contracts) during the Bidding and Tendering Phase	262
Table 4.33	Checklist for Bid Evaluation	265
Table 4.34	Pre-Qualification Questionnaires (PQQ) for Selecting Contractor	266
Table 4.35	Major Risk Factors Affecting Contractor	270
Table 4.36	Responsibilities of Various Participants (Design-Bid-Build Type of Contracting System) During the Construction Phase	276
Table 4.37	Responsibilities of Supervision Consultant	278
Table 4.38	Subcontractor Prequalification Questionnaire	282

Table 4.39	Contents of Contractor's Quality Control Plan	298
Table 4.40	Matrix for Site Administration and Communication	300
Table 4.41	Guidelines to Prepare a Communication Matrix	305
Table 4.42	Contents of Contractor's Communication Management Plan	306
Table 4.43	List of Project Control Documents	307
Table 4.44	Contents of Contract Management Plan	313
Table 4.45	Contents of Contractor's HSE Plan	315
Table 4.46	Consultant's Checklist for Smooth Functioning of a Project	332
Table 4.47	Monitoring and Controlling Plan References for Construction Projects	336
Table 4.48	Causes of Changes in a Construction Project	342
Table 4.49	Monthly Progress Report	351
Table 4.50	Contents of Progress Report (Consultant)	353
Table 4.51	Responsibility for Site Quality Control	359
Table 4.52	Analysis for Communication Matrix	368
Table 4.53	Communication Matrix	369
Table 4.54	Typical Categories of Risks in Construction Projects	375
Table 4.55	Responsibilities of Various Participants during Testing, Commissioning, and Handover Phase	398
Table 4.56	Typical Responsibilities of Supervision Consultant during Project Testing, Commissioning, and Handover Phase	398
Table 4.57	Major Items for Testing and Commissioning of Equipment	401
Table 4.58	Project Closeout Documents	406
Table 4.59	Punch List	413
Table 4.60	Project Close-out Checklist	414

Foreword

My first introduction to Dr. Rumane came at the 2021 Advancing Construction Quality Conference in Denver, Colorado, where we both were in attendance as presenters, representing the ASQ Design and Construction Division, for whom I was Chair-Elect at the time. I had the pleasure to attend Dr. Rumane's conference workshop titled "Where Do I Start?: Introducing a New Quality Management System to Your Company." I found it to be an insightful, structured, and well-thought-out workshop that resonated across all levels of experience for construction quality professionals active in implementing and/or improving their organization's quality program.

After seeing Dr. Rumane speak at that conference, I decided to take a journey into reading a couple of his previously published books, and it is no wonder he has been revered as a credible industry expert on construction project management and quality. Dr. Rumane has a keen ability to leverage his industry experience and education, to provide readers with useable frameworks for practical deployment methods of quality auditing practices, and various quality tools and philosophies, including how to integrate these into the aspects of managing construction projects throughout the entire project lifecycle.

This book, *Quality Management: How to Achieve Sustainability in Projects*, is no exception, as Dr. Rumane provides detailed roadmaps for ensuring the successful implementation of quality and sustainability concepts, with insights into practical integration methodologies that are understandable at all levels. Dr. Rumane begins the book by taking the reader through a journey of understanding the pillars of sustainability from project conceptualization, all the way through project turnover and beyond, while explaining the synergies between quality and sustainability, and the role quality plays in achieving sustainability. As the book progresses, readers will be provided with a litany of information on the use of quality tools and philosophies, from basic elementary concepts like PDCA and Process Mapping, to the implementation of much more advanced concepts in Six Sigma and Quality Function Deployment (QFD) House of Quality, while gaining a deeper understanding of the relationship between quality and sustainability, and how to achieve both at various stages throughout the project lifecycle.

Environmental, social, and governance (ESG) has emerged as a critical aspect of society and the business world across the globe, impacting all industries, and ESG is increasingly influencing the way we design and deliver construction projects through a major focus on sustainability. This book will be a critical tool in helping construction professionals at all levels, from junior-level associates to top-level executives, better understand how to integrate quality best practices through the various stages of construction projects, to achieve their goals in sustainability, and should be considered for incorporation into any construction education curriculum, as well as the personal library of any seasoned construction professional.

C. Collin Sutt, CMQ/OE, ASQ DCD
Chair – Excellence & Quality Consultant

Dr. Sutt is a recognized industry thought leader, who has written for publications such as Quality Magazine, and has been selected to chair and present at numerous industry conferences. Collin is an active member of the Insurance and Risk Committee for the Construction Quality Executives Council (CQEC) and has also had the honor of being selected to serve as a Baldrige Examiner on the Quality Texas Foundation (QTF) Board of Examiners for 2019/2020 and 2021/2022 for Texas's state-level Malcolm Baldrige Performance Excellence Award program.

Preface

The construction industry makes an important contribution to the competitiveness and prosperity of the economy.

Construction plays an important role in our drive to promote sustainable growth and development.

Buildings are responsible for almost half of carbon emissions, half of our water consumption, about one-third of landfill waste, and one-quarter of all raw materials used in the economy.

Numerous UN meetings, such as:

- The first United Nations Conference on Human Development in Stockholm, Sweden, 1972
- The Brundtland Commission on Environment and Development in 1987
- The 1992 Earth Summit in Rio de Janeiro, Brazil
- The 2002 Earth Summit in Johannesburg, South Africa
- The 2005 World Summit in New York City, USA
- The UN 2012 Sustainable Development Conference in Rio de Janeiro, Brazil
- The UN 2015 Sustainable Development Summit in New York City, USA
- The 2021 Climate Change Conference COP26 (Conference of Parties) in Glasgow, UK

have emphasized "sustainability," whether it be a sustainable environment, sustainable economic development, sustainable agricultural and rural development, and so on.

Sustainable development, as defined by the Brundtland Commission, is "Development that meets the needs of the present without compromising the ability of future generations to meet their own needs."

Sustainability focuses on:

- Economical factors
- Environmental issues
- Social requirements

Sustainable construction is designing and developing a construction project(s) that focuses on:

1. Economical
2. Environmental Protection
3. Social

Construction projects involve many participants, comprising Owner, Designer, Contractor, and many other professionals from construction-related industries.

Each of these participants is involved in implementing quality in construction projects.

These participants are both influenced by and dependent on each other in addition to "other players" involved in the construction process.

Participation involvement of the owner/client, designer, and contractors at different stages of the project is required to develop sustainable construction.

Construction project development has three major elements:

1. Study
2. Design
3. Construction

In practice, there are several types of tools, techniques, and methods which are used as quality improvement tools and have a variety of applications in manufacturing, process, and the construction industry. However, not all of these tools are used in construction projects due to the nature of construction projects, which are customized and non-repetitive. The use of various quality principles, tools, techniques, and methods by all the participants at different stages of the construction project will improve the process and ensure completion of the construction project, making the project qualitative, competitive, and economical.

The book discusses the sustainability framework, an overview of the quality of construction projects, the most common quality tools, and techniques used in construction projects for sustainability, and management of sustainability from the inception of the project to handover of the projects (study stage, design stage, and construction stage) to make the project sustainable.

For the benefit of project professionals who are not familiar with quality tools that are used at various activities of the project stages, the book contains a chapter covering different categories of quality tools that are useful to achieve sustainability in projects.

For the sake of proper understanding, the book is divided into four chapters and each chapter is divided into a number of sections covering sustainability requirement processes at each of the phases of construction projects.

Chapter 1 is about sustainability. It discusses the definition of sustainability and the elements (pillars) of sustainability, namely, economical, environmental, and social, in the project. It covers brief details about the sustainability framework. The chapter also discusses the major activities to be considered for development of a sustainable project. These activities are listed under the sustainable elements (pillars), economical, environmental, and social. The chapter lists in brief the contribution of sustainable projects to achieve some of the UN Sustainability Development Goals Agenda 2030.

Chapter 2 presents an overview of the quality of construction projects. It covers brief descriptions of the quality of construction projects.

Chapter 3 is about quality tools and their uses in construction projects. It elaborates applications of quality tools and techniques for achieving sustainability in the project.

Chapter 4 is about the management of quality for sustainability in the project stages (study stage, design stage, bidding and tendering, construction stage, and testing and commissioning). The relevant sustainability requirements that are essential to establish the scope of project works are considered to develop project design, construction documents, and construction phase processes in order to meet the requirements to achieve sustainability in the project.

The book, I am certain, will meet the requirements of construction professionals, quality professionals, project owners, students, and academics, and satisfy their needs.

Acknowledgments

Share the knowledge with others is the motto of this book.
  Many of my colleagues and friends extended help during the preparation of the book by arranging reference material, so many thanks to all of them for their support.
  I thank the publisher and the authors, whose writings are included in this book, for extending their support by allowing me to reprint their material.
  I thank the reviewers from various professional organizations for their valuable input to improve my writing. I thank members of ASQ Audit Division, ASQ Design & Construction Division, The Institution of Engineers (India), IEI, Kuwait Chapter, Kuwait Society of Engineers, and ASQ Kuwait Section, for their support in bringing out this book.
  I thank Collin Sutt, Chair, ASQ (DCD) for his nicely worded thought-provoking Foreword, and support and best wishes.
  I thank Cindy Renee Carelli, Executive Editor of CRC Press, the Senior Editorial Assistant, and other CRC staff for their support and contributions to make this construction-related book a reality.
  I thank Mr. Cliff Moser, former Chair, ASQ Design and Construction Division, Mr. Raymond R. Crawford, former Chair, ASQ Design and Construction Division. I thank Dr Adedeji Badiru of Airforce Institute of Technology for their best wishes all the time.
  I thank Engr Adel Kharafi, former Chairman, Kuwait Society of Engineers, and former President of the World Federation of Engineering Organizations (WFEO) for his good wishes at all times, and Engr Ahmad Alkandari, Director, Kuwait Municipality for his support and good wishes. I thank Engr. Ahmad Almershed, former Undersecretary, MSNA, Kuwait, Dr. Ayed Alamri, President, Saudi Quality Council, KSA, and Dr Fadel Safer, Former Minister of Public Works, Kuwait for his support and good wishes. I thank Engr Faisal D. Alatal, Chairman, the Kuwait Society of Engineers and President of Federation of Arab Engineers for his support and good wishes. I thank Prof. Mohammed Aichouni, University of Hail, KSA, Dr Mohammad Ben Salamah, Chair, ASQ Kuwait Section, and Dr N. N. Murthy of Jagruti Kiran Consultants for their support and good wishes at all times. I thank Dr Othman Alshamrani, Imam Abdulrehman Bin Faisal University, KSA, Ms. Rima Al Awadhi, former Chair, ASQ Kuwait Section and Team Leader at Kuwait Oil Company, Engr Wael Aljasem of Kuwait Project Management Society, and Engr Yaseen Farraj, former Director, Ministry of Public Works Kuwait, for their support and good wishes.
  I thank Nabila Rumane of Morgan Stanley, India, and Mr Bashir Ibrahim Parkar of Dar SSH International for their valuable input and support.
  I extend my thanks to Dr Ted Coleman, Professor and Department Chair, California State University, San Bernardino and former Chancellor KW University, for his unending support.

My special thanks go to H.E. Sheikh Rakan Nayef Jaber Al Sabah for his support and good wishes.

I thank my well-wishers whose inspiration encouraged me to complete this book.

Most of the data discussed in this book are from the author's practical and professional experience and are accurate to the best of the author's knowledge and ability. However, if any discrepancies are observed in the presentation, I would appreciate being made aware of them.

The contributions of my son Ataullah, my daughter Farzeen, and daughter-in-law Masum are worth mentioning here. They encouraged me and helped me in my preparatory work to achieve the final product. I thank my brothers, sisters, and all my family members for their support, encouragement, and good wishes at all times.

Abdul Razzak Rumane

About the Author

Abdul Razzak Rumane Ph.D. is a Chartered Quality Professional Fellow of The Chartered Quality Institute (UK) and a certified consultant engineer in the field of electrical engineering and project management. He earned a Bachelor of Engineering (Electrical) degree from Marathwada University (now Dr Babasaheb Ambedkar Marathwada University), India in 1972 and completed his Doctor of Philosophy (Ph.D.) from Kennedy Western University, USA (now Warren National University) in 2005. His dissertation topic was "Quality Engineering Applications in Construction Projects."

Dr Rumane has been honored by The International University of Ministry and Education, Missouri, USA by being awarded in 2021 the honorary degree of Doctor of Philosophy (D.Phil.) for his expertise in quality management, and by The Yorker International University, USA, who awarded him with an Honorary Doctorate of Engineering in 2007. The Albert Schweitzer International Foundation honored him with the gold medal for "Outstanding Contribution in the Field of Quality in Construction Projects," while the World Quality Congress awarded him the "Global Award for Excellence in Quality Management and Leadership."

Dr Rumane is an accomplished Engineer. He is associated with a number of professional organizations. He has attended many international conferences and has made technical presentations at various conferences. Dr Rumane is the author of seven practical books, entitled *Quality Management in Construction Projects* (first edition 2010), *Quality Tools for Managing Construction Projects* (2013), *Quality Management in Construction Projects* (second edition 2017), *Quality Management in Oil and Gas Projects* (2021), *Risk Management Applications Used to Sustain Quality in Projects: A Practical Guide* (2022), as well as the being the editor of *Handbook of Construction Management: Scope, Schedule, and Cost Control* (2016). All of these books are published by CRC Press (a Taylor & Francis Group Company), USA. A sixth authored book, entitled *Quality Auditing in Construction Projects: A Handbook* (2019) is published by Routledge, UK (a Taylor & Francis Group Company). His book, *Quality Management in Construction Projects*, has been translated into the Korean language.

Presently, he is Treasurer of ASQ Kuwait Section and was Chair of the International Liaison Committee, ASQ (Design and Construction Division) for 2022. He served as Secretary ASQ GC in 2019 and Secretary ASQ LMC, Kuwait in 2017–2018. He was honorary Chairman of The Institution of Engineers (India), Kuwait chapter, for 2016–2017, 2013–2014, and 2005–2007.

Dr. Rumane's professional career exceeds 50 years, including 10 years in manufacturing industries and over 40 years in construction projects. Presently, he is associated with SIJJEEL Co., Kuwait, as Advisor and Director, Construction Management.

Abbreviations

AACE	American Association of Cost Engineers
ASCE	American Society of Civil Engineers
ASHRAE	American Society of Heating, Refrigeration and Air-conditioning Engineers
ASQ	American Society for Quality
BMS	Building Management System
CII	Construction Industry Institute
CDM	Construction (Design and Management)
CEN	European Committee for Standardization
CMAA	Construction Management Association of America
CSC	Construction Specifications, Canada
CSI	Construction Specification Institute
FEED	Front End Engineering Design
FIDIC	Federation Internationale des Ingenieurs-Counceils
HAZID	Hazard Identification
HAZOP	Hazard and Operability
HSE	Health, Safety and Environment
ICE	Institute of Civil Engineers (U.K.)
IEC	International Electrotechnical Commission
IEEE	Institute of Electrical and Electronics Engineers
IoT	Internet of Things
IP	Ingress Protection
ISO	International Organization for Standardization
OH&S	Occupational Health and Safety
PHSER	Procedure for Project HSE Review
PCM	Planning and Control Manager
PMC	Project Management Consultant
PMI	Project Management Institute
PMBOK	Project Management Book of Knowledge
QMS	Quality Management System
QS	Quantity Surveyor
RE	Resident Engineer
TIC	Total Investment Cost

Synonyms

Owner	Client, Employer
Consultant	Architect/ Engineer (A/E), Designer, Design Professionals, Designer, Consulting Engineers, Supervision Professional, Specialist Consultant
Engineer	Resident Project Representative
Engineer's Representative	Resident Engineer
Project Charter	Terms of Reference (TOR), Client Brief, Definitive Project Brief
Project Manager	Construction Manager
Contractor	Constructor, Builder, EPC Contractor
Quantity Surveyor	Cost Estimator, Contract Attorney, Cost Engineer, Cost and Works Superintendent
Main Contractor	General Contractor

1 Sustainability Framework

1.1 SUSTAINABILITY

Sustainability is basically the ability to meet our own needs, without compromising the ability of future generations to meet their needs. Sustainability is the human preservation of the environment, whether economically or socially, through personal responsibility, management of resources, and maintenance of environmental assets.

Sustainability focuses on:

1. Economical factors
2. Environmental issues
3. Social requirements

Figure 1.1 illustrates the model of sustainability.

Adedeji Badiru (2021) stated the multidimensionality of sustainability as follows:

> Different experts, researchers, and practitioners have differing definitions of sustainability. The lack of a consistent view is probably the reason that many sustainability programs have not taken root as expected. The following systems-based definitions are essential for the purpose of the theme of this book.
>
> Sustainability – The human preservation of the environment, whether economically or socially, through responsibility, management of resources, and maintenance of physical infrastructure.
> Recycling – The reprocessing of materials already used into new materials or products to prevent waste and protect the environment by reducing energy, air pollution, water pollution, and emissions.
> Reducing Waste – Reducing the amount of unwanted materials technologically and socially to economically benefit the environment.
> Reuse – To use again in a different circumstance after processing.
> Energy – The capacity of a physical system to perform work through heat, kinetics, mechanical systems, light, or electrical means.
> Sustainable Energy – The condition of energy that meets the needs of the present without compromising the ability of future generations to meet their needs.
> Environment – The setting, surroundings, or conditions in which living objects operate.
> Ecosystem – Interacting organisms and their physical means of living in a biological community.
> Humanity – Human nature and civilization.
> Global – Relation to the world, Earth, or planet.
> Global Warming – The gradual increase in temperature of the Earth and oceans predicted to be from pollution and the inconsideration of the environment.

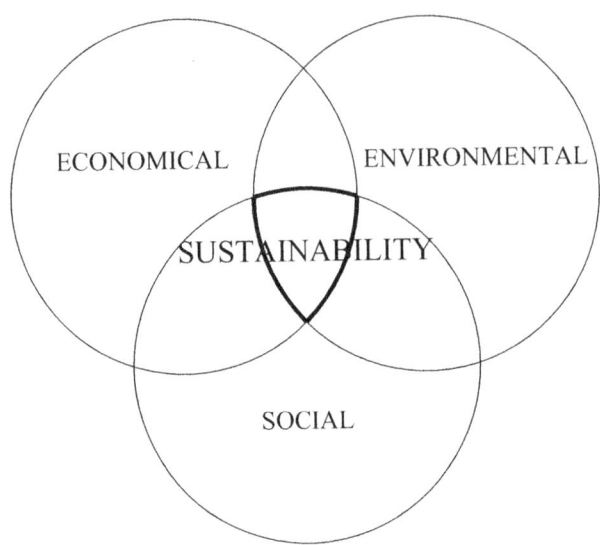

FIGURE 1.1 Sustainability.

- Greenhouse – Solar radiation entrapment caused by atmospheric gases caused by pollution, which allows sunlight to pass through and be remitted as heat radiation back from the Earth's surface.
- Quality Management – The key for a system to ensure the end goal is met through a desired level of excellence to be competitive in the business.
- Total Quality Management – Continuous improvement in products and processes by increasing the quality and reducing the defects through management methodologies.
- Systems – A network of things interacting together.
- Biodiversity – The natural variation among living organisms in particular ecosystems.
- Climate – Temperature, precipitation, and wind characteristics and conditions.
- Natural Resources – Materials or matter in nature such as minerals, freshwater, forests, or abundant land that can be used for economic benefits.
- Forecasting – The prediction or estimation of future events performed normally due to trending.
- Agriculture – Farming and manipulation of soils to harvest and grow crops while raising livestock.
- Human Impact – The impacts from human beings that affect biophysical environments, biodiversity, and any other environments.
- Public Health – The science and art of preventing diseases while prolonging life.
- Project Management – Planning, managing, monitoring, and controlling projects utilizing feedback and knowledge of tools and techniques.
- Ethics – A system of moral values that makes one perform the right conduct.
- Education – Education or training performed through knowledge and studying.
- Policies – Plans or courses of action.
- Physics – The science of matter and energy and the interaction of the two.

Sustainability Framework

Cooperation – The process of two or more beings working together toward the same goal.

Coordination – Organization of different elements so that there is cooperation for effective processes.

Planetary – Relating to the Earth as a planet.

Systems – Detailed methods or procedures established to transmit a specific activity or to perform a responsibility.

Theory – A set of assumptions based on accepted facts that provide rational explanations of cause and effect relationships among groups.

Community Service – Voluntary work intended to assist others in a particular area.

Enterprise – A business or company with resources.

Biosphere – The actual regions where living organisms occupy or reside.

Biophysical – The science dealing with the application of the physics of biological processes.

Sustainable Development – The concept of needs mostly from the idea of limitations imposed by society and technology to prevent the environment from meeting present and future needs.

Society – The collective group of people living together in a particular region with the same customs, laws, or organizations.

Economy – The prosperity, possessions, and resources of a country or region in terms of production and consumption of goods and services.

Energy Efficient Coding – Codes that set minimum requirements for energy-efficient design and construction for new and renovated buildings that impact energy usage.

Fenestration – The arrangement of windows and doors in a building to help operate with lower heating and cooling losses.

Figure 1.2 illustrates the major elements of sustainability.

1.2 DEVELOPMENT OF SUSTAINABLE PROJECTS

The construction project quality is the fulfillment of the owner's needs as per the defined scope within the schedule and budget. The quality of the construction project consists of mainly three attributes:

1. Scope
2. Schedule
3. Budget

Sustainability focuses on:

1. Economical factors
2. Environmental issues
3. Social requirements

In order to achieve sustainability in the quality of projects, it is essential that all the sustainability requirements are considered in each stage of the project while developing a sustainable project.

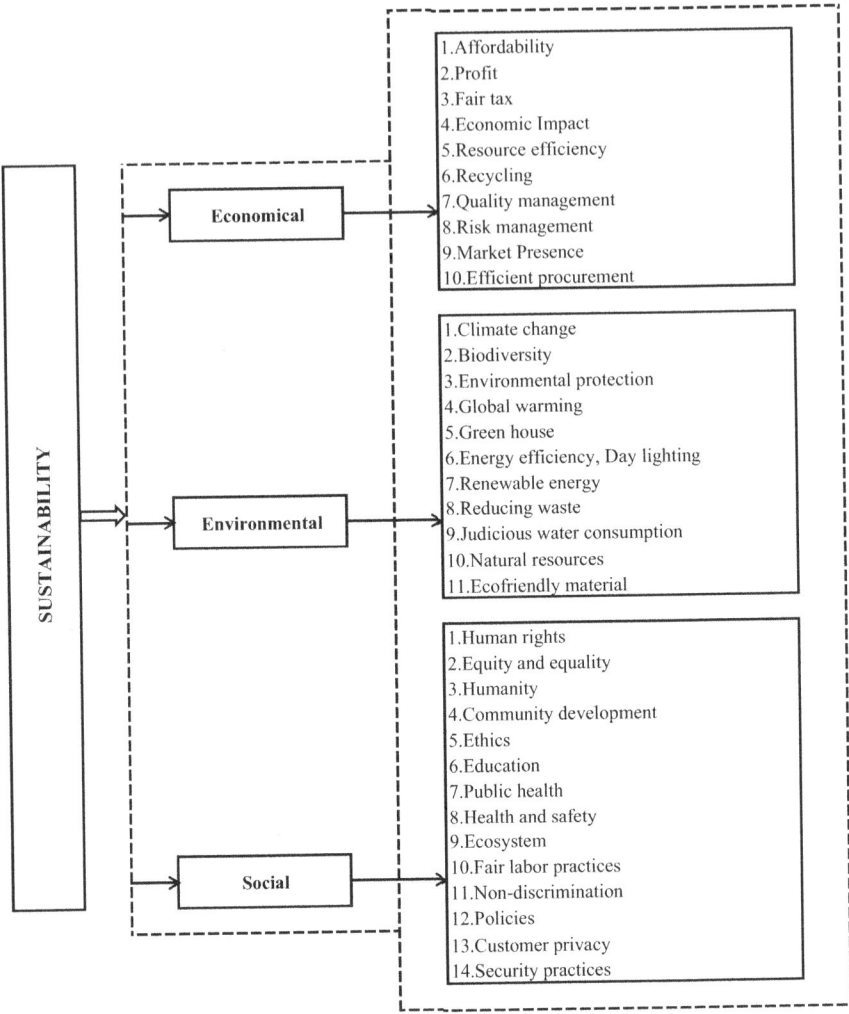

FIGURE 1.2 Sustainability elements.

1.2.1 Sustainability Framework

The Sustainability Framework (SF) encompasses the normative, operational (safeguards), and management, and aims by:

- Balancing flexibility and accountability
- Adding value to existing procedures and policies
- Applying to all types of United Nations (UN) activities
- Strengthening monitoring, evaluation, and transparency
- Advancing internal systematic approaches (not mainstreaming one issue or the other, or new strategies)

Sustainability Framework

Sustainability frameworks have become common practice among organizations aiming at disclosing their contribution to sustainable development and enhancing their credibility with stakeholders. Sustainability frameworks aid companies in selecting and implementing sustainability initiatives by developing and applying a sustainability strategy to achieve, among other things, competitive advantage and to leverage efficient business operations, support better corporate decision-making, and long-term value creation through transparency.

The following are major sustainability frameworks that have been helpful in developing a sustainable project.

1.2.1.1 Sustainable Development Goals (SDGs)

SDGs refer to the initiative passed by the United Nations in 2015 and present the foundation of the 2030 Agenda. The SDG framework comprises 17 specific goals and is aimed at ending poverty, protecting the planet, and ensuring prosperity for all by 2030. These 17 goals follow and build on the Millennium Development Goals, while including new areas such as climate change, economic inequality, innovation, sustainable consumption, and peace and justice, among other priorities. The SDGs provide a great source of inspiration in setting internal goals and strategies.

Table 1.1 lists UN Sustainability Development Goals Agenda 2030 for Sustainable Development

1.2.1.2 ISO 26000 Social Responsibility

The International Standard ISO 26000 is a guideline that supports organizations in implementing social responsibility. ISO 26000 standards were released in 2010 and were developed by groups of experts and national delegations from all over the world. The standard guideline was developed with the involvement of relevant interest groups and 450 experts from almost 100 countries. ISO 26000 is thus regarded as an internationally recognized reference framework for [corporate] social responsibility and contributes to all three aspects of sustainable development – economy, environment, and society. The standard is aimed not only at companies but also at organizations of all kinds. This distinguishes ISO 26000, for example, from the OECD (Organization for Economic Cooperation and Development) Guidelines for Multinational Enterprises or the ILO (International Labor Organization) Tripartite Declaration on Multinational Enterprises, which refer specifically to only business organizations.

The standard describes how organizations can meet their responsibilities in the areas of governance, human rights, labor practices, environmental protection, fair operating and business practices, consumer concerns, and community involvement and development. So, it is a framework for sustainability management.

1.2.1.3 ISO 14000 Environmental Management System

ISO 14000 is a series of international standards that has been developed to incorporate environmental aspects into business operations and product standards. ISO 14001 is a specific standard in the series for a management system that incorporates

TABLE 1.1
UN Sustainability Development Goals Agenda 2030

Serial Number	Goal	Description
1	No poverty	End poverty in all its forms everywhere
2	Zero hunger	End hunger, achieve food security and improved nutrition, and promote sustainable agriculture
3	Good health and well-being	Ensure healthy lives and promote well-being for all at all ages
4	Quality education	Ensure inclusive and equitable quality education and promote lifelong learning opportunities for all
5	Gender equality	Achieve gender equality and empower all women and girls
6	Clean water and sanitation	Ensure availability and sustainable management of water and sanitation for all
7	Affordable and clean energy	Ensure access to affordable, reliable, sustainable, and clean energy for all
8	Decent work and economic growth	Promote sustained, inclusive, and sustainable economic growth, full and productive employment, and decent work for all
9	Industry, innovation, and infrastructure	Build resilient infrastructure, promote inclusive and sustainable industrialization, and foster innovation
10	Reduced inequality	Reduce inequality within and among countries
11	Sustainable cities and communities	Make cities and human settlements inclusive, safe, resilient, and sustainable
12	Responsible consumption and production	Ensure sustainable consumption production patterns
13	Climate action	Take urgent action to combat climate change and its impacts
14	Life below water	Conserve and sustainably use the oceans, seas, and marine resources for sustainable development
15	Life on land	Protect, restore, and promote sustainable use of terrestrial ecosystems, sustainably manage forests, combat desertification, halt and reverse land degradation, and halt biodiversity loss
16	Peace, justice, and institutions	Promote peaceful, and inclusive societies for sustainable development, provide access to justice for all, and build effective, accountable, and inclusive institutions at all levels
17	Partnership for the goals	Strengthen the means of implementation and revitalize the global partnership for sustainable development

a set of interrelated elements designed to minimize harmful effects on the environment due to the activities performed by an organization and to achieve continual improvement of its environmental performance. ISO 14000 outlines how to put an effective environmental system in place. ISO 14001 incorporates a quality management system (QMS) philosophy, terminology, and requirement structure similar to that of ISO 9001, and thus achieves system compatibility.

Sustainability Framework

1.2.1.4 CDP (Carbon Disclosure Project)

The CDP framework is solely concentrated on the environment. Since 2002, the CDP has been supporting organizations to disclose their environmental impacts. The purpose of the CDP framework is to encourage organizations to manage, measure, and reduce their impact on the environment while providing high-quality information to the market.

1.2.1.5 GRI (Global Reporting Initiative)

The Global Reporting Initiative (GRI) is an independent international organization that has pioneered sustainability reporting since 1997. The GRI is the most widely used framework for sustainability reporting.

1.2.1.6 IIRC (International Integrated Reporting Council)

The IIRC was established in 2011. The primary purpose of an integrated report is to explain to providers of financial capital how an organization creates value over time. IIRC focuses on the organization's short-term and long-term value creation process, based on financial, manufactured, intellectual, human, social and relationship, that reflect the interconnections between environmental, social, governance and financial factors in decisions that affect long-term performance and condition, making clear the link between sustainability and economic value, and provide the necessary framework for environmental and social factors to be taken into account systematically in reporting and decision-making

1.2.1.7 SASB (Sustainability Accounting Standards Board)

SASB standards were established in 2011 using the definition of "materiality" applied under the US federal securities laws. SASB standards are industry-specific standards aiming at covering solely information that is financially material at the organization level.

1.2.1.8 UNGC (United Nations Global Compact)

The UN Global Compact was launched in 2000 and represents a platform for the development, implementation, and disclosure of responsible and sustainable corporate policies and practices. The UN Global Compact is a framework, based on ten principles, that asks organizations to embrace and enact a set of core values in the areas of human rights, labor, environment, and anti-corruption.

1.2.1.9 AccountAbility (Accountability AA1000 Series of Standards)

AccountAbility's AA1000 series was launched in 1996 and represents principle-based standards aimed at helping organizations to become more accountable, responsible, and sustainable. These standards address issues affecting governance, business models, and organizational strategy, and provide operational guidance on sustainability assurance and stakeholder engagement. Accountability AA1000 Series is composed of:

- The AA1000 Accountability Principles Standard (AA1000APS)
- The AA1000 Assurance Standard (AA1000AS)
- The AA1000 Stakeholder Engagement Standard (AA1000SES)

Quality Management

Each of the frameworks referred to above has the same goal of supporting better corporate decision-making and long-term value creation through transparency. The specific framework can be selected to meet the organization's goal to develop a specific project.

1.2.2 Sustainable Project Development Process

Figure 1.3 illustrates the Plan-Do-Check-Act (PDCA) cycle for the development of a sustainable project. Chapter 4 in the current book discusses in detail the stages of project development, taking into account the sustainability requirement at each of these stages.

1.3 ELEMENTS OF SUSTAINABLE PROJECT

Sustainable projects have three major elements. These are as follows:

1. Economical
2. Environmental
3. Social

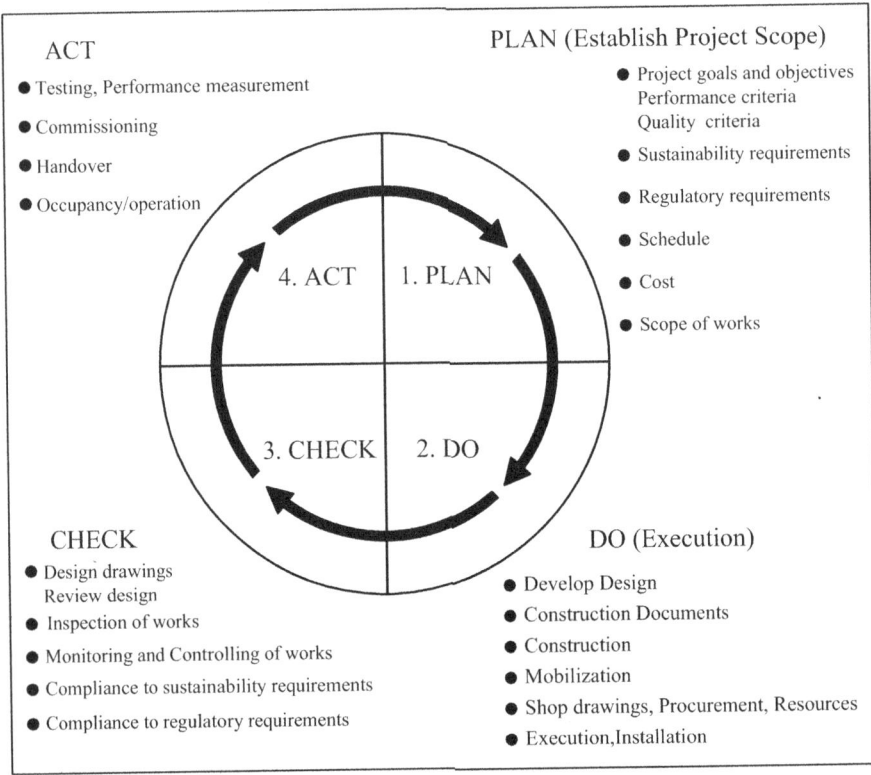

FIGURE 1.3 PDCA cycle for project development.

Sustainability Framework

When developing a sustainable project, a sustainable framework and the activities of the major elements must be considered.

The following are the main criteria for selecting a sustainability framework to develop a sustainable project:

- Business need
- Organization's core business activity
- Project goals and objectives
- Organization's vision and strategic plan
- Justification or business case for the project
- Need to be SMART (Specific, Measurable, Achievable, Realistic, Time bound)
- Stakeholder's project goals and objectives
- Type of industry
- Size of the project
- Location of the project
- Innovation
- Technical and functional capability
- Maturity level (lifecycle span)
- Suitable for usage as per owner's requirements
- Additional value to existing policies
- Additional value to existing procedures
- Performance requirements
- Quality criteria
- Regulatory requirements
- Economic benefits
- Energy efficiency
- Green building
- Diversity
- Reduction in environmental impact
- Water conservation
- Waste reduction
- Pollution control
- Local resources
- Key indicators to measure sustainability of project
- Green rating

The three major elements (also known as pillars) of sustainability, namely economic, environmental, and social, have the following major items which need to be considered when developing sustainability in the quality of a project.

1.3.1 Economical

- Project benefits
- Durability
- Overall project schedule
- Project cost

- Quality management and performance criteria
- Potential risks
- Regulatory requirements
- Flowchart to develop project design
- Interdisciplinary coordination
- Flowchart for construction activities
- Supply chain management system
- Detailed calculation of design activities
- Schematics of systems
- Energy-efficient equipment, systems
- Alternate energy sources
- Monitoring of design
- Monitoring of construction
- Monitoring and control of schedule
- Overrun of schedule
- Monitoring and control of cost
- Overrun of cost
- Design review procedure
- Modern technology
- Economical construction methods
- Modular construction
- Prevention of reworks
- Sequencing of execution activities
- Inspection of executed/installed work
- Ease of maintenance
- Cost effectiveness over the entire lifecycle of the project

1.3.2 Environmental

- Aesthetic
- Landscape matching with surrounding area
- Plantation in project area
- Environmental impact
- Ecofriendly
- Natural resources
- Ecofriendly material
- Material compatibility with the environment
- Air quality
- Water conservation
- Comfortable lighting
- Day lighting control
- Low-noise equipment (noise control)
- Waste reduction
- Stormwater management
- High-performance facility
- Better services to the user of the facility

Sustainability Framework

1.3.3 Social

- Training and development of team members
- Local resources
- Integration with local community
- Diversity and inclusion
- Health, safety, and the environment
- High-performance facility

1.4 SUSTAINABILITY DEVELOPMENT GOALS AND SUSTAINABLE PROJECT

A sustainable project has the capability to influence and contribute to achieving some of the goals of the UN Sustainability Development Goals Agenda 2030.

Table 1.2 illustrates the project contribution to the related goals.

TABLE 1.2
UN Sustainability Development Goals Agenda 2030 and Sustainable Projects

	UN Goal	Project Contribution
Serial Number	Goal	Description

1	No poverty	• Employment generation opportunities during:
		   • Construction period
		   • Operation and maintenance
		 • Employment to local community
		 • Supporting farmers and improving agriculture productivity through community support

2	Zero hunger	• Employment opportunities will result in well-being of community • Nutrition programs for vulnerable communities
3	Good health and well being	• Improved health services • Setting OPD medical units in rural areas of operations • Combating malnutrition • Health insurance for employees and their families
4	Quality education	• Training and development for local team members • Training for digital inclusion • Supporting education for children of local employees – rural youth empowerment
5	Gender equality	• Working opportunities without any segregation • Promoting diversity and inclusion • Equal pay across workforce • Diversity in senior management
6	Clean water and sanitation	• Environmental issues are taken care while developing sustainable project • Safe water discharge • Effluent treatment plant (ETP) to reuse wastewater in operations • Judicial water withdrawal and consumption

(Continued)

TABLE 1.2 CONTINUED
UN Sustainability Development Goals Agenda 2030 and Sustainable Projects

UN Goal		Project Contribution
Serial Number	Goal	Description
7	Affordable and clean energy	• Economical factors and environmental issues are taken care while developing sustainable projects that will improve the existing situation • Adopting renewable projects • Using energy-efficient equipment • Less energy-intensive operations
8	Decent work and economic growth	• Economical factors and social requirements are taken care while developing sustainable project that will improve the existing situation • Promote local employment • Safe working conditions
9	Industry, innovation, and infrastructure	• Construction projects are non-repetitive and complex. Each project by itself is an innovative project. The project can be designed to fulfill the latest requirements by incorporating the latest technology • Invest in research and development • Adopt the latest cost-effective technologies
10	Reduced inequality	• Participation in international organizations to develop sustainable projects • Promoting diversity, equity, and inclusion in the organization • Public commitment to equality
11	Sustainable cities and communities	• Sustainable projects will result in sustainable cities and communities • Designing green buildings • Knowledge hub for promoting sustainability initiatives • Clean up campaigns
12	Responsible consumption and production	• Sustainable projects will help reduce water consumption and reduce energy consumption • Using sustainable and renewable raw materials for construction
13	Climate action	• Sustainable projects will reduce impact on climate • Invest in renewable technologies and replace high emission fuels • Reduce GHG emissions across value chain from procuring raw materials to customer use
14	Life below water	• Marine related sustainable projects can achieve sustainability in marine resources
15	Life on land	• Life on land will improve with development of sustainable projects • Plantation of trees • Preserve biodiversity especially endangered species
16	Peace, justice, and institutions	• Sustainable projects will increase harmony, peace, and justice among all • Form institutions to conduct sustainable activities
17	Partnership for the goals	• Sustainable project developers can participate and be a partner to achieve goals. They can have a partnership with the UN Economic and Social Council (ECOSOC) • Global partnerships with organizations like the United Nations, etc. to address global challenges and achieve the goals

2 Overview of Quality in Construction Projects

2.1 CONSTRUCTION PROJECTS

Construction involves translating the owner's goals and objectives, by the contractor, to build the facility as stipulated in the contract documents, plans, and specifications on schedule and within budget.

Construction has a history of several thousand years. The first shelters were built from stone or mud and the materials were collected from the forests to provide protection against cold, wind, rain, and snow. These buildings were primarily for residential purposes, although some may have had commercial functions.

During the New Stone Age, people introduced dried bricks, wall construction, metalworking, and irrigation. Gradually, people developed the skills to construct villages and cities and considerable skills in building were acquired. This can be seen from the great civilization in different parts of the world some 4,000–5,000 years ago. In Greek settlements, which were constructed about 2000 BC, the buildings were made of mud on timber frames. Later, temples and theaters were built from marble. Some 1,500–2,000 years ago, Rome became the leading center of world culture, which extended to construction.

Marcus Pollo was the first century BC military and civil engineer, who published a book in Rome. This was the world's first major publication on architecture and construction, and it dealt with building materials, the styles and design of building types, the construction process, building physics, astronomy, and building machines.

During the Medieval Age (476–1492), improvements in agriculture and artisanal productivity, and the exploration and consequent broadening of commerce took place, so that, in the late Middle Ages, building construction became a major industry. Craftsmen were given training and education in order to develop skills and to raise their status. At this time, Guilds were responsible for managing quality.

The fifteenth century brought a "renaissance" or renewal in architecture, building, and science. Significant changes occurred during the seventeenth century and thereafter due to the increasing transformation of construction and urban habitats.

The scientific revolution of the seventeenth and eighteenth centuries gave birth to the great Industrial Revolution of the eighteenth century. After some delay, construction followed these developments in the nineteenth century.

The first half of the twentieth century witnessed the construction industry become an important sector throughout the world, employing many workers. During this period, skyscrapers, long-span dams, shells, and bridges were developed to satisfy new requirements and marked the continuing progress of construction techniques. The provision of services, such as heating, air conditioning, electrical lighting, mains water, and elevators, to buildings became common. The twentieth century has seen the transformation of the construction and building industry into a major economic sector. During the second half of the twentieth century, the construction industry began to industrialize, introducing mechanization, prefabrication, and system building. The design of building services systems changed considerably in the last 20 years of the twentieth century. Construction projects are constantly increasing in technological complexity. In addition, the requirements of construction clients are on the increase, and, as a result, construction projects must meet various performance standards. Therefore, to ensure the adequacy of the client brief, which addresses the numerous complex client/user needs, it has become the responsibility of the designer to evaluate the requirements in terms of activities and their relationship, and to follow health, safety, and environmental regulations while designing any building.

Residential and commercial, traditional A/E type of construction projects accounts for an estimated 25% of the annual construction volume. Building construction is a labor-intensive endeavor. Every construction project has some elements that are unique. No two construction or research and development projects are alike. Though it is clear that some construction projects are usually more routine than R & D projects, some degree of customization is a characteristic of all projects.

There are several types of projects. Table 2.1 illustrates the various types of construction projects.

A project is a temporary endeavor undertaken to create a unique product or service. "Temporary" means that every project has a definite beginning and a definite end. "Unique" means that the product or service is different in some distinguishing way from all similar products or services. Projects are often critical components of the performing organization's business strategy. Examples of projects include:

- Developing a new product or service.
- Effecting a change in structure, staffing, or style of an organization.
- Designing a new transportation vehicle/aircraft.
- Developing or acquiring a new or modified information system.
- Running a campaign for political office.
- Implementing a new business procedure or process.
- Constructing a building or facilities.

The duration of a project is finite, projects are not ongoing efforts and the project ceases when its declared objectives have been achieved. Some of the characteristics of projects, for example, are that they are:

1. Performed by people.
2. Constrained by limited resources.
3. Planned, executed, and controlled.

Based on various definitions, a project can be defined as follows:

"A project is a plan or program performed by people with assigned resources to achieve an objective within a finite duration."

TABLE 2.1
Types of Construction Projects

1	Process Type Projects		
1.1	Liquid chemical plants		
1.2	Liquid/solid plants		
1.3	Solid process plants		
1.4	Petrochemical plants		
1.5	Petroleum refineries		
2	Non-Process Type Projects		
2.1	Power plants		
2.2	Manufacturing plants		
2.3	Support facilities		
2.4	Miscellaneous (R&D) projects		
2.5	Civil construction projects	Residential construction	Family homes, Multi unit town houses, Garden, Apartments, Condominiums, High-rise apartments, Villas.
2.6	Commercial A/E projects	Building construction (institutional and commercial)	Schools, Universities, Hospitals, Commercial office complexes, Shopping malls, Banks, Theaters, Stadiums, Government buildings, Warehouses, Recreation centers, Amusement parks, Holiday resorts, Neighborhood centers.
		Industrial construction	Petroleum refineries, Petroleum plants, Power plants, Heavy manufacturing plants, Steel mills, Chemical processing plants.
		Heavy engineering	Dams, Tunnels, Bridges, Highways, Railways, Airports, Urban rapid transit system, Ports, Harbors, Power lines, and Communication networks.
		Environmental	Water treatment and clean water distribution, Sanitary and sewage system, Waste management.

(Categories of Civil construction projects and Commercial A/E projects)

Source: Abdul Razzak Rumane (2013). Quality Tools for Managing Construction, CRC Press, Florida. Projects. Reprinted with permission of Taylor & Francis Group

Construction projects comprise a cross-section of many different participants. These participants both influence and depend on each other in addition to "other players" involved in the construction process. Figure 2.1 illustrates the concept of traditional construction project organization.

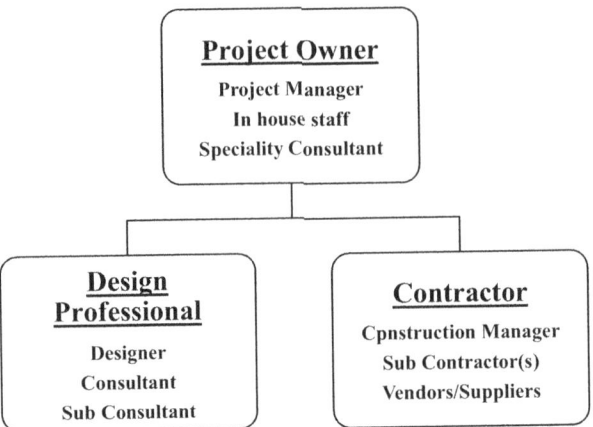

FIGURE 2.1 Concept of traditional construction project organization.

Traditional construction projects involve three main groups of participants. These are:

1. Owner – A person or an organization that initiates and sanctions a project. He/she requests the need of the facility and is responsible for arranging the financial resources needed for the creation of the facility.
2. Designer (A/E) – Consists of Architects/Engineers (A/E) or consultants. They are the owner's appointed entity accountable to convert the owner's conception and needs to design the specific facility with detailed directions through drawings and specifications within the economic objectives. They are responsible for the design of the project and, in certain cases, for supervision of the construction process.
3. Contractor – A construction firm engaged by the owner to complete the specific facility by providing the necessary staff, workforce, materials, equipment, tools, and other accessories to the satisfaction of the owner/end-user in compliance with the contract documents. The contractor is responsible for implementing the project activities and to achieve the owner's objectives.

Construction projects are executed based on a predetermined set of goals and objectives. Under traditional construction projects, the owner heads the team, designating a project manager. The project manager is a person/member of the owner's staff or an independently hired person/firm who has overall or principal responsibility for the management of the project as a whole.

Complex and major construction projects face many challenges, such as delays, changes, disputes, and accidents at the site, and therefore projects need to be efficiently managed from the beginning to the end to meet the intended use and the owner's expectations. The owner/client may not have the necessary staff/resources in-house to manage the planning, design, and construction of the project to achieve the desired results. Therefore, in such cases, owners engage a professional firm (a construction manager) that is trained in the management of construction processes, to assist in developing bid documents and overseeing and coordinating projects for the owner. The

Overview of Quality in Construction Projects

basic construction management concept is that the owner assigns a contract to a firm that is knowledgeable and capable of coordinating all the aspects of the project to meet the intended use of the project by the owner. In the construction management type of construction project, the consultant (architect/engineer) prepares the complete design drawings and contract documents, then the project is put up for competitive bids and the contract is awarded to the most competitive bidder (contractor). The owner hires a third party (construction manager) to oversee and coordinate the construction.

Construction projects are mainly capital investment projects. They are customized and non-repetitive in nature. Construction projects have become more complex and technical, and the relationships and the contractual grouping of those who are involved are also more complex and contractually varied. The products used in construction projects are expensive, complex, immovable, and long-lived. Generally, a construction project is composed of building materials (civil), electro-mechanical items, finishing items, and equipment. These are normally produced by other construction-related industries/manufacturers. These industries produce products according to their own quality management practices, complying with certain quality standards or against specific requirements for a particular project. The owner of the construction project or his representative has no direct control over these companies unless he/his representative/appointed contractor commits to buying their product for use in their facility. These organizations may have their own quality management program. In manufacturing or service industries, the quality management of all in-house-manufactured products is performed by the manufacturer's own team or under the control of the same organization, having jurisdiction over their manufacturing plants at different locations. Quality management of vendor-supplied items/products is carried out as stipulated in the purchasing contract, according to the quality control specifications of the buyer.

2.2 QUALITY DEFINITION FOR CONSTRUCTION PROJECTS

Quality has different meanings to different people. The definition of quality, relating to manufacturing, processes, and service industries, is as follows:

- Meeting the customer's needs.
- Customer satisfaction.
- Fitness for use.
- Conforming to requirements.
- Degree of excellence at an acceptable price.

The International Organization for Standardization (ISO) defines quality as "the totality of characteristics of an entity that bears on its ability to satisfy stated or implied needs."

However, the definition of quality for construction projects is different from that of manufacturing or services industries as the product is not repetitive, but a unique piece of work with specific requirements.

Quality in construction project is not only the quality of the products and equipment used in the construction of the facility/project – it is the total management approach to complete the facility. The quality of construction depends mainly upon the control of construction, which is the primary responsibility of the contractor.

Quality in manufacturing passes through a series of processes. Materials and labor are inputted through a series of processes from which a product is obtained. The output is monitored by inspection and testing at various stages of production. Any non-conforming product identified is repaired, reworked, or scrapped and proper steps are taken to eliminate problem causes. Statistical process control methods are used to reduce variability and increase the efficiency of the process. In construction projects, the scenario is not the same. If anything goes wrong, the non-conforming work is very difficult to rectify and remedial actions are sometimes not possible.

Quality management in construction projects is different from that in manufacturing projects. Quality in construction projects is not only the quality of products and equipment used in the construction, but it is also the total management approach to completing the facility according to the scope of works to the satisfaction of the customer/owner to be completed within the specified schedule and within agreed upon budget to meet the owner's defined purpose. The nature of the contracts between the parties plays a dominant role in the quality system required from the project and the responsibility for achieving them must therefore be specified in the project documents. The documents include plans, specifications, schedules, bill of quantities, and so on. Quality control in construction typically involves ensuring compliance with minimum standards of material and workmanship in order to ensure the performance of the facility according to the design. These minimum standards are contained in the specification documents. For the purpose of ensuring compliance, random samples and statistical methods are commonly used as the basis for accepting or rejecting work completed and batches of materials. Rejection of a batch is based on non-conformance or violation of the relevant design specifications.

Based on the above quality of construction projects can be defined as follows. Construction project quality is a fulfillment of the owner's needs as per the defined scope of works within the specified schedule and budget to satisfy the requirements of the owner/user requirements. The phenomenon of these three components is known as the "Construction Project Trilogy" and is illustrated in Figure 2.2

Thus, the quality of construction projects can evolve as follows:

1) A properly defined scope of work.
2) Owner, Project Manager, Design Team Leader, Consultant, and Constructor's Manager are responsible to implement the quality.
3) Continuous improvement can be achieved at different levels as follows:
    (a) Owner – specify the latest needs.
    (b) Designer – specification to include the latest high-quality materials, products, and equipment
    (c) Constructor – use the latest construction equipment to build the facility
4) Establishment of performance measures.
    (a) Owner:
        (I) To review and ensure that designer has prepared the contract documents which satisfy his needs

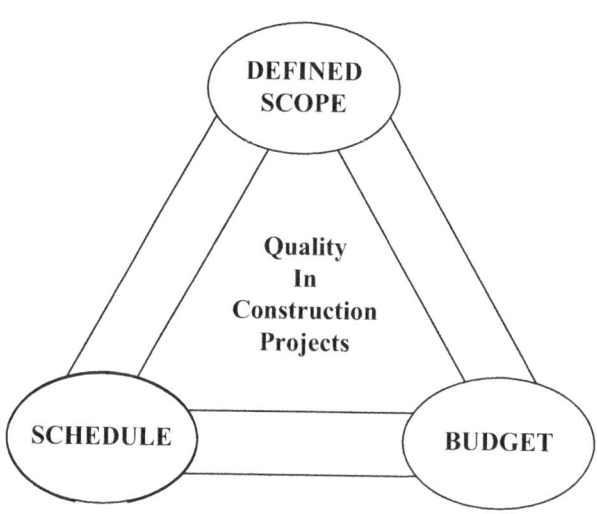

FIGURE 2.2 Construction project quality trilogy.

      (II) To check the progress of work to ensure compliance with the contract documents
- (b) Consultant:

      (I) As a consultant designer, to include the owner's requirements explicitly and clearly defined in the contact documents

      (II) As a supervision consultant, to supervise the contractor's work according to the contract documents and the specified standards
- (c) Contractor: To construct the facility as specified and use the materials, products, and equipment that satisfy the specified requirements

5) Team Approach: Every member of the project team should know the principles of Total Quality Management (TQM), knowing that TQM involves collaborative efforts, and everybody should participate in all the functional areas to improve the quality of project work. They should know that it is a collective effort by all the participants to achieve project quality
6) Training and Education: The consultant and contractor should have customized training plans for their management, engineers, supervisors, office staff, technicians, and laborers
7) Establish Leadership: Organizational leadership should be established to achieve the specified quality, and encourage and help the staff and laborers to understand the quality to be achieved for the project.

These definitions, when applied to construction projects, relate to the contract specifications or the owner/end-user requirements to be constructed in such a way that the construction of the facility is suitable for the owner's use or that it meets the owner's requirements. Quality in construction is achieved through complex interaction of many participants in the facility/project's development process.

The quality plan for construction projects is part of the overall project documentation, consisting of:

1. Well-defined specifications for all the materials, products, components, and equipment to be used to construct the facility.
2. Detailed construction drawings.
3. Detailed work procedure.
4. Details of the quality standards and codes to be complied.
5. Manpower and other resources to be used for the project.
6. Project completion schedule.
7. Cost of the project.

The participation of all three parties at different levels of the construction phases is required to develop the quality system and the application of quality tools and techniques. With the application of various quality principles, tools, and methods by all the participants at different stages of a construction project, rework can be reduced, resulting in savings in the project cost and making the project qualitative and economical. This will ensure the completion of the construction and make the project as qualitative, competitive, and economical as possible.

2.2.1 Quality Principles of Construction Projects

Construction projects are mainly capital investment projects. They are customized and non-repetitive in nature. Construction projects have become more complex and technical, and the relationships and the contractual grouping of those who are involved are also more complex and contractually varied. Quality in construction is achieved through the complex interaction of many participants in the facility's development process.

Construction projects comprise a cross-section of many different participants. These participants both influence and depend on each other, in addition to "other players" involved in the construction process. Traditional construction projects involve three main groups. These are as follows;

1. Owner
2. Designer (Consultant)
3. Contractor

Participation of all three parties at different levels of the construction phases is required to develop a quality system and the application of quality tools and techniques. In construction projects, the project owner engages supervision consultant (a project manager, construction manager as applicable). The PM/CM plays an important role in the development of construction project. Construction project quality has three main components. These are:

1. Scope
2. Schedule
3. Cost (Budget)

Overview of Quality in Construction Projects

In order to achieve a successful project based on owner/end-user satisfaction, project documents need to be formulated in such a way that the construction of the project is suitable for the use of the owner or end-user or to meet the owner's requirements. An ISO document has listed eight quality management principles (CLIPSCFM) on which the quality management system standards of the revised ISO 9000:2000 series are based. These are as follows:

Principle 1 – Customer focus
Principle 2 – Leadership
Principle 3 – Involvement of people
Principle 4 – Process approach
Principle 5 – Systems approach to management
Principle 6 – Continual improvement
Principle 7 – Factual approach to design making
Principle 8 – Mutual beneficial supplier relationship

Table 2.2 summarizes the quality principles that are applicable to construction projects.

TABLE 2.2
Quality Principles of Construction Projects

Principle	Construction Project Quality Principle
Principle 1 (Customer Focus)	1.1 Designer or Consultant is responsible for providing the owner's requirements explicitly and clearly, defining the standards of the end-products and their compliance in the contract documents.
	1.2 Engineering design should include the process, process equipment, engineering system requirements clearly and without any ambiguity for ease of operation
	1.3 The project and end-products should satisfy the owner's needs and requirements and be suitable for intended usage
Principle 2 (Leadership)	2.1 Owner, Designer, Consultant, and Contractor are fully responsible for the application of the quality management system to meet customer requirements and strive to exceed customer expectations by complying with the defined scope of work in the contract documents
	2.2 Every member of the project team should exert collaborative efforts in all the functional areas to improve the quality of the project
Principle 3 (Involvement/ Engagement of People)	3.1 Each member of the project team should participate and be fully involved according to their abilities in all the functional areas by adhering to a team approach and coordination to continuously improve the quality of the project
Principle 4 (Process Approach)	4.1 The contractor is employed to build the facility as stipulated in the contract documents, plan, and specifications as per the approved schedule and within the agreed-upon budget to meet the owner's objectives

(Continued)

TABLE 2.2 CONTINUED
Quality Principles of Construction Projects

Principle	Construction Project Quality Principle
	4.2 The contractor should study all the documents during the tendering/bidding stage and submit his proposal, taking into consideration all the requirements specified in the contract documents and identifying, understanding, and managing the interrelated processes as a system to achieve the specified product output
	4.3 The contractor is responsible for providing all the resources, manpower, materials, equipment, etc., to build the facility as per the specifications to produce the specified products
	4.4 The contractor checks executed/installed works to confirm that works have been performed/executed as specified, using specified/approved materials, approved shop drawings, installation methods and specified references, codes and standards to meet the intended use
Principle 5 (Systems Approach to Management)	5.1 The contractor prepares the contractor's quality control plan (CQCP) and follows the same to ensure meeting the performance standards specified in the contract documents.
	5.2 The method of payments (work progress, material, equipment, etc.) to be clearly defined in the contract documents. Rate analysis of the Bill of Quantities (BOQ) or the Bill of Materials (BOM) item to be agreed before signing of the contract
	5.3 The contract documents should include a clause to settle any dispute arising during construction stage
Principle 6 (Continual Improvement)	6.1 The contractor shall follow the submittal procedure specified in the contract documents for detailed design, procurements, check lists, inspection and testing procedures as per the communication matrix. Review the contents of transmittals and executed works prior to submission for approval
Principle 7 (Factual Approach to Evidence-Based Design Making)	7.1 The contractor shall follow an agreed-upon quality assurance and quality control plan. The consultant, PMC shall be responsible for overseeing compliance with the contract documents and specified standards and codes
	7.2 Contractor is responsible for constructing the facility to produce the products as specified and use the material, products, systems, equipment, and methods which satisfy the specified requirements (Factual Approach to Design Making)
Principle 8 (Mutual Beneficial Relationship) Relationship Management	8.1 The contractor/all team members should participate and put collective efforts into performing the works as per the agreed-upon construction program and handover of the project as per the contracted schedule to meet the owner's requirements
	8.2 All team members should focus on participative management and strong operational accountability at the individual contributory level to follow the principles of Total Quality Management

Source: Abdul Razzak Rumane(2021). Quality Management in Oil and Gas Projects, CRC Press, Florida,. Reprinted with permission of Taylor & Francis Group

2.3 QUALITY MANAGEMENT SYSTEM (QMS)

ISO 9000 quality management system standards are a tested framework for taking a systematic approach to managing the business process so that organizations turn out products or services conforming to customer satisfaction. The typical ISO quality management system is structured on four levels, usually portrayed as a pyramid. Figure 2.3 illustrates the QMS documentation pyramid.

At the top of the pyramid is the quality policy, which sets out what management requires its staff to do in order to ensure operation of the quality management system. Beneath the quality policy is the quality manual, which details the work to be done. Beneath the quality manual are work instructions, procedures, and records. The number of manuals containing work instructions or procedures is based on the size and complexity of the organization. The procedures discuss mainly the following;

- What is to be done?
- How is it done?
- How does one know that it has been done properly (for example, by inspecting, testing, or measuring)?
- What is to be done if there are problems (for example, failure)

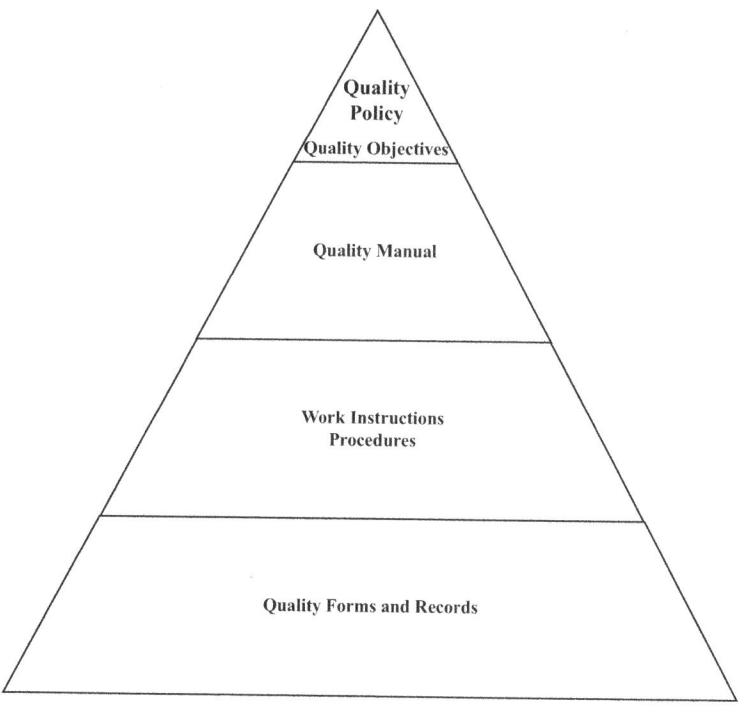

FIGURE 2.3 QMS documentation pyramid.

The bottom level of the pyramid contains forms and records that are used to capture the history of routine events and activities.

The ISO 9000 quality management system requires documentation that includes a quality manual and quality procedures as well as work instructions and quality records. All documentation (including quality records) must be controlled according to a document control procedure. The structure of the quality management system depends largely on the management structure in the organization.

ISO 9001:2015 identifies certain minimum requirements that all quality management systems must meet to ensure customer satisfaction. ISO 9001:2015 specifies requirements for a quality management system when an organization:

- Needs to demonstrate its ability to consistently provide products that meet customer and applicable regulatory requirements.
- Aims to enhance customer satisfaction through the effective application of the system, including processes for continual improvement of the system and the ensurance of conformity for customer and applicable regulatory requirements.

A quality system has to cover all the activities leading to turning out the final product or service. The quality system depends entirely on the scope of operation of the organization and particular circumstances such as the number of employees, the type of organization, and the physical size of the premises of the organization. The quality manual is the document that identifies and describes the quality management system, by:

1. Identifying the process (activities and necessary elements) needed for a quality management system.
2. Determining the sequence and interaction of these processes and how they fit together to accomplish quality goals.
3. Determining how these processes are effectively operated and controlled.
4. Measuring, monitoring, and analyzing these processes and implementing actions necessary to correct the process and achieve continual improvement.
5. Ensuring that all information is available to support the operation and monitoring of the process.
6. Displaying the most options, thus helping to make the right management system.

ISO 9001:2015 requirements fall into the following sections (clauses):

1. Scope
2. Normative references
3. Terms and definitions
4. Context of Organization/Quality Management System
5. Leadership
6. Planning for Quality Management System
7. Support

Overview of Quality in Construction Projects

8. Operation
9. Performance evaluation
10. Improvement

ISO 9000:2015 has ten clauses, compared with eight clauses in ISO 9000:2008. The most important is that the revised QMS focuses on risk-based thinking which has to be considered from the beginning and throughout the life cycle of the project.

In the construction industry, a contractor may be working at any time for a number of projects of varied nature. Each of these projects has its own contract documents to implement project quality which require the contractor to submit a contractor's quality control plan to ensure that specific requirements of the project are considered to meet the client's requirements. Therefore, when preparing a quality management system at a corporate level, the organization has to take into account tailor-made requirements for the projects; the manual should be prepared accordingly.

The QMS documentation pyramid includes quality policy and objectives, a quality manual, and quality procedures, as well as work instructions and quality records. The manual is developed taking into consideration the following:

1. Eight principles (CLIPSCFM) of QMS, as defined by ISO Technical Committee TC 176
2. All the related and applicable documents produced, taking into consideration ten sections/clauses listed under ISO 9001:2015 to ensure that the manual is in compliance with ISO 9001:2015. Figure 2.4 illustrates the relationship between QMS principles and ISO sections.

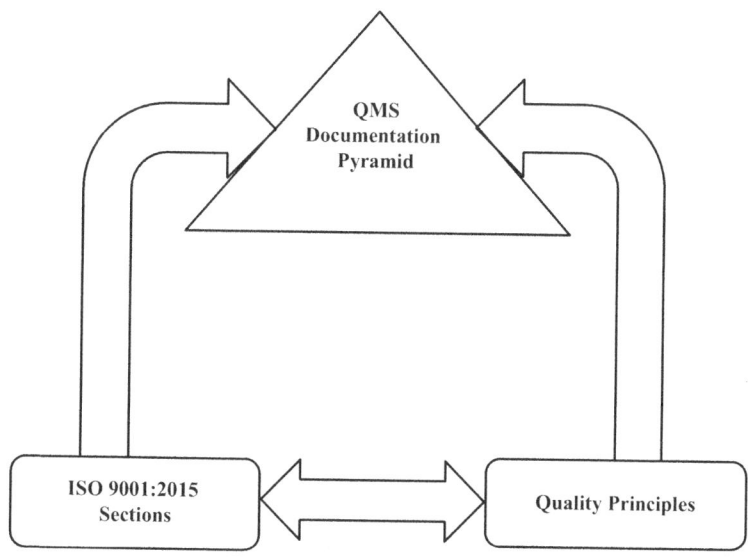

FIGURE 2.4 QMS documentation model.

2.3.1 Integrated Quality Management System

The integrated quality management system (IQMS) is the integration and proper coordination of functional elements of quality to achieve efficiency and effectiveness in implementation and maintaining an organization's quality management system to meet customer requirements and satisfaction. IQMS consists of any element or activity which has an effect on quality. Customer satisfaction is the goal of quality objectives.

During the past three decades, many programs have been implemented for organizational improvements. In the 1980s, programs such as statistical process control, various quality tools, and total quality management were implemented. In the 1990s, the most popular quality management framework, ISO 9000, came into being which resulted in improved productivity, cost reduction, improved duration, improved quality, and increased customer satisfaction.

With globalization and competition, it became necessary for organizations to improve continuously to achieve the highest performance and greatest competitive advantage.

In the 1980s, the major challenge facing most organizations was to improve quality. In the 1990s, the major challenge was to improve faster by restructuring and re-engineering all operations.

In today's competitive global environment, organizations are facing many challenges due to an increase in customer demand for higher performance requirements at a competitive cost. They are finding that their survival in the competitive market is increasingly in doubt. To achieve a competitive advantage, effective quality improvement is critical.

Processes and systems are essential for the performance and expansion of any organization. ISO 9000 is an excellent tool to develop a strong foundation for good processes and systems. The ISO 9000 quality management system is accepted worldwide and ISO 9000 certification has global recognition.

The Integrated Quality Management system is developed by merging recommendations and specifications from ISO 9000 (Quality Management System), ISO 14000 (Environmental Management System), and OHSAS 18000 (Occupational Health and Safety Management), together with other contract documents. If an organization has a certified Quality Management System (ISO 9000), it can build an IQMS system by adding the environmental, health, safety, and other requirements of management system standards.

The benefits of implementing an IQMS are:

- Reduced duplication and therefore cost.
- Improved resource allocation.
- Standardized process.
- Elimination of conflicting responsibilities and relationships.
- Greater consistency.
- Improved communication.
- Reduced risk and increased profitability.
- Facilitation of training development.
- Simplify document maintenance.
- Reduced record keeping.
- Ease of managing legal and other requirements.

Overview of Quality in Construction Projects

Construction projects are unique and non-repetitive in nature and have their own quality requirements which can be developed by integration of project specifications and the organization's quality management system. Normally, quality management system manuals consist of procedures to develop a project quality control plan, taking into consideration contract specifications. This plan is known as the Contractor's Quality Control Plan (CQCP). Certain projects require value engineering studies to be undertaken during the construction phase. The contractor is required to include the same while developing the CQCP. This plan is termed the Integrated Quality Management System (IQMS) for construction projects. The contractor has to implement a quality system to ensure that the construction is carried out in accordance with the specification details and the approved COQP. Figure 2.5 illustrates a flowchart for the development of IQMS for construction projects.

2.4 QUALITY MANAGEMENT IN CONSTRUCTION PROJECTS

Quality management is an organization-wide approach to understanding customer needs and delivering the solutions to satisfy the customer. Quality management involves management and implementation of a quality system to achieve customer satisfaction at the lowest overall cost to the organization while continuing to improve the process. A quality system is a framework for quality management. It embraces the organization structure, policies, procedures, and processes needed to implement a quality management system.

Quality management in construction projects is different to that in manufacturing.

Quality in construction projects is not only the quality of products and equipment used in the construction but is also the total management approach to completing the facility according to the scope of works to the satisfaction of the customer/owner within the budget and within the specified schedule to meet the owner's defined purpose. Quality management in construction addresses the management of both the project and the project's product and all its components. It also involves incorporation of changes or improvements, if needed. Construction project quality is fulfillment of owner's needs as per the defined scope of works, as per the specified schedule, and within the budget to satisfy the requirements of the owner/user.

The quality management system in construction projects mainly consists of:

- Quality management (planning to plan quality).
- Quality assurance (to perform quality assurance).
- Quality control (to perform quality control).

2.4.1 Quality Plan

The quality plan for construction projects is part of the overall project documentation, addressing and describing the procedures to manage construction quality and project deliverables. The quality plan consists mainly of:

- Stakeholders' quality requirements.
- Well-defined specification for all the materials, products, components, and equipment to be used to construct the facility.

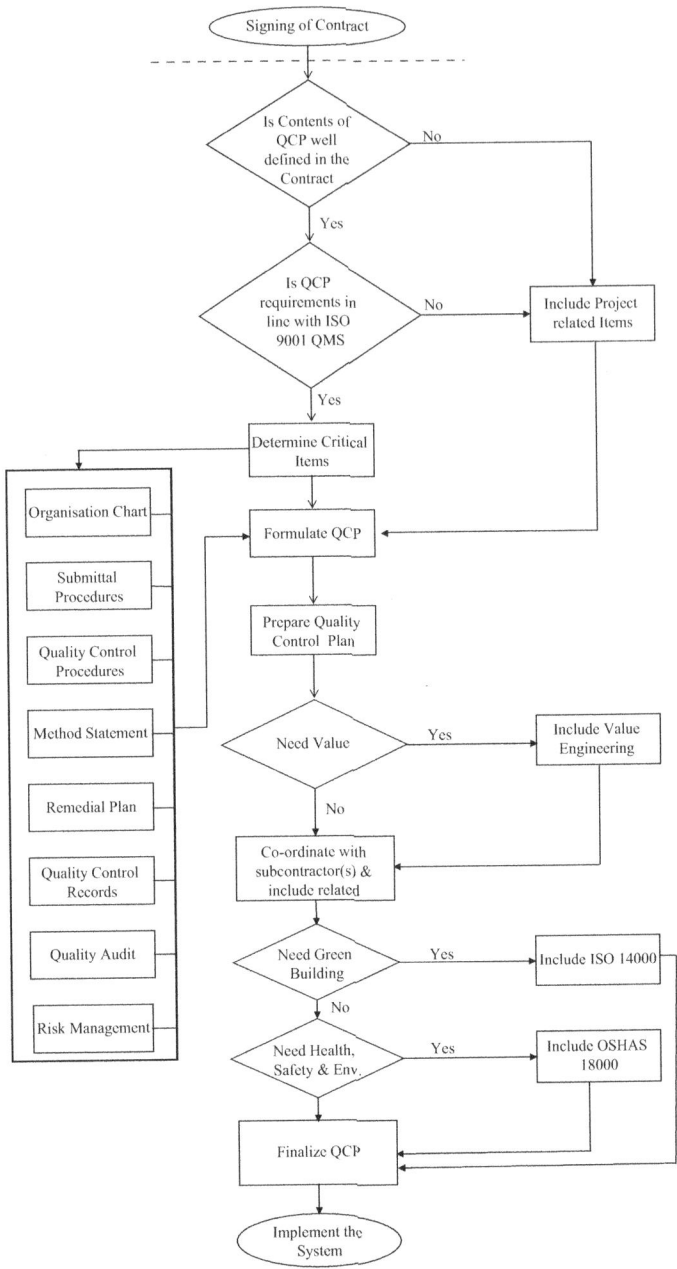

FIGURE 2.5 Flowchart diagram for development of IQMS.

Overview of Quality in Construction Projects

- Detailed construction drawings.
- Detailed work procedure.
- Details of the quality standards and codes to be complied.
- Regulatory requirements.
- Manpower and other resources to be used for the project.
- Project completion schedule.
- Cost of the project.

2.4.2 Quality Assurance

Quality assurance in construction projects covers all the activities performed by the design team, the contractor, and the quality controller/auditor (supervision staff) to meet the owner's objectives as specified and to ensure and guarantee that the project/facility is fully functional to the satisfaction of the owner/end-user. Auditing is part of the quality assurance function.

2.4.3 Quality Control

Quality control in construction projects is performed at every stage through the use of various control charts, diagrams, check lists, etc., and can be defined as:

- Checking of executed/installed works to confirm that the works have been performed/executed as specified, using specified/approved materials, installation methods, and specified references, codes, standards to meet the intended use.
- Planning, monitoring, and controlling project schedule.
- Controlling budget.

Construction projects involve the owner, designer (consultant), and contractor. In order to achieve the project objectives, the designer, as well as the contractor, has to develop project quality management plans. The designer's quality management plan should be based on the owner's project objectives whereas the contractor's plan should take into consideration requirements of the contract documents.

The following sections discuss in brief the quality management processes (quality planning, quality assurance, and quality control) and the activities to be performed during the following main stages of construction projects:

1. Study stage.
2. Design stage (conceptual design, preliminary design, design development, construction documents).
3. Bidding and tendering stage
4. Construction stage (construction, testing, commissioning, and handover)

2.4.4 QUALITY DURING THE STUDY STAGE

2.4.4.1 Quality Plan
Table 2.3 lists the major quality planning activities to be performed during the study stage.

2.4.4.2 Quality Assurance
Table 2.4 lists the major quality assurance activities to be performed during the study stage.

TABLE 2.3
Major Quality Planning Activities During the Study Stage

Serial Number	Activities
1	Needs identification
2	Perform feasibility study
3	Project goals and objectives
4	Identify potential risks associated with the project
5	Develop the design basis
6	Establish project delivery and contracting system
7	Estimate the resources
8	Estimate the timescale for completion of the project
9	Estimate the cost of the project
10	Establish the statutory requirements
11	Select the suitable project delivery system
12	Select the contracting/pricing most appropriate for the benefit of the owner.

TABLE 2.4
Major Quality Assurance Activities During the Study Stage

Serial Number	Activities
1	Needs assessment
2	Needs analysis
3	Feasibility assessment
4	Develop the project initiation documents on the SMART concept
5	Conduct the cost/benefit analysis
6	Outline the key requirements that will drive design trade-off and identify which requirements are critical
7	Confirm the availability of resources
8	Identify the issues, sustainability, impacts, and potential approvals
9	Identify the project delivery system

Overview of Quality in Construction Projects

2.4.4.3 Quality Control

Table 2.5 lists the major quality control activities to be performed during the study stage.

2.4.5 QUALITY DURING THE DESIGN STAGE

2.4.5.1 Quality Plan

Table 2.6 lists the major quality planning activities to be performed during the design stage

2.4.5.2 Quality Assurance

Table 2.7 lists the major quality assurance activities to be performed during the design stage

2.4.5.3 Quality Control

Table 2.8 lists the major quality control activities to be performed during the design stage

2.4.6 QUALITY DURING THE BIDDING AND TENDERING STAGE

The following are the main quality management-related activities to be considered:

- Tendering procedure/process.
- Management of submissions/bids.
- Management of project budget.

TABLE 2.5
Major Quality Control Activities During the Study Stage

Serial Number	Activities
1	Needs statement
2	Feasibility statement
3	Establish the project schedule
4	Determine the project budget
5	Establish the authority approval process
6	Establish how the participants, owner, designer (A/E), and contractor will be involved in constructing the project/facility
7	Identify the type of contracting system
8	Identify the project team
9	Develop the project charter

TABLE 2.6
Major Quality Planning Activities During the Design Stage

Serial Number	Activities
1	**Conceptual Design**
1.1	The owner's requirements
1.2	Quality standards and codes to be complied with
1.3	Regulatory requirements
1.4	Design review procedure
1.5	Drawings review procedure
1.6	Document review procedure
1.7	Quality management during all the phases of the project life cycle
2	**Schematic/Preliminary Design**
2.1	Establish the owner's requirements
2.2	Determine the number of drawings to be produced
2.3	Establish the scope of the work
2.4	Identify the quality standards and codes to be complied with
2.5	Establish the design criteria
2.6	Identify the regulatory requirements
2.7	Identify the requirements listed under TOR
2.8	Establish the quality organization with the responsibility matrix
2.9	Develop the design (drawings and documents) review procedure
2.10	Establish the submittal plan
2.11	Establish the design review procedure
3	**Design Development/Detail Design**
3.1	Review the comments on the schematic design
3.2	Determine the number of drawings to be produced
3.3	Establish the scope of work for the preparation of the detail design
3.4	Identify the requirements listed under TOR
3.5	Identify the quality standards and codes to be complied with
3.6	Establish the design criteria
3.7	Identify the regulatory requirements
3.8	Identify the environmental requirements
3.9	Establish the quality organization with the responsibility matrix
3.10	Develop the design (drawings and documents) review procedure
3.11	Establish the submittal plan
3.12	Establish the design review procedure
4	**Construction Documents**
4.1	Review the comments on the design development package
4.2	Determine the number of drawings to be produced
4.3	Establish the scope of the work for preparation of the construction documents
4.4	Identify the requirements listed under TOR
4.5	Identify the quality standards and codes to be complied with
4.6	Identify the regulatory requirements
4.7	Identify the environmental requirements
4.8	Establish the quality organization with the responsibility matrix
4.9	Develop the review procedure for the working drawings produced
4.10	Develop the review procedure for the specifications and contract documents
4.11	Establish the submittal plan for the construction documents

TABLE 2.7
Major Quality Assurance Activities During the Design Stage

Serial Number	Activities
1	**Conceptual Design**
1.1	Prepare the concept design
1.2	Reports
1.3	Model
1.4	Project schedule
1.5	Project cost
2	**Schematic/Preliminary Design**
2.1	Collect the data
2.2	Investigate the site conditions
2.3	Prepare the preliminary drawings
2.4	Prepare the outline specifications
2.5	Ensure functional and technical compatibility
2.6	Coordinate with all disciplines
2.7	Select the material to meet the owner's objectives
3	**Design Development/Detail Design**
3.1	Collect the data
3.2	Investigate the site conditions
3.3	Prepare the design drawings
3.4	Prepare the detailed specifications
3.5	Prepare the contract documents
3.6	Prepare the Bill of Quantities
3.7	Ensure the functional and technical compatibility
3.8	Ensure the design is constructible
3.9	Ensure the operational objectives are met
3.10	Ensure the drawings are fully coordinate with all disciplines
3.11	Ensure the design is cost-effective
3.12	Ensure the selected/recommended material meets the owner's objectives
3.13	Ensure that the design fully meets the owner's objectives/goals
4	**Construction Documents**
4.1	Prepare the working drawings
4.2	Prepare the detailed specifications
4.3	Prepare the contract documents
4.4	Prepare the Bill of Quantities and Schedule of Rates
4.5	Ensure the functional and technical compatibility
4.6	Ensure the design is constructible
4.7	Ensure the operational objectives are met
4.8	Ensure the drawings are fully coordinate with all disciplines
4.9	Ensure the design is cost-effective
4.10	Prepare the working drawings
4.11	Prepare the detailed specifications
4.12	Prepare the contract documents

TABLE 2.8
Major Quality Control Activities During the Design Stage

Serial Number	Activities
1	**Conceptual Design**
1.1	Conformity to the owner's requirements
1.2	Conformity to the requirements listed under TOR
1.3	Regulatory compliance
2	**Schematic/Preliminary Design**
2.1	Check the design drawings
2.2	Check the specifications/contract documents
2.3	Check for regulatory compliance
2.4	Check the preliminary schedule
2.5	Check the cost of the project (preliminary cost)
3	**Design Development/Detail Design**
3.1	Check the quality of the design drawings
3.2	Check the accuracy and correctness of the design
3.3	Verify the Bill of Quantities for correctness according to the design drawings and specifications
3.4	Check the specifications
3.5	Check the contract documents
3.6	Check for regulatory compliance
3.7	Check the project schedule
3.8	Check the project cost
3.9	Check the interdisciplinary requirements
3.10	Check the required number of drawings prepared
4	**Construction Documents**
4.1	Check the quality of the design drawings
4.2	Check the accuracy and correctness of the design
4.3	Verify the Bill of Quantities for correctness as per the design drawings and specifications
4.4	Check the specifications
4.5	Check the contract documents
4.6	Check for regulatory compliance
4.7	Check the project schedule
4.8	Check the project cost
4.9	Check the interdisciplinary requirements
4.10	Check the required number of drawings prepared
4.11	Check the quality of the design drawings
4.12	Check the accuracy and correctness of –he design
4.13	Verify the Bill of Quantities for correctness as per the design drawings and specifications

2.4.7 Quality During the Construction Stage

2.4.7.1 Quality Plan

Develop the Contractor's Quality Control Plan (CQCP)
Table 2.9 lists the contents of the Contractor's Quality Control Plan

TABLE 2.9
Contents of the Contractor's Quality Control Plan

Serial Number	Description
1.0	Introduction
2.0	Description of project
3.0	Quality control (QC) organization
4.0	Qualification of QC staff
5.0	Responsibilities of QC personnel
6.0	Procedure for submittals
6.1	Submissions of subcontractor(s)
6.2	Submissions of shop drawings
6.3	Submissions of materials
6.4	Modification request
6.5	Construction program
7.0	Quality control procedure
7.1	Procurement
7.2	Inspection of site activities (checklists)
7.3	Inspection and testing procedure for systems
7.4	Off-site manufacturing, inspection, and testing
7.5	Procedure for laboratory testing of material
7.6	Inspection of material received at site
7.7	Protection of works
7.8	Material storage and handling
8.0	Method statement for various installation activities
9.0	Project-specific procedures
10.0	Risk management
11.0	Quality control records
12.0	Company's quality manual and procedures
13.0	Periodic testing
14.0	Quality updating program
15.0	Quality auditing program
16.0	Testing, commissioning, and handover
17.0	Health, safety, and the environment (HSE)

2.4.7.2 Quality Assurance

The contractor's quality assurance activities mainly consist of the following:

1. Selecting the materials, systems fully compliant with contract specifications, and installing the approved material, systems only.
2. Preparing the shop drawings detailing all the requirements included in the working drawings and installing/executing the works as per the approved shop drawings.
3. Installing the works, materials, and systems as per the specified method statement and as per the recommendations from the manufacturer of the products.

2.4.7.3 Quality Control

The construction project quality control process is a part of the contract documents which provide details about specific quality practices, resources, and activities relevant to the project. On a construction site, inspection and testing is carried out at three stages during the construction period to ensure quality compliance.

1. During the construction process. This is carried out with the checklist request submitted by the contractor for testing of ongoing works before proceeding to the next step.
2. Receipt of material, equipment or services. This is performed by a Material Inspection Request submitted by the contractor to the consultant upon receipt of the material.
3. Before final delivery or commissioning and handover.

3 Quality Tools for Sustainability

3.1 INTRODUCTION

The quality tools are the charts, check sheets, diagrams, graphs, techniques, and methods which are used to create an idea, develop planning, analyze the cause, analyze the process, foster evaluation, and create a wide variety of situations for continuous quality improvement. Applications of tools enhance the chances of success and help maintain consistency and accuracy, increase efficiency, and process improvement.

3.2 CATEGORIZATION OF TOOLS

There are several types of tools, techniques, and methods in practice, which are used as quality improvement tools and have a variety of applications in the manufacturing and process industry. However, not all of these tools are used in construction projects due to the nature of these projects, which are customized and non-repetitive. Some of these quality management tools that can be used to achieve sustainability in projects are listed under the following broader categories:

1. Classic Quality tools
2. Management and planning tools
3. Process analysis tools
4. Process improvement tools
5. Innovation and creative tools
6. Lean tools
7. Cost of quality
8. Quality function deployment
9. Six Sigma
10. Triz

Brief descriptions of each of these tools are given below.

3.2.1 CLASSIC QUALITY TOOLS

Classic quality tools have a long history. These tools are listed in Table 3.1.

TABLE 3.1
Classic Quality Tools

Sr. No.	Name of Quality Tool	Usage
Tool 1	Cause-and-Effect diagram	To identify possible cause-and-effect relationships in processes.
Tool 2	Check sheet	To provide a record of quality. How often does it occur?
Tool 3	Control Chart	A device in Statistical Process Control to determine whether or not the process is stable
Tool 4	Flowchart	Used for graphical representation of a process in sequential order.
Tool 5	Histogram	Graphs used to display frequency of various ranges of values of a quantity.
Tool 6	Pareto Chart	Used to identify the most significant cause or problem.
Tool 7	Pie Chart	Used to show classes or a group of data in proportion to the whole dataset.
Tool 8	Run Chart	Used to show measurement against time in a graphical manner with a reference line to show the average of the data.
Tool 9	Scatter diagram	To determine whether there is a correlation between two factors.
Tool 10	Stratification	Used to show the pattern of data collected from different sources.

Source: Abdul Razzak Rumane (2013). Quality Tools for Managing Construction Projects. Reprinted with permission of Taylor & Francis Group.

All of these tools have been in use since World War II. Some of these tools date prior to 1920. The approach includes both quantitative and qualitative aspects, which, taken together, focus on company-wide quality.

A brief definition of these quality tools is as follows: (values shown in the figures are indicative only).

3.2.1.1 Cause-and-Effect Diagram

The Cause and Effect Diagram is also called the Ishikawa Diagram, after its developer Kaoru Ishikawa, or the Fishbone Diagram. It is used to identify possible causes and effects in the process. It is used to explore all the potential or real causes that result in a single output. The causes are organized and displayed in a graphical manner to reflect their level of importance or details. It is a graphical display of multiple causes with a particular effect. The causes are organized and arranged mainly into four categories:

1. Machine
2. Manpower
3. Material
4. Method

Quality Tools for Sustainability

The effect or problem being investigated is shown at the end of the horizontal arrow. Potential causes are shown as labeled arrows entering the main cause arrow. Each arrow may have a number of other arrows entering as the principal cause or factor are reduced to their sub-causes. Figure 3.1 illustrates an example of a cause-and-effect diagram for rejection of executed (installed) false ceiling works by the supervision engineer for not complying with contract specifications.

3.2.1.2 Check Sheet

A check sheet is a structured list, prepared from the collected data, to indicate how often each item occurs. It is an organized way of collecting and structuring data. The purpose of a check sheet is to collect the facts in the most efficient manner. Data are collected and ordered by adding tally or check marks against predetermined categories of items or measurements. Figure 3.2 illustrates the check sheet for check list approval record for wire bundles.

3.2.1.3 Control Chart

A control chart is the fundamental tool of statistical process control. The control chart is a graph used to analyze how a process behaves over time and to show whether the process is stable or is being affected by special causes of variation, creating an

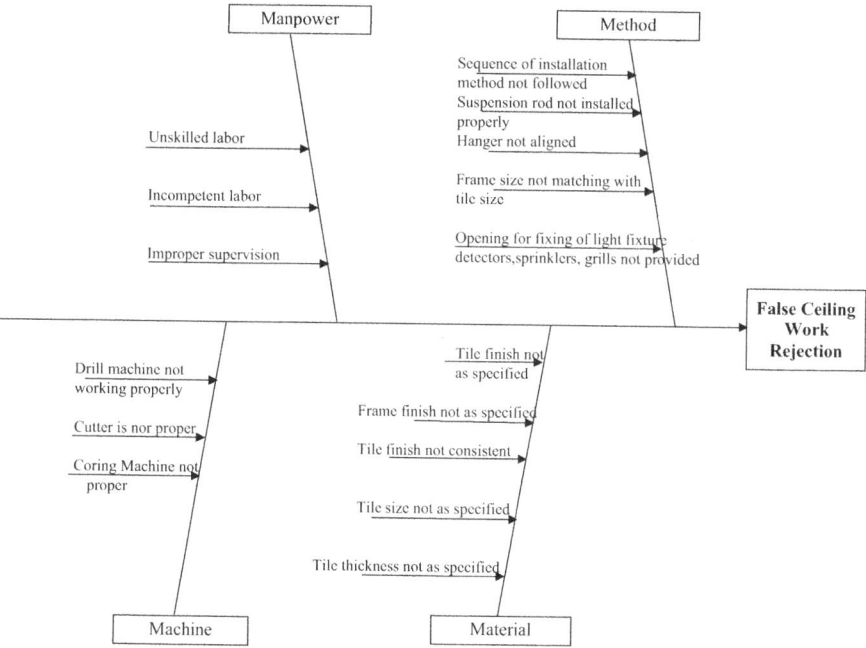

FIGURE 3.1 Cause and effect diagram for false ceiling rejection. Source: Abdul Razzak Rumane. (2013). *Quality Tools for Managing Construction Projects*, CRC Press, Florida. Reprinted with permission of Taylor & Francis Group.

out-of-control condition. It is used to determine whether the process is stable or varies between predictable limits. It can be employed to distinguish between the existence of a stable pattern of variation and the occurrence of an unstable pattern. Control charts make it easy to see both special and common cause variation in a process. There are many types of control charts. Each is designed for a specific kind of process or data. Figure 3.3 illustrates a control chart for the distribution of Air Handling Unit (AHU) air.

Approval Record for Wire Bundles

	Approved	Not-Approved	Total	% Not Approved
1.5 mm² Wire	/⊬/⊬///	///	50	4
2.5 mm² Wire	/⊬/⊬/⊬/ /⊬///	///	85	6
4.0 mm² Wire	/⊬/⊬/⊬/ /⊬///	///	25	3
6.0 mm² Wire	/⊬///	///	15	2
10.0 mm² Wire	/⊬///	///	10	2

FIGURE 3.2 Check sheet. Source: Abdul Razzak Rumane. (2013). *Quality Tools for Managing Construction Projects*. CRC Press, Florida. Reprinted with permission of Taylor & Francis Group.

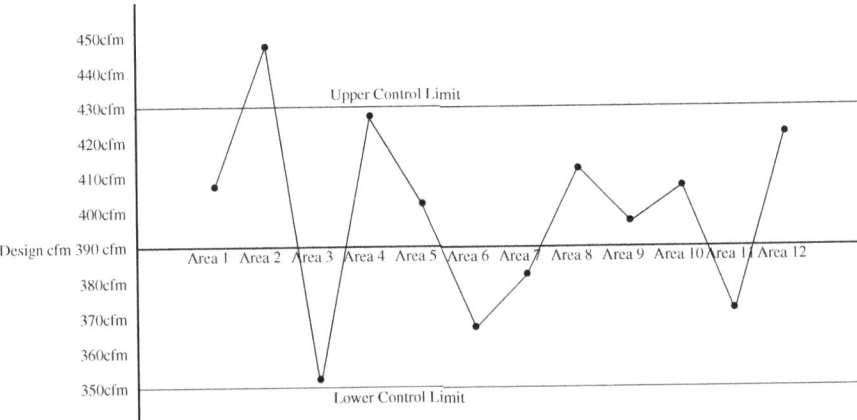

FIGURE 3.3 Control chart for air handling unit air distribution (cfm). Source: Abdul Razzak Rumane. (2013). *Quality Tools for Managing Construction Projects*, CRC Press, Florida. Reprinted with permission of Taylor & Francis Group.

Quality Tools for Sustainability

3.2.1.4 Flowchart

A flowchart is a pictorial tool that is used for the representation of a process in sequential order. Flow chart uses graphic symbols to depict the nature and flow of the steps in a process. It helps to show whether the steps of a process are logical, uncovering the problems, or miscommunications, defining the boundaries of the process and developing a common base of knowledge about the process. The flow of steps is indicated with arrows connecting the symbols. Flowcharts can be applied at all stages of the project life cycle. Figure 3.4 illustrates a flowchart for contractor's staff approval in construction projects.

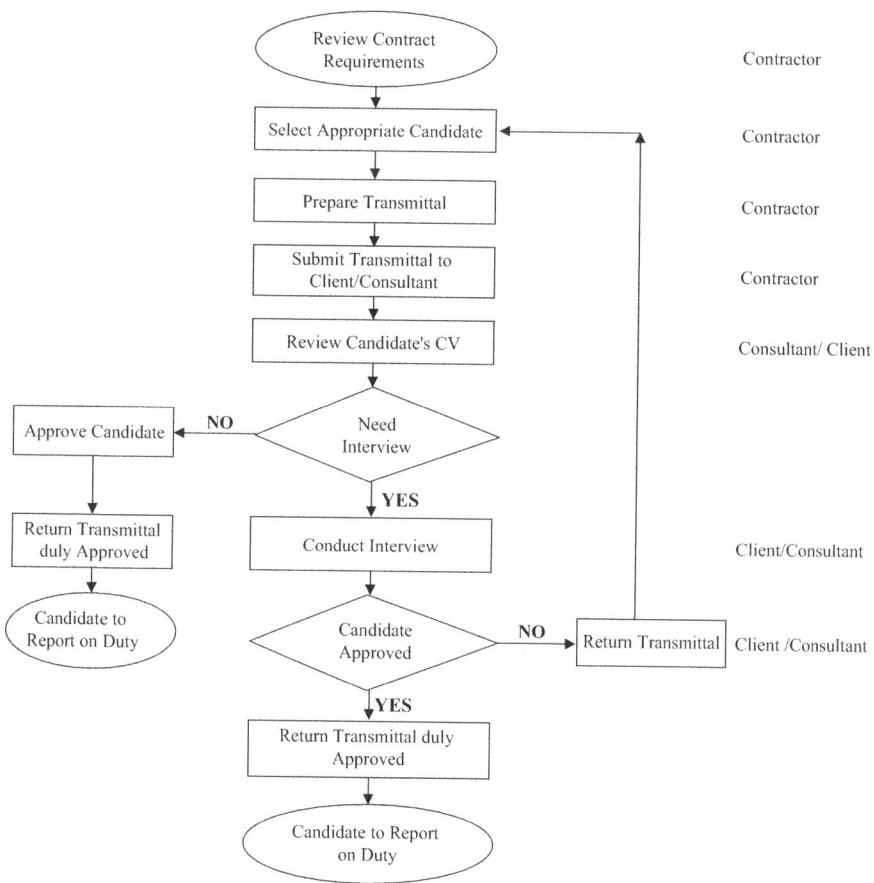

FIGURE 3.4 Flowchart for contractor's staff approval. Source: Abdul Razzak Rumane. (2013). *Quality Tools for Managing Construction Projects*, CRC Press, Florida. Reprinted with permission of Taylor & Francis Group.

3.2.1.5 Histogram

A histogram is a pictorial representation of the frequency distribution of the data. It is created by grouping the data points into cells and displaying how frequently different values occur in the dataset. Figure 3.5 illustrates a histogram for employee reporting time.

3.2.1.6 Pareto Chart

The Pareto chart is named after Vilfredo Pareto, a nineteenth-century Italian economist, who postulated that a large share (80%) of wealth is owned by a small percentage (20%) of the population. A Pareto chart is a bar chart having a series of bars whose heights reflect the frequency of occurrence. Pareto charts are used to display the Pareto Principle in action, arranging data so that the few vital factors that are causing most of the problems reveal themselves. The bars are arranged in descending order of height from left to right. Pareto charts are used to identify those factors which have the greatest cumulative effect on the system, and thus less-significant factors can be screened out from the process. Pareto charts can be used at various stages in a quality improvement program to determine which step to take next. Figure 3.6 illustrates a Pareto chart for division of the cost of a construction project.

3.2.1.7 Pie Chart

A pie chart is a circle divided into wedges to depict proportions of data or information in order to understand how they make up the whole. The entire pie chart represents all of the data, while each slice or wedge represents a different class or group within the whole. The sum of the portions of the entire circle or pie add up to 100%. Figure 3.7 illustrates the contents of a contractor's site staff at a construction project site.

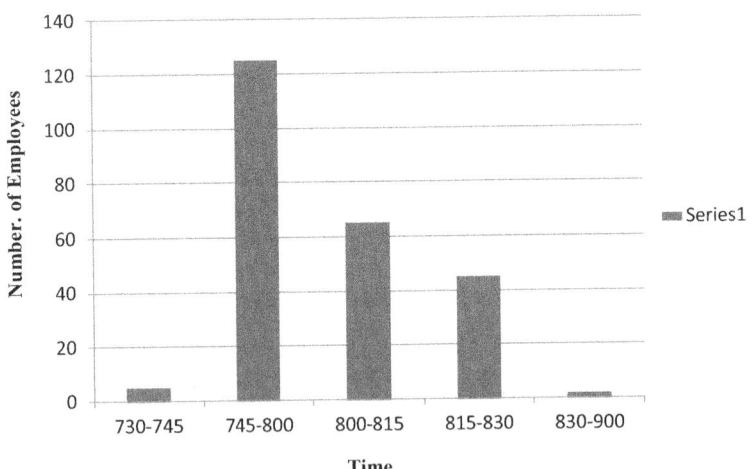

FIGURE 3.5 Histogram of employee reporting. Source: Abdul Razzak Rumane. (2013). *Quality Tools for Managing Construction Projects*, CRC Press, Florida. Reprinted with permission of Taylor & Francis Group.

Quality Tools for Sustainability

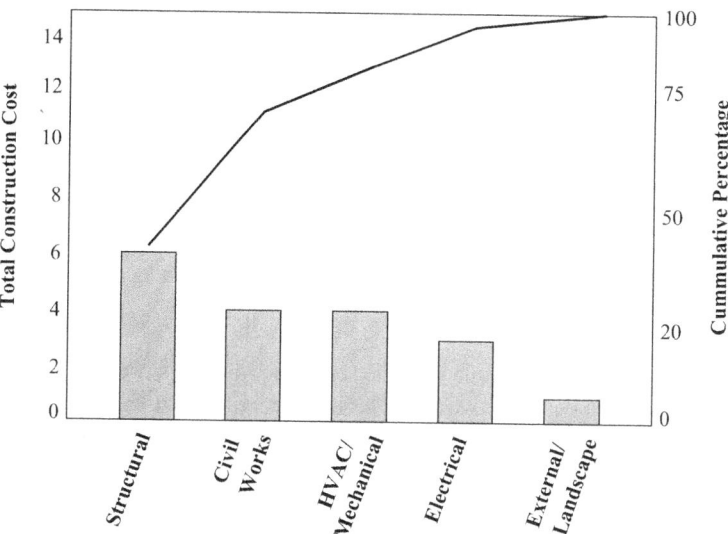

FIGURE 3.6 Pareto analysis for construction cost. Source: Abdul Razzak Rumane. (2013). *Quality Tools for Managing Construction Projects*, CRC Press, Florida. Reprinted with permission of Taylor & Francis Group.

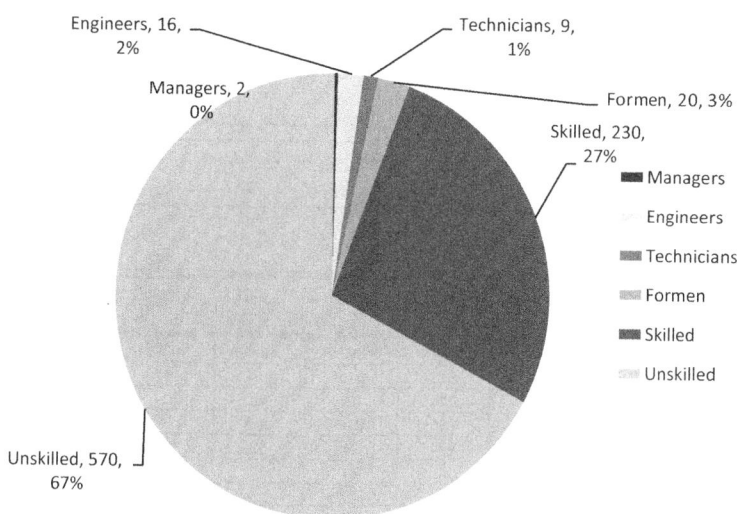

FIGURE 3.7 Pie chart for site staff. Source: Abdul Razzak Rumane. (2013). *Quality Tools for Managing Construction Projects*, CRC Press, Florida. Reprinted with permission of Taylor & Francis Group.

3.2.1.8 Run Chart

A run chart is a graph plotted by showing measurement (data) against time. Run charts are used to know the trends or changes in a process variation above the average over time and also to determine whether the pattern can be attributed to common causes of variation or if special causes of variation are present. A run chart is also used to monitor process performance. Run charts can be used to track improvements that have been put in place, checking to determine their success. Figure 3.8 illustrates a run chart for weekly manpower for different trades of a project. It is similar to a control chart but does not show control limits

3.2.1.9 Scatter Diagram

A scatter diagram is a plot of one variable versus another. It is used to investigate the possible relationship between two variables that both relate to the same event. It helps to show how one variable changes with respect to another. It can be used to identify the potential root cause of problems and to evaluate cause-and-effect relationships. Figure 3.9 illustrates a scatter diagram for beam quantity of various length.

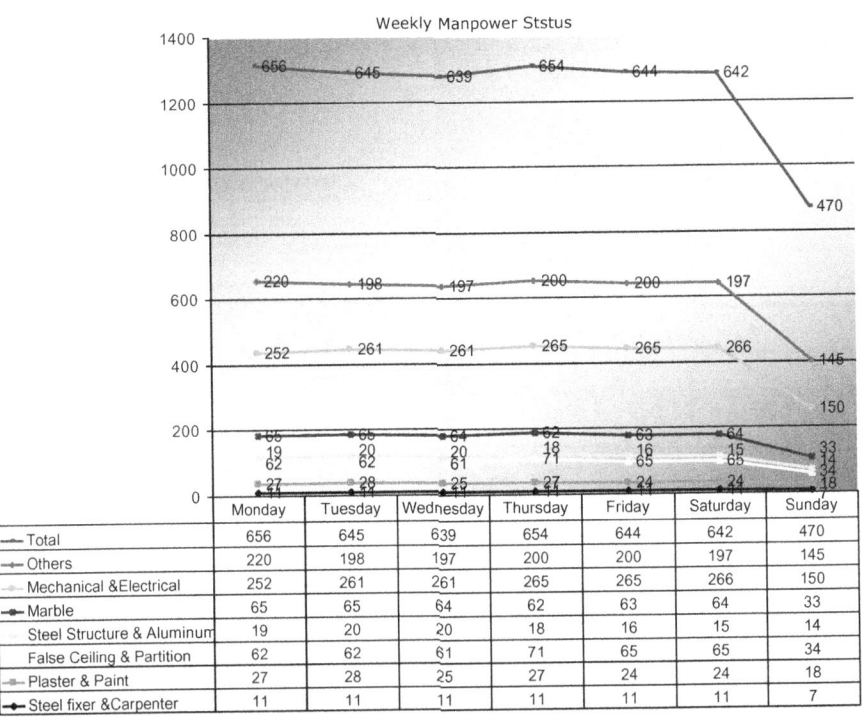

FIGURE 3.8 Run chart for manpower. Source: Abdul Razzak Rumane. (2013). *Quality Tools for Managing Construction Projects*, CRC Press, Florida. Reprinted with permission of Taylor & Francis Group.

Quality Tools for Sustainability

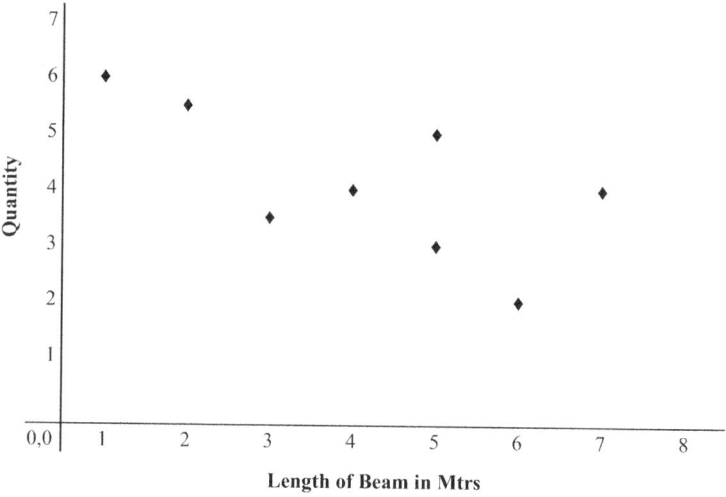

FIGURE 3.9 Scatter diagram. Source: Abdul Razzak Rumane. (2013). *Quality Tools for Managing Construction Projects*, CRC Press, Florida. Reprinted with permission of Taylor & Francis Group.

3.2.1.10 Stratification:

Stratification is a graphical representation of data collected from different samples. Figure 3.10 shows a stratification diagram for cable drums.

3.2.2 MANAGEMENT AND PLANNING TOOLS:

Seven management tools are popular in addition to quality classic tools. These tools are listed in Table 3.2.

These tools are focused on managing and planning quality improvement activities. Brief definitions of these tools is as follows: (values shown in the figures are indicative only)

3.2.2.1 Activity Network Diagram

The activity network diagram (AND) is a graphical representation chart showing the interrelationship among the activities (tasks) associated with a project. An activity network diagram was developed by the US Department of Defense. It was first used as a management tool for military projects. It was adapted as an educational tool for business managers. In activity network diagrams, each activity is represented by one and only one arrow in the network, and is associated with an estimated time to perform the activity. The activity network diagram analyzes the sequences of tasks necessary to complete the project. The direction of the arrow specifies the order in which the events must occur. The event represents a point in time that indicates the

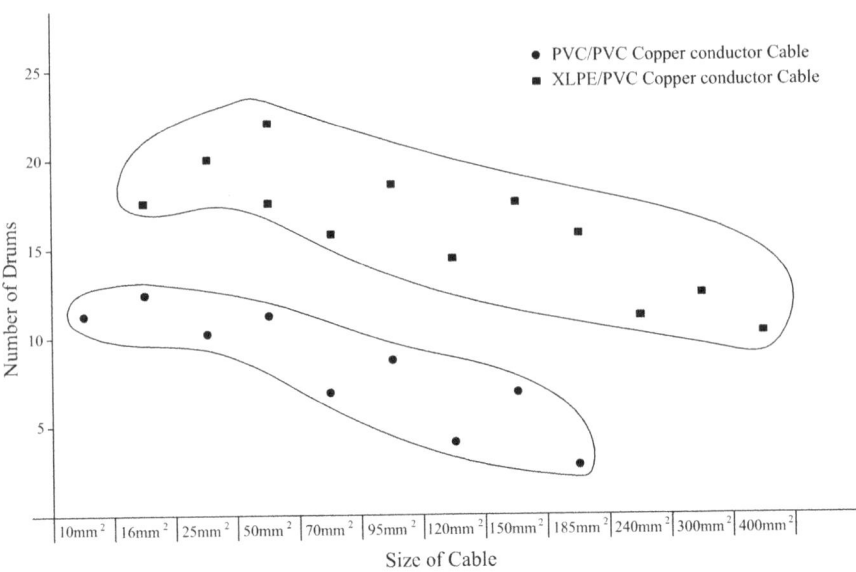

FIGURE 3.10 Stratification chart. Source: Abdul Razzak Rumane. (2013). *Quality Tools for Managing Construction Projects*, CRC Press, Florida. Reprinted with permission of Taylor & Francis Group.

TABLE 3.2
Management and Planning Tools

Sr. No.	Name of Quality Tool	Usage
Tool 1	Activity Network diagram (Arrow diagram)/Critical Path Method	Used when scheduling or monitoring a task is complex or lengthy and has schedule constraints.
Tool 2	Affinity diagram	Used to organize a large group of items into smaller categories that are easier to understand and deal with.
Tool 3	Interrelationship digraph (Relationship diagram)	Used to show a logical relationship between ideas, process, cause and effect.
Tool 4	Matrix diagram	Used to analyze the correlations between two or more groups of information.
Tool 5	Prioritization matrix	Used to choose one or two options from several options which have important criteria.
Tool 6	Process decision program chart	Used to help with contingency plans.
Tool 7	Tree diagram	Used to break down or stratify ideas into a progressively more detailed step.

Source: Abdul Razzak Rumane (2013). Quality Tools for Managing Construction Projects. Reprinted with permission of Taylor & Francis Group.

Quality Tools for Sustainability

completion of one or more activities and the beginning of new ones. Figure 3.11 illustrates an activity network diagram.

There are two kinds of network diagrams: the "Activity-on-Arrow" (A-O-A) network diagram and the "Activity-on-Node" (A-O-N) network diagram.

Arrow diagrams or activity-on-arrows (A-O-A) is a diagramming method to represent the activities on arrows and connect them at nodes (circles) to show the dependencies. With the A-O-A method, the detailed information about each activity is placed on an arrow or as footnotes at the bottom.

Figure 3.12 illustrates the use of an arrow diagramming method for concrete foundation work.

Activities originating from a certain event cannot start until the activities terminating the same event have been completed. This is known as the precedence relations. These relationships are drawn using the Precedence Diagramming Method (PDM). The PDM technique is also referred to as "Activity-on-Node" (A-O-N) because it shows the activities in a node (box) with the arrows showing dependencies. An "Activity-on-Node" network diagram has the activity information written in a small box that is a node of the diagram. Arrows connect the boxes to show the logical relationships between pairs of activities.

In networking diagrams, all activities are related in some direct way and may be further constrained by indirect relationships. The following are direct logical relationships or dependencies among project-related activities:

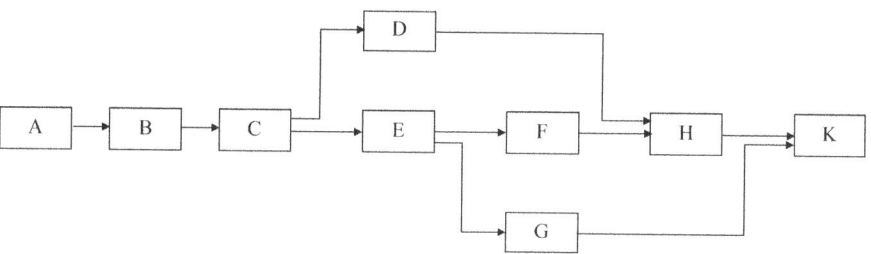

FIGURE 3.11 Activity network diagram. Source: Abdul Razzak Rumane. (2013). *Quality Tools for Managing Construction Projects*, CRC Press, Florida. Reprinted with permission of Taylor & Francis Group.

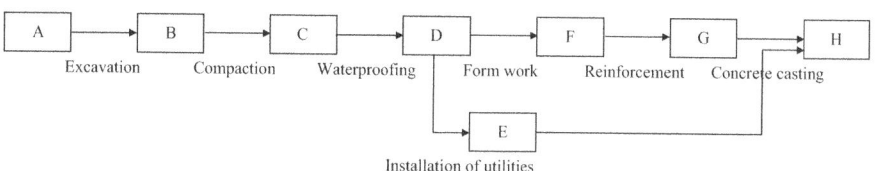

FIGURE 3.12 Arrow diagramming method for concrete foundation. Source: Abdul Razzak Rumane. (2013). *Quality Tools for Managing Construction Projects*, CRC Press, Florida. Reprinted with permission of Taylor & Francis Group.

1. Finish-to-Start ... activity A must finish before activity B can begin.
2. Start-to-Start ... activity A must begin before activity B can begin.
3. Start-to-Finish ... activity A must begin before activity B can finish.
4. Finish-to-Finish ... activity A must finish before activity B can finish.

Apart from these, there are other dependencies such as:

1. Mandatory
2. Discretionary
3. External

Figure 3.13 illustrates dependency relationship diagrams and Figure 3.14 illustrates the precedence diagramming method.

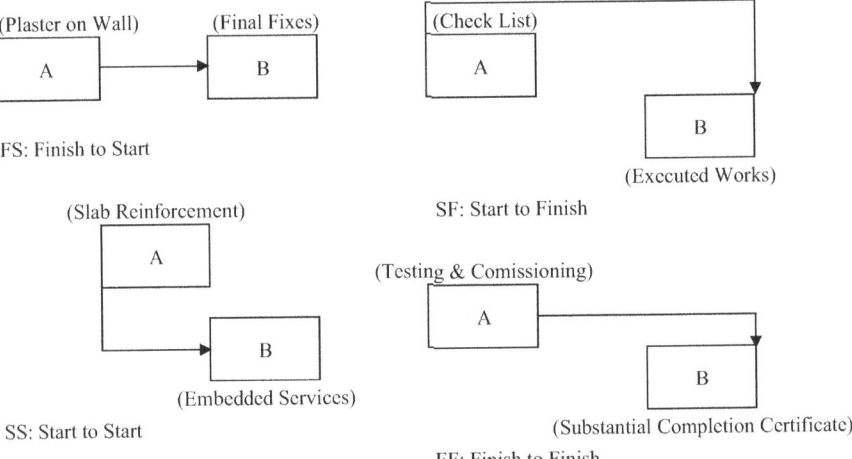

FIGURE 3.13 Dependency relationship diagram. Source: Abdul Razzak Rumane. (2013). *Quality Tools for Managing Construction Projects*, CRC Press, Florida. Reprinted with permission of Taylor & Francis Group.

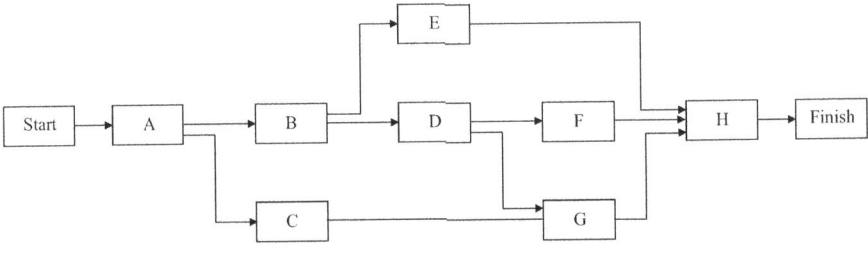

FIGURE 3.14 PDM diagramming method. Source: Abdul Razzak Rumane. (2013). *Quality Tools for Managing Construction Projects*, CRC Press, Florida. Reprinted with permission of Taylor & Francis Group.

Quality Tools for Sustainability

Activity network diagram (AND) or arrow diagram is a tool used for detailed planning, for analyzing the schedule during execution and for controlling a complex or large-scale project. A network diagram uses nodes and arrows. Date information is added to each activity node.

A Critical Path Method (CPM) chart is an expanded activity network diagram, showing an estimated time to complete each activity and connecting these activities based on the task to be performed. A critical path is a sequence of interrelated predecessor/successor activities that determines the minimum completion time for a project. The duration of a critical path is the sum of the activities' duration along the path. The activities in the critical path have the least scheduling flexibility. Any delays along the critical path would imply that additional time would be required to complete the project. There may be more than one critical path among all the project activities, so completion of the entire project could be affected due to delaying activity along any of the critical paths. Table 3.3 illustrates the activity relationship for a substation project and Figure 3.15 presents a Critical Path Method (CPM) diagram for the construction of a substation.

The activity network diagram is also known as Program Evaluation and Review Technique (PERT). PERT is used to schedule, organize, and coordinate tasks within a project.

PERT planning involves the following steps:

1. Identify a specific activity.
2. Identify the milestones of each activity.
3. Determine the proper sequence of each activity.
4. Construct a network diagram.
5. Estimate the time required to complete each activity. The three-point estimation method, using the following formula, can be used to determine the approximate estimated time for each activity:
    Estimated time = (Optimistic + 4 × most likely + pessimistic)/6
6. Compute the Early Start (ES), Early Finish (EF), Late Start (LS), and Late Finish (LF) times for each activity in the network.
7. Determine the critical path
8. Identify the critical path for possible schedule compression
9. Evaluate the diagram for milestones and target dates in the overall project.

Figure 3.16 shows a Gantt chart.

3.2.2.2 Affinity Diagram

An affinity diagram is a tool that gathers a large group of ideas/items and organizes them into a smaller grouping based on their natural relationships. An affinity diagram is a refinement of brainstorming ideas into somewhat smaller groups which can be dealt with more easily to satisfy the team members. The affinity process is often used to group ideas generated by brainstorming. An affinity diagram is created according to the following steps:

TABLE 3.3
Activities to Construct a Substation Building

Activity Number	Description of Activity	Duration in Days	Preceding Activity (ies)
1	Start	0	
2	Mobilization	21	1
3	Preparation of site	15	2
4	Staff approval	15	2
5	Material approval	15	2,4
6	Shop drawing approval	15	2,5
7	Procurement (structural work)	15	5
8	Procurement (pipes, ducts, sleeves)	7	5
9	Procurement (HT switchgear, transformer)	60	5
10	Procurement (Civil, MEP, Furnishing)	30	5
11	Excavation	4	3,6
12	Blinding concrete	4	5,7,11
13	Raft foundation	7	5,6,7,12
14	Utility services (Embedded)	4	8,12
15	Concrete (Floor)	1	6,13,14
16	Trenches	7	13
17	Embedded services/ducts in trench	2	8,13
18	Concrete (Transformer Area)	1	16,17
19	Walls and Columns	7	15,18
20	Form work for slab	3	6,7,19
21	Reinforcement	2	7,20
22	Embedded services	1	8,20
23	Concrete (roof slab)	1	21,22
24	Masonry work	14	6,10,23
25	Installation of equipment	7	9,18,24
26	Installation of electromechanical items	14	6,10,23,24
27	Installation of ventilation system	4	10,23,24
28	Finishes	7	10,24
29	Installation of final fixes	3	10,28
30	Furnishing	2	10,28
31	Testing of equipment	2	25
32	Testing of HVAC, firefighting, electrical system	4	25,26,27
33	Handing over	1	29,30,31,32
34	End	0	

Note: The duration is indicative only to understand the sequencing and relationship between activities.
Source: Abdul Razzak Rumane (2013). Quality Tools for Managing Construction Projects. Reprinted with permission of Taylor & Francis Group.

Quality Tools for Sustainability 51

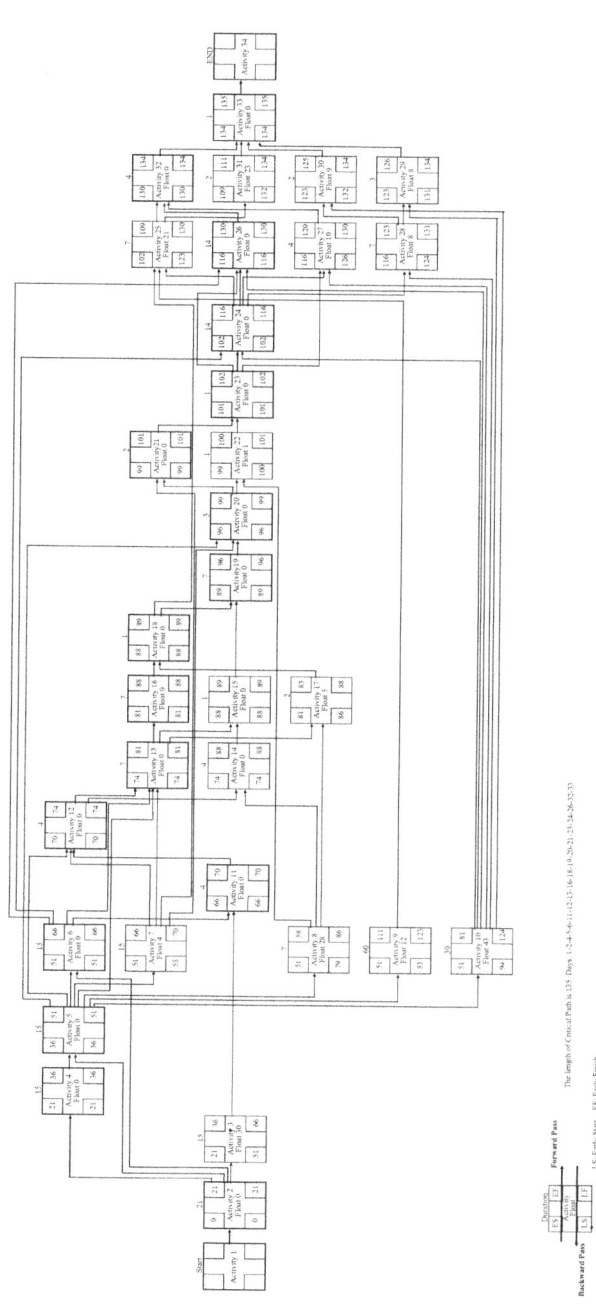

FIGURE 3.15 Critical path method. Source: Abdul Razzak Rumane. (2013). *Quality Tools for Managing Construction Projects*, CRC Press, Florida. Reprinted with permission of Taylor & Francis Group.

Quality Management

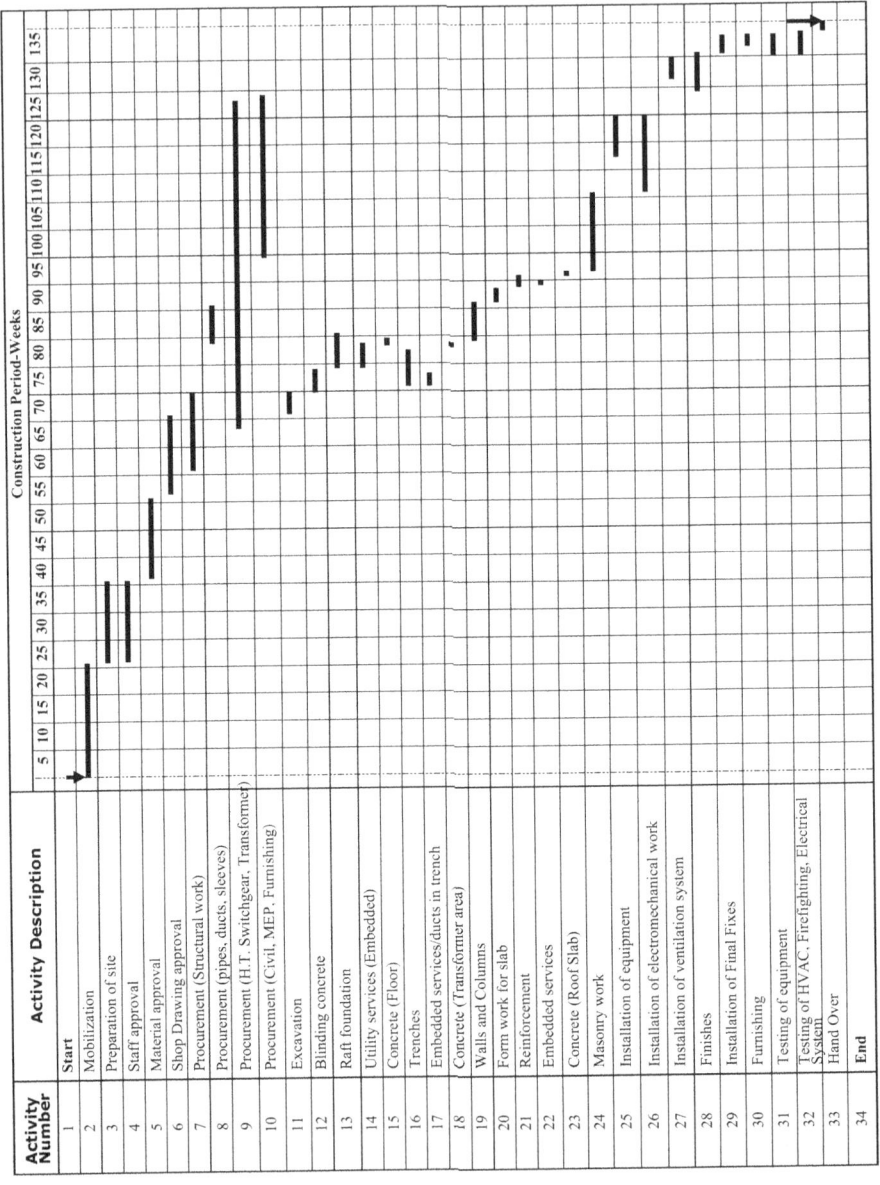

Quality Tools for Sustainability

1. Generate ideas and list the ideas without critiquing.
2. Display the ideas in a random manner.
3. Sort the ideas and place them into multiple groups.
4. Continue until smaller groups satisfy all the members.
5. Draw the affinity diagram.

Figure 3.17 shows an affinity diagram for a concrete slab.

3.2.2.3 Interrelationship Digraph

The interrelationship digraph ("di-" is for directional) is an analysis tool that allows the team members to identify logical cause-and-effect relationships between the ideas. It is drawn to show all the different relationships between factors, areas, or processes. They make it easy to pick out the factors in a situation which are the ones which are driving many of the other symptoms or factors. Whereas affinity diagrams organize and arrange the ideas into groups, interrelationship digraphs identify problems in defining the ways in which ideas influence one another. An interrelation digraph is used to identify cause-and-effect relationship with the help of directional arrows among critical issues. The number of arrows coming into the node determines the outcome (Key Indicator) whereas the outgoing arrows determine the cause (Driver) of the issue.

Figure 3.18 represents an interrelationship digraph for causes of bridge collapse.

3.2.2.4 Matrix Diagram

A matrix diagram is constructed to systematically analyze the correlations between two or more groups of items or ideas. The matrix diagram can be shaped in the following ways:

1. L-shaped
2. T-shaped
3. X-shaped
4. C-shaped
5. Inverted Y-shaped
6. Roof-shaped

Each shape has its own purpose:

1. An L-shaped matrix is used to show interrelationships between two groups or processes.
2. A T-shaped matrix is used to show the relationship among three groups. For example, consider there are three groups, A, B, and C. In a T-shaped matrix, group A and B are each related to group C whereas group A and B are not related to each other.
3. An X-shaped matrix is used to show relationship among four groups. Each group is related to two other groups in a circular fashion.

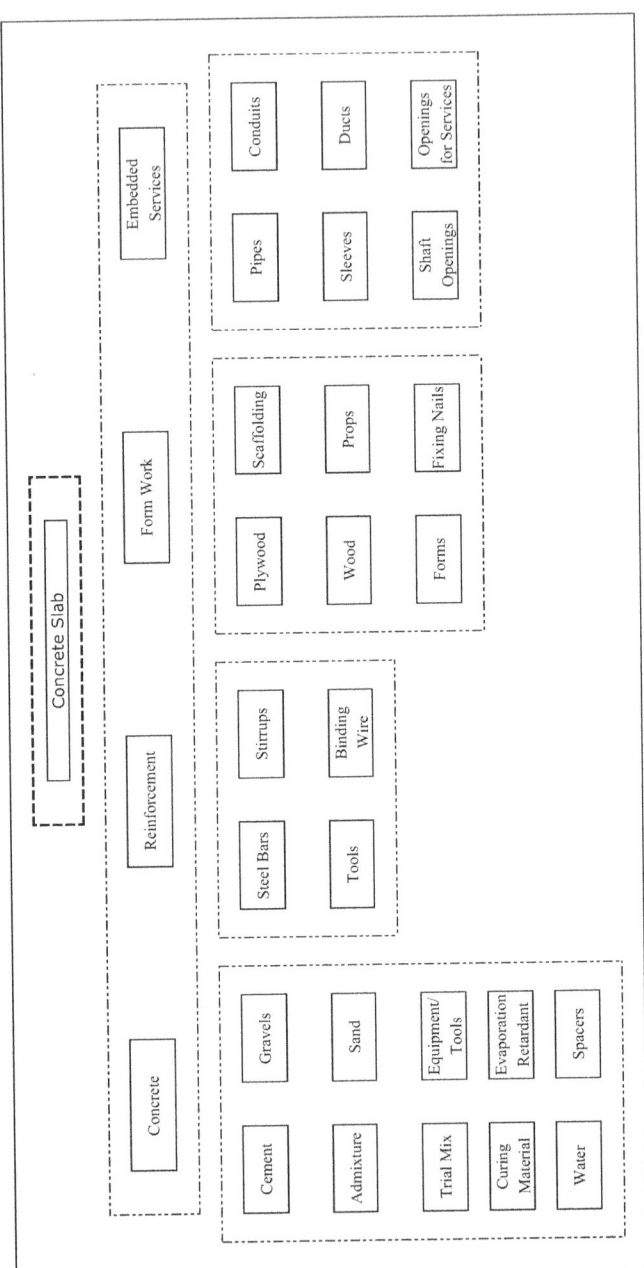

FIGURE 3.17 Affinity diagram for concrete slab. Source: Abdul Razzak Rumane. (2013). *Quality Tools for Managing Construction Projects*, CRC Press, Florida. Reprinted with permission of Taylor & Francis Group.

Quality Tools for Sustainability

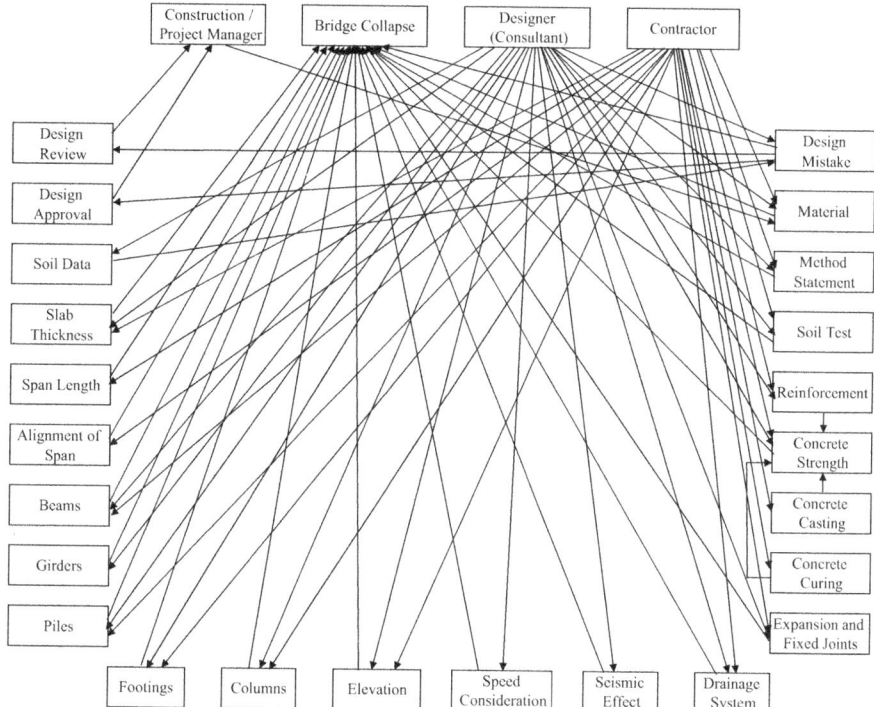

FIGURE 3.18 Interelationship digraph. Source: Abdul Razzak Rumane. (2013). *Quality Tools for Managing Construction Projects*, CRC Press, Florida. CRC Press, Florida. Reprinted with permission of Taylor & Francis Group.

4. A C-shaped matrix interrelates three groups of process or ideas in three-dimensional ways.
5. An inverted Y-shaped matrix is used to show relationships between three groups. Each group is related with other groups in a circular fashion
6. A roof-shaped matrix relates one group of items to itself. A roof-shaped matrix is used with L- or T-shaped matrices.

Table 3.4 presents an L-shaped matrix.
Figure 3.19 presents a T-shaped matrix.
Figure 3.20 presents a roof-shaped matrix.

3.2.2.5 Prioritization Matrix

The prioritization matrix assists in choosing between several options in order of importance and priority. It helps decision-makers determine the order of importance, considering the relative merit of each of the activities or goals being considered. A

TABLE 3.4
L-Shaped Matrix

Customer Requirements of Distribution Boards

Serial Number	Component Details	Customer A	Customer B	Customer C
1	Isolator	1	1	-
2	Molded Case Circuit Breaker (MCCB)	-		1
3	HRC Fuse	1	1	-
4	Earth Leakage Circuit Breaker	2	1	2
5	Miniature Circuit Breaker (MCB)	18	12	18
6	Single Bus Bar	-	1	-
7	Double Bus Bar	2	-	2
8	Enclosure with Lock	Yes	No	Yes
9	Surface Mounting	Yes	No	Yes
10	Flush Mounting	-	Yes	

Source: Abdul Razzak Rumane (2013). Quality Tools for Managing Construction Projects. Reprinted with permission of Taylor & Francis Group.

Manufacturing Plant	Products				
Customer					
International				#	#
European Manufacturer	≠	#	#		
Local Plant	#	≠	≠		
# Large Capacity ≠ Small Capacity	600 KVA Transformer	1000 KVA Transformer	1250 KVA Transformer	1600 KVA Transformer	2000 KVA Transformer
ABC Company	#	≠	≠		
XYZ Company	≠	#	#		
Others				#	#

FIGURE 3.19 T-shaped matrix. Source: Abdul Razzak Rumane. (2013). *Quality Tools for Managing Construction Projects*, CRC Press, Florida. Reprinted with permission of Taylor & Francis Group.

prioritization matrix focuses the attention of team members on those key issues and options which are more important for the organization or project.

Figure 3.21 presents a prioritization matrix.

3.2.2.6 Process Decision Program

The process decision program is a technique used to help prepare contingency plans. The process decision program systematically identifies what might go wrong in a project plan or project schedule and describe specific actions to be taken to prevent

Quality Tools for Sustainability

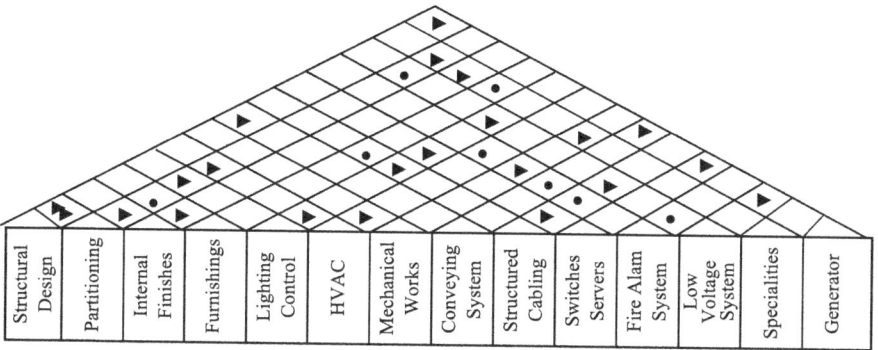

FIGURE 3.20 Roof shaped matrix. Source: Abdul Razzak Rumane. (2013). *Quality Tools for Managing Construction Projects*, CRC Press, Florida. Reprinted with permission of Taylor & Francis Group.

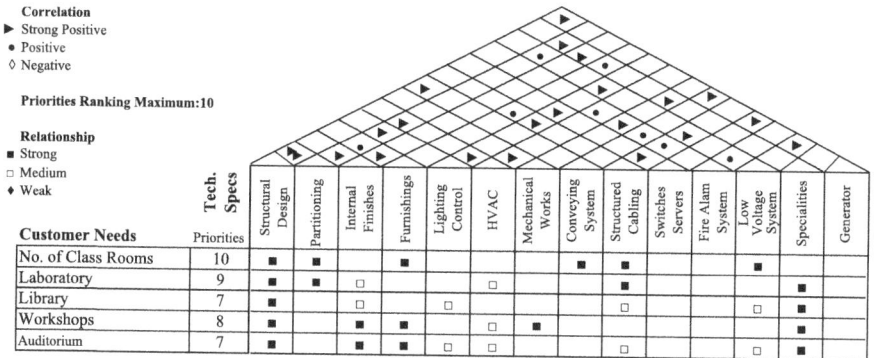

FIGURE 3.21 Prioritization matrix. Source: Abdul Razzak Rumane. (2013). *Quality Tools for Managing Construction Projects*, CRC Press, Florida. Reprinted with permission of Taylor & Francis Group.

the problems from occurring in the first place, and to mitigate or avoid the impact of the problems if they occur.

Figure 3.22 presents a process decision program for submission of contract documents.

3.2.2.7 Tree Diagram

Tree diagrams are used to break down or stratify ideas progressively into more detailed steps. A tree diagram breaks broader ideas into specific details and helps make decision-making easier to select the alternative. It is used to figure out all the various tasks that must be undertaken to achieve a given objective.

Figure 3.23 presents a tree diagram for water in the storage tank.

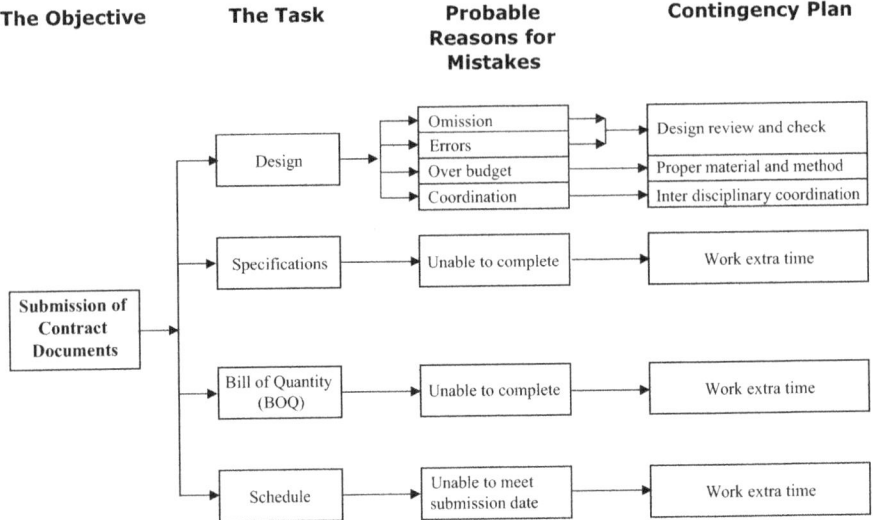

FIGURE 3.22 Process decision diagram chart. Source: Abdul Razzak Rumane. (2013). *Quality Tools for Managing Construction Projects*, CRC Press, Florida. Reprinted with permission of Taylor & Francis Group.

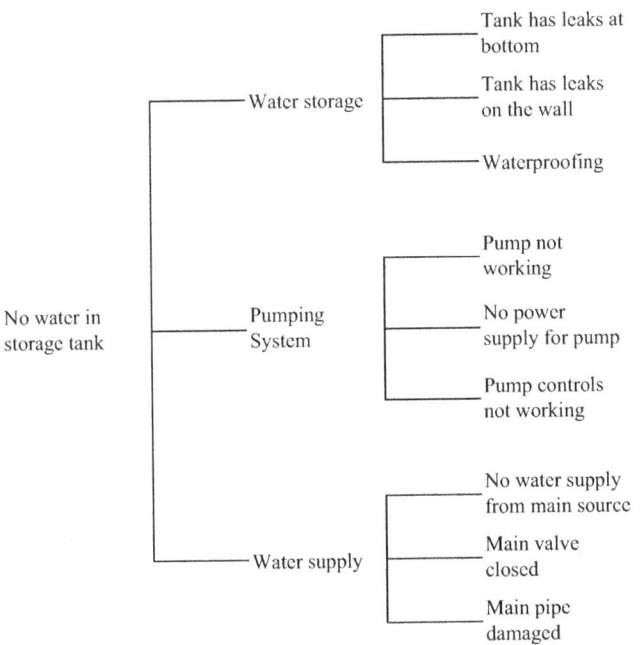

FIGURE 3.23 Tree diagram for no water in storage tank. Source: Abdul Razzak Rumane. (2013). *Quality Tools for Managing Construction Projects*, CRC Press, Florida. Reprinted with permission of Taylor & Francis Group.

Quality Tools for Sustainability

3.2.3 Process Analysis Tools

Table 3.5 presents process analysis tools.

3.2.3.1 Benchmarking:

Benchmarking is the process of measuring the actual performance of the organization's products, processes, and services and comparing it to the best-known industry standards to assist the organization in improving the performance of their products, processes, and services. Benchmarking involves analyzing an existing situation, identifying and measuring factors critical to the success of the product or services, comparing them with other businesses, analyzing the results and implementing an action plan to achieve better performance. The following is the process for benchmarking:

1. Collect internal and external data on the work, process, method, product characteristics, and system selected for benchmarking.
2. Analyze the data to identify performance gaps and determine causes and differences.
3. Prepare an action plan to improve the process in order to meet or exceed the best practices in the industry.
4. Search for the best practices among market leaders, competitors, and non-competitors that lead to their superior performance.
5. Improve the performance by implementing these practices.

Figure 3.24 presents the benchmarking process.

TABLE 3.5
Process Analysis Tools

Sr. No.	Name of Quality Tool	Usage
Tool 1	Bench marking	To identify best practices in the industry and improve the process or project.
Tool 2	Cause and effect	To identify possible cause and its effect on the process
Tool 3	Cost of quality	To identify hidden or indirect costs affecting the overall cost of the product/project.
Tool 4	Critical to quality	To identify quality features or characteristics most important to the client.
Tool 5	Failure Mode and Effects Analysis (FEMA)	To identify and classify failures according to their effect.
Tool 6	5 Why Analysis	Used to analyze and solve any problem where the root cause is unknown.
Tool 7	5w2H	The questions used to understand why the things happen the way they do.
Tool 8	Process mapping/ Flowcharting	It is a technique used for designing, analyzing, and communicating work processes.

Source: Abdul Razzak Rumane (2013). Quality Tools for Managing Construction Projects. Reprinted with permission of Taylor & Francis Group.

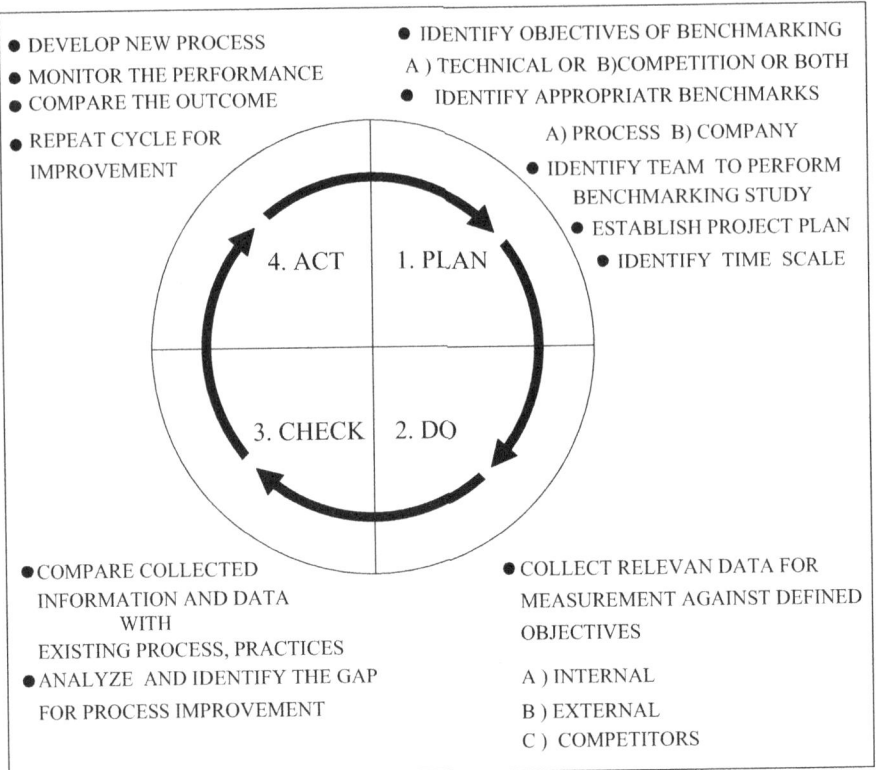

FIGURE 3.24 Benchmarking process. Source: Abdul Razzak Rumane. (2013). *Quality Tools for Managing Construction Projects*, CRC Press, Florida. Reprinted with permission of Taylor & Francis Group.

3.2.3.2 Cause-and-Effect Diagram

This is one of the quality classic tools. It is used to analyze the cause and effect of defects or non-conformance and their effect on the process due to these causes.

Figure 3.25 presents a cause-and-effect diagram for rejection of masonry work.

3.2.3.3 Cost of Quality

The cost of quality is discussed in detail under Section 3.2.7 of this chapter.

3.2.3.4 Critical to Quality

Critical to quality is a significant step in the design process of a product or service to identify the expectation of the customer/client to fulfill their needs and requirements.

3.2.3.5 Failure Mode and Effects Analysis (FMEA)

The goal of failure mode and effects analysis is to identify all the possible failures in the design of a product, process, and service ,and their effects on the product, process, and service. Its aim is to reduce the risk of failure and improve the process.

Quality Tools for Sustainability

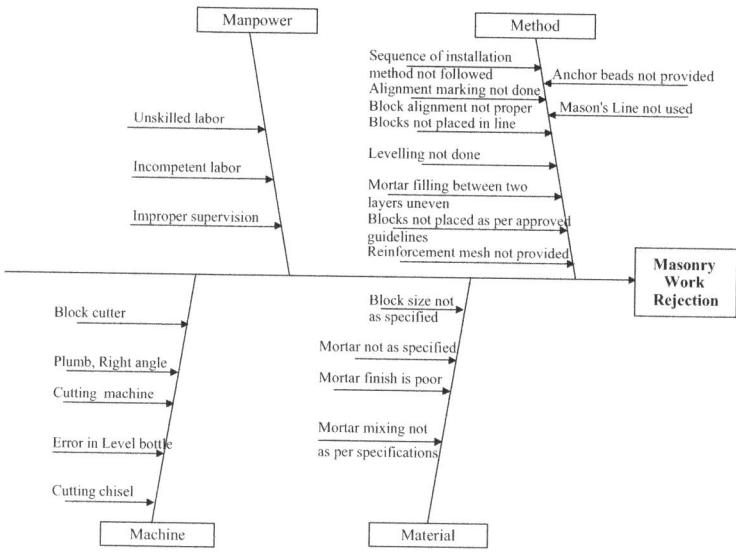

FIGURE 3.25 Cause and effect diagram for masonary work. Source: Abdul Razzak Rumane. (2013). *Quality Tools for Managing Construction Projects*, CRC Press, Florida. Reprinted with permission of Taylor & Francis Group.

Figure 3.26 presents the failure mode and effects analysis process and Figure 3.27 illustrates an example of a form used to record FMEA readings.

3.2.3.6 Five Whys Analysis

Five whys analysis is used to analyze and solve any problem where the root cause is unknown.

Table 3.6 presents a five whys analysis chart for the burning of cable.

3.2.3.7 5W2H

5W2H is about asking the questions to better understand a process or problem.

The five Ws are:

1. Why
2. What
3. When
4. Where
5. Who

and the two Hs are:

1. How
2. How much

Table 3.7 presents 5W2H for slab collapse.

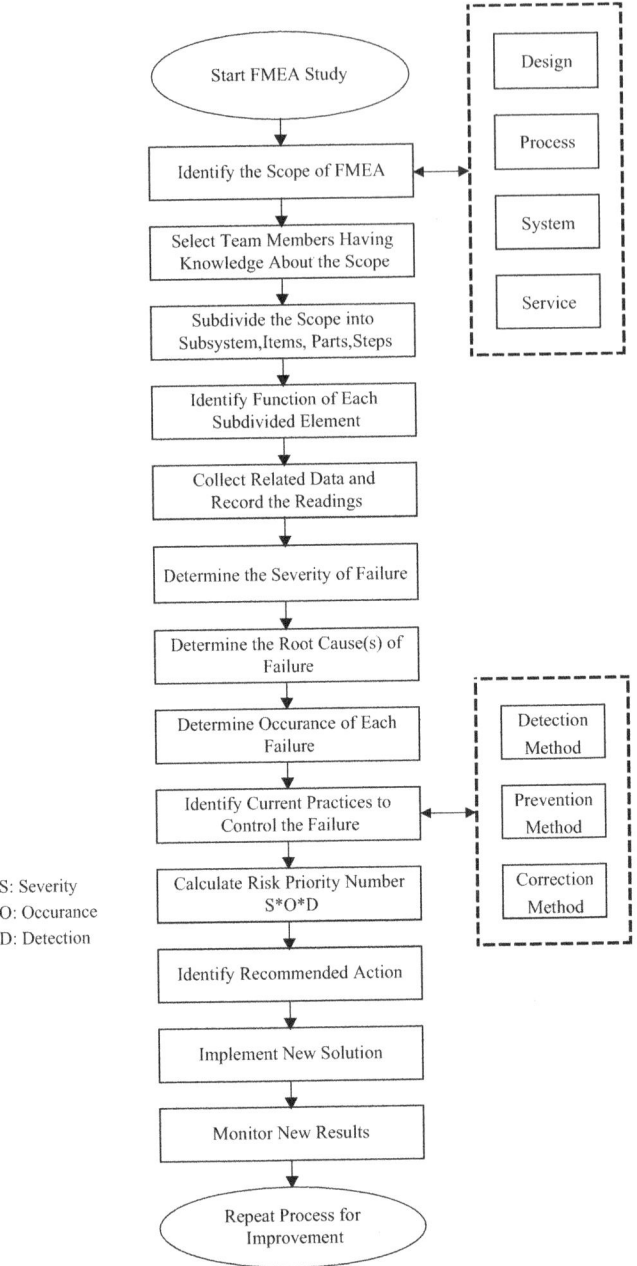

FIGURE 3.26 Failure mode and effects analysis (FMEA) process. Source: Abdul Razzak Rumane. (2013). *Quality Tools for Managing Construction Projects*, CRC Press, Florida. Reprinted with permission of Taylor & Francis Group.

Quality Tools for Sustainability

Failure Mode and Effect Analysis

Product Name:
Drawing Reference:
Revision:

Team Members
1.
2.
3.
4.

EXAMPLE ANALYSIS

Operation Number	Scope Description	Subdivided Elements	Failure Mode	Effects of Failure	Severity Rating	Cause of Failure	Occurrence Rating	Current Practice of Controls – Detection	Current Practice of Controls – Prevention	Current Practice of Controls – Correction	Detection Rating	Current Status of Product – S	O	D	RPN	Recommended Corrective Action	Action By	Action Taken Date	Revise Status – S	O	D	RPN
1	Emergency generator system	Operation	Generator failed to start	1. No lights 2. Life support equipment stopped functioning 3. No power supply for IT System 4. Water supply pumps stopped 5. No power supply for lift operation equipment will not operate 6. Fire mode 5. HVAC system stopped	7 10 9 6 4	1. No signal from ATS 2. Automatic starting system failed 3. Low Battery voltage for Starter motor 4. Circuit breaker in off position 5. No diesel in day tank		1. Through BMS 2. No regular check 3. Manual 4. Manual	1. Not in practice	1. Manual						1. Regular check of starting system 2. Check starter regularly 3. Check diesel level regularly. Interface level indicator with BMS 4. Check breaker position regularly. Interface with BMS	Mintenance Engineer					

Legend:-
RPN: Risk Priority Number
S: Severity
O: Occurance
D: Dietection

ATS: Automatic Transfer Switch
BMS: Building Management System

FIGURE 3.27 FMEA recording form. Source: Abdul Razzak Rumane. (2013). *Quality Tools for Managing Construction Projects*, CRC Press, Florida. Reprinted with permission of Taylor & Francis Group.

TABLE 3.6
5Why Analysis for Cable Burning

Serial Number	Why	Related Analyzing Question
1	Why	Why did the cable burn?
2	Why	Why did the earth leakage relay not trip?
3	Why	Why did the circuit breaker not trip?
4	Why	Why was the poor insulation of the cable not noticed?
5	Why	Why was the undersize rating of the breaker, with respect to current-carrying capacity of the cable, not noticed?

Source: Abdul Razzak Rumane (2013). Quality Tools for Managing Construction Projects. Reprinted with permission of Taylor & Francis Group.

TABLE 3.7
5W2H Analysis for Slab Collapse

Serial Number	Why/What/Who	Related Analyzing Question
1	Why	Why did the slab collapse?
2	What	What is the reason for the collapse?
3	Who	Who is responsible?
4	Where	Where is the mistake?
5	When	When did the slab collapse?
6	How many	How many persons were affected (injured or died)?
7	How much	How much was the loss in terms of cost and time?

Source: Abdul Razzak Rumane (2013). Quality Tools for Managing Construction Projects. Reprinted with permission of Taylor & Francis Group.

3.2.3.8 Process Mapping/Flowcharting

Process mapping/flowcharting is a graphical representation of the workflow, giving a clear understanding of a process or services of parallel processes. Process mapping/flowcharting are techniques that can be employed not only to process visual representations of the production processes, but processes related to other departments.

Figure 3.28 presents a process mapping/flowcharting diagram for approval of the variation order.

3.2.4 Process Improvement Tools

Table 3.8 lists process improvement tools.

Quality Tools for Sustainability

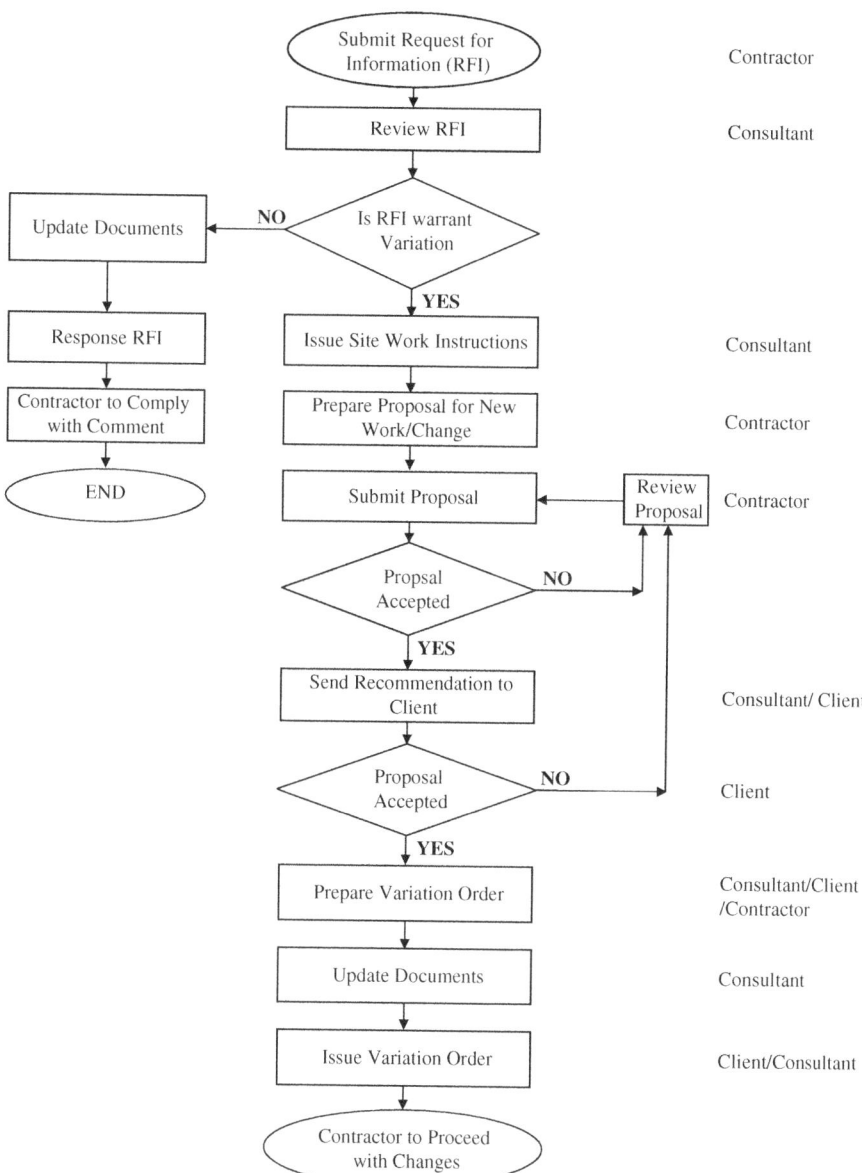

FIGURE 3.28 Process mapping/flowcharting for approval of variation order. Source: Abdul Razzak Rumane. (2013). *Quality Tools for Managing Construction Projects*, CRC Press, Florida. Reprinted with permission of Taylor & Francis Group.

3.2.4.1 Root Cause Analysis

This tool is used to analyze the root causes of problems. The analysis is generally performed by using an Ishikawa diagram or cause-and-effect diagram.

Figure 3.29 presents root cause analysis for the rejection of executed marble work.

TABLE 3.8
Process Improvement Tools

Sr. No.	Name of Quality Tool	Usage
Tool 1	Root cause analysis	To identify the root causes of the problem
Tool 2	PDCA cycle	Used to plan for improvement followed by putting it into action
Tool 3	SIPOC analysis	Used to identify supplier-input-process-output-customer (SIPOC) relationship
Tool 4	Six Sigma DMAIC	Used as an analytic tool for improvement
Tool 5	Failure Mode and Effects Analysis (FMEA)	To identify and classify failures according to the effect and prevent or reduce failure
Tool 6	Statistical process control	Used to study how the process changes over time

Source: Abdul Razzak Rumane (2013). Quality Tools for Managing Construction Projects. Reprinted with permission of Taylor & Francis Group.

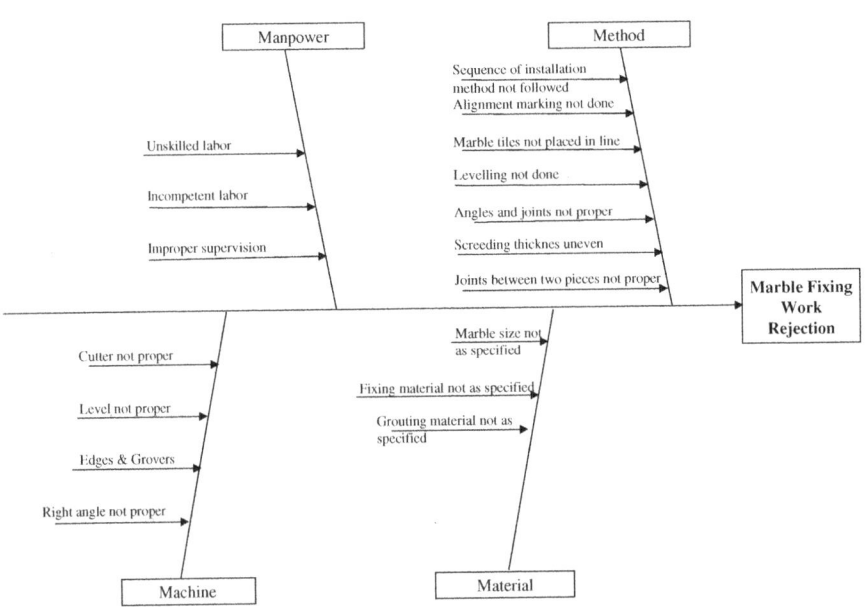

FIGURE 3.29 Root cause analysis for rejection of executed marble work. Source: Abdul Razzak Rumane. (2013). *Quality Tools for Managing Construction Projects*, CRC Press, Florida. Reprinted with permission of Taylor & Francis Group.

Quality Tools for Sustainability

3.2.4.2 PDCA Cycle

Plan-Do-Check-Act (P-D-C-A) is mainly used for continuous improvement. It consists of a four-step model for carrying changes. The PDCA cycle model can be developed as a process improvement tool to reduce the cost of quality.

Figure 3.30 presents the PDCA cycle for preparation of shop drawings.

3.2.4.3 SIPOC Analysis

This analysis is used to identify the Supplier-Input-Process-Output-Customer (SIPOC) relationship. The purpose of the SIPOC analysis is to show the process flow by defining and documenting the suppliers, inputs, process steps, outputs, and customers.

Table 3.9 presents an SIPOC analysis for an electrical panel.

3.2.4.4 Six Sigma DMAIC

Six Sigma is discussed in detail under Section 3.2.9 of this chapter.

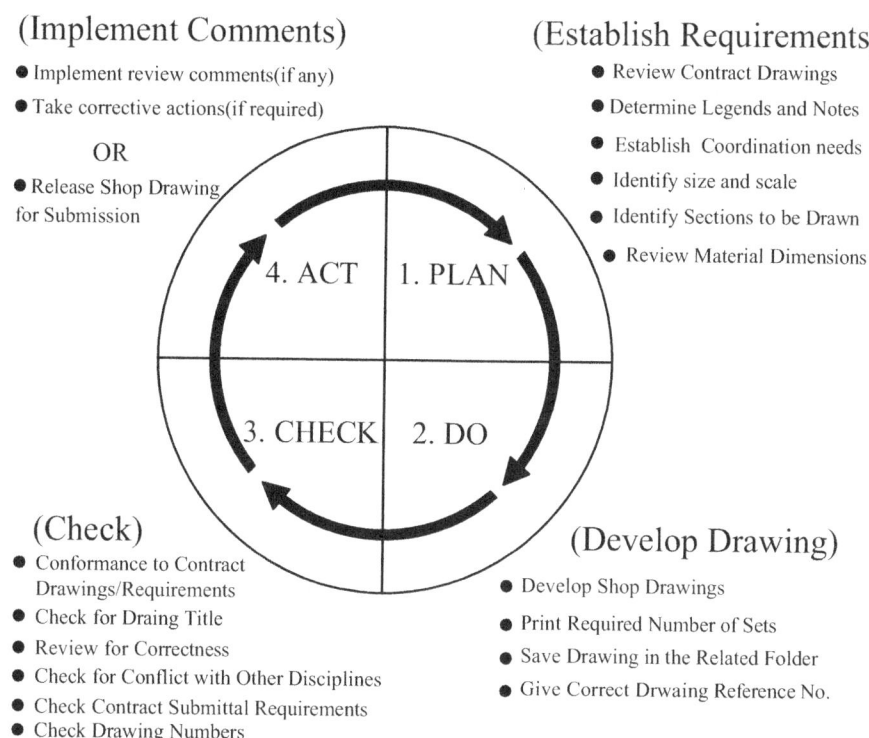

FIGURE 3.30 PDCA cycle for preparation of shop drawing. Source: Abdul Razzak Rumane. (2013). *Quality Tools for Managing Construction Projects*, CRC Press, Florida. Reprinted with permission of Taylor & Francis Group.

TABLE 3.9
SIPOC Analysis for an Electrical Panel

Who are the suppliers?	What are the suppliers providing?	What is the process?	What is the output of process?	Who are the customers?
Supplier	Inputs	Process	Outputs	Customer
Electrical Panel Builder/ Assembler	Main Low-Tension Panel Main Switchboards Distribution Boards Starter Panels Control Panels	Electrical Installation Work	Electrical Distribution Network	Power supply for project

Source: Abdul Razzak Rumane (2013). Quality Tools for Managing Construction Projects. Reprinted with permission of Taylor & Francis Group.

3.2.4.5 Failure Mode and Effects Analysis (FMEA)

Failure mode and effects analysis is also used as a process improvement tool. It identifies all the possible failures in the design of a product, process, or service and their effects on the product, process, or service. Its aim is to reduce the risk of failure and improve the process. Figure 3.27 under Section 3.2.3.5 illustrates the failure mode and effects analysis process.

3.2.4.6 Statistical Process Control

Statistical process control (SPC) is a quantitative approach based on the measurement of process control. Dr Walter A. Shewhart developed the control charts as early as 1924. Statistical process control charts are used for identification of common cause and special (or assignable) cause variations. assisting in the diagnosis of quality problems. SPC charts reveal whether a process is "under control" – stable and exhibiting only random variation – or "out of control" and needing attention. Control chart is one of the key tools of SPC. It is used to monitor processes that are not under control, using measured ranges. There are two types of process control charts:

1. Variable charts
2. Attributes charts

Variable charts relate to variable measurement, such as length, width, temperature, weight, etc.

Attributes charts relate to the characteristics possessed (or not possessed) by the process or the product.

Figure 3.31 presents statistical control process charts for generator frequency.

Quality Tools for Sustainability

3.2.5 Innovation and Creative Tools

Table 3.10 illustrates innovative and creative tools.

3.2.5.1 Brainstorming

Brainstorming is listing all the ideas put forward by a group in response to a given question or problem. It is a process of creating ideas by "storming" some objective. In 1939, a team, led by advertising executive Alex Osborn, coined the term

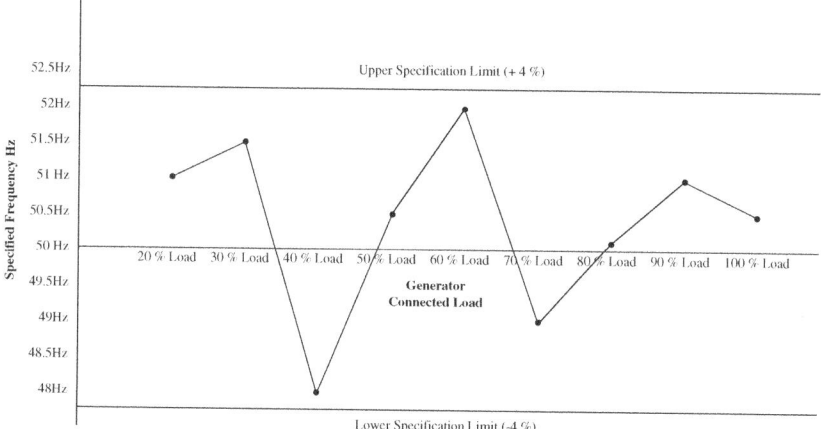

FIGURE 3.31 Statistical process control chart for generator frequency. Source: Abdul Razzak Rumane. (2013). *Quality Tools for Managing Construction Projects*, CRC Press, Florida. Reprinted with permission of Taylor & Francis Group.

TABLE 3.10
Innovative and Creative Tools

Sr. No.	Name of Quality Tool	Usage
Tool 1	Brainstorming	Used to generate multiple ideas
Tool 2	Delphi technique	Used to get ideas from a select group of experts
Tool 3	5W2H	The questions which are used to understand why the things happen the way they do
Tool 4	Mind mapping	Used to create a visual representation of many issues that can help get a better understanding of the situation
Tool 5	Nominal group technique	Used to enhance brainstorming by ranking the most useful ideas
Tool 6	Six Sigma-DMADV	Used primarily for the invention and innovation of modified or new products, services, or processes
Tool 7	TRIZ	Used to provide systematic methods and tools for analysis and innovative problem solving

Source: Abdul Razzak Rumane (2013). Quality Tools for Managing Construction Projects. Reprinted with permission of Taylor & Francis Group.

"brainstorm." According to Osborn, a brainstorm means using the brain to storm a creative problem. Classical brainstorming is the most well-known and often-used technique for idea generation in a short period of time. It is based on the fundamental principles of deferment of judgment, and that quantity breeds quality. It involves questions such as:

- Does the item have any design features that are not necessary?
- Can two or more parts be combined together?
- How can we cut down the weight?
- Are these nonstandard parts that can be eliminated?

There are four rules for successful brainstorming:

1. Criticism is ruled out.
2. Freewheeling is welcomed.
3. Quantity is wanted.
4. Contribution and improvements are sought.

A classical brainstorming session has the following basic steps:

- Preparation: The participants are selected, and a preliminary statement of the problem is circulated.
- Brainstorming: A warm-up session, with simple unrelated problems, is conducted, the relevant problem and the four rules of brainstorming are presented, and ideas are generated and recorded using checklists and other techniques, if necessary.
- Evaluation: The ideas are evaluated relative to the problem.

Generally, a brainstorming group should consist of four to seven people, although some suggest larger groups.

Figure 3.32 presents the brainstorming process.

3.2.5.2 The Delphi Technique

The Delphi technique is intended to determine a consensus among experts on a subject. The goal of the Delphi technique is to pick the brains of experts in the subject area, treating them as contributors to create ideas. It is a measure and method for consensus building by using questionnaires and obtaining responses from the panel of experts in the selected subjects. The Delphi technique employs multiple iterations designed to develop consensus opinion about the specific subject. The selected expert group answers questions from a facilitator. The responses are summarized and further circulated for group comments to reach the consensus. The iteration/feedback process allows the team members to reassess their initial judgment and change or modify the earlier suggestions.

Figure 3.33 presents the Delphi Technique Process.

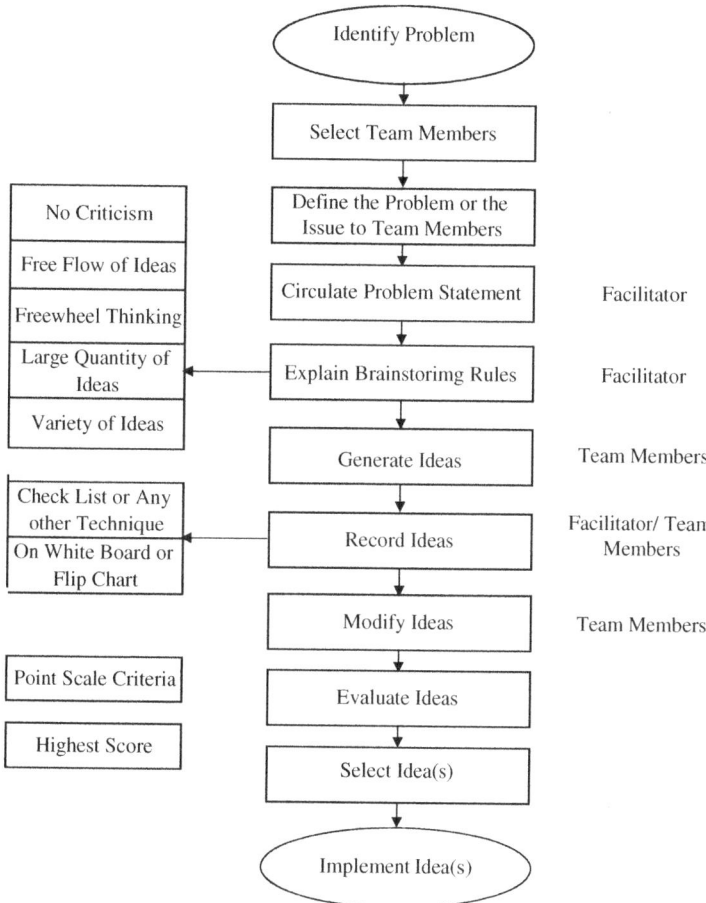

FIGURE 3.32 Brainstorming process. Source: Abdul Razzak Rumane. (2013). *Quality Tools for Managing Construction Projects*, CRC Press, Florida. Reprinted with permission of Taylor & Francis Group.

4.2.5.3 5W2H

5W2H is also known as the Innovation and Creative Tool. 5W2H is about asking the questions to better understand a process or problem.

The five Ws are:

1. Why
2. What
3. When
4. Where
5. Who

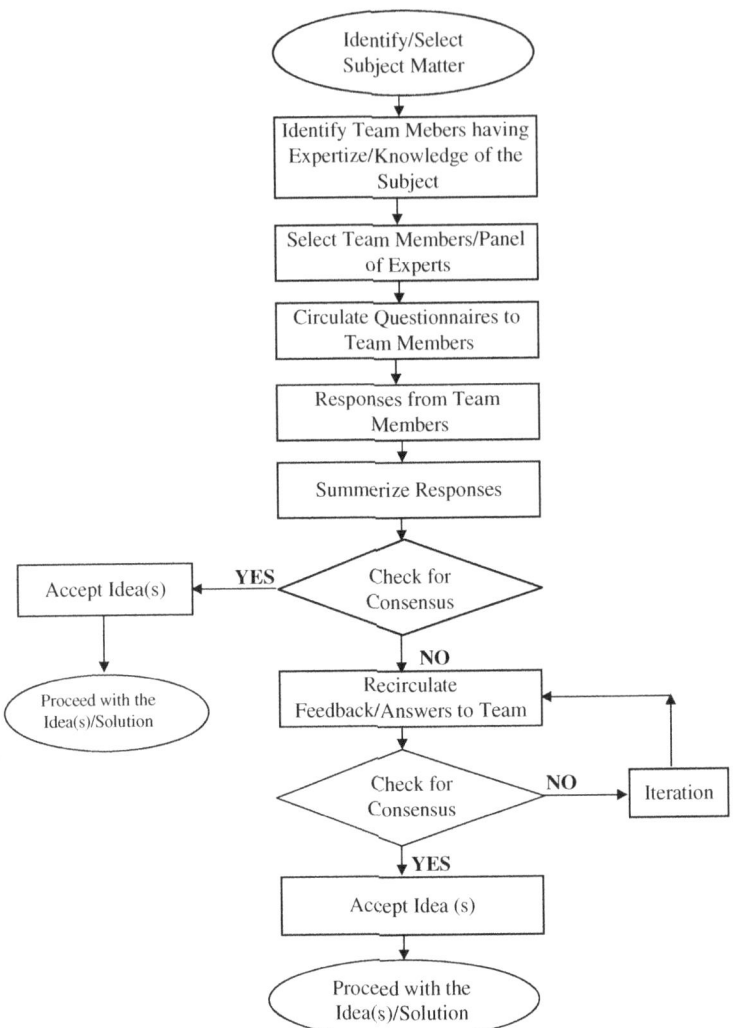

FIGURE 3.33 Delphi technique process. Source: Abdul Razzak Rumane. (2013). *Quality Tools for Managing Construction Projects*, CRC Press, Florida. Reprinted with permission of Taylor & Francis Group.

and the two Hs are:

3. How
4. How much

Table 3.11 presents 5W2H for the development of a new product.

TABLE 3.11
5W2H Analysis of a New Product

Serial Number	Why/What/Who	Related Analyzing Question
1	Why	Why make a new product?
2	What	What advantage will it have over other similar products?
3	Who	Who will be the customers for this product
4	Where	Where can we market the product?
5	When	When will the product will be ready for sale?
6	How many	How many pieces will be produced/sold per year?
7	How much	How much market share we will get for this product?

Source: Abdul Razzak Ruman. (2013). Quality Tools for Managing Construction Projects. Reprinted with permission of Taylor & Francis Group.

3.2.5.4 Mind Mapping

Mind mapping is a graphical representation of ideas which can help achieve a greater understanding of the situation and create the solution or improve the task. Figure 3.34 presents a mind mapping sketch to improve site safety.

3.2.5.5 Nominal Group Technique

The nominal group technique (NGT) involves a structural group meeting designed to incorporate individual ideas and judgments into a group consensus. By correctly applying the NGT, it is possible for groups of people (preferably five to ten) to generate alternatives or other ideas for improving the competitiveness of the firm. The technique can be used to obtain group thinking (consensus) on a wide range of topics. The technique, when properly applied, draws on the creativity of the individual participants, while reducing two undesirable effects of most group meetings, namely:

a. The dominance of one or more participants, and
b. The suppression of conflict ideas.

The basic format of an NGT session is as follows:

- Individual silent generation of ideas.
- Individual round-robin feedback and recording of the ideas.
- Group's clarification of each idea.
- Individual voting and ranking to prioritize ideas.
- Discussion of the group consensus results.

74 Quality Management

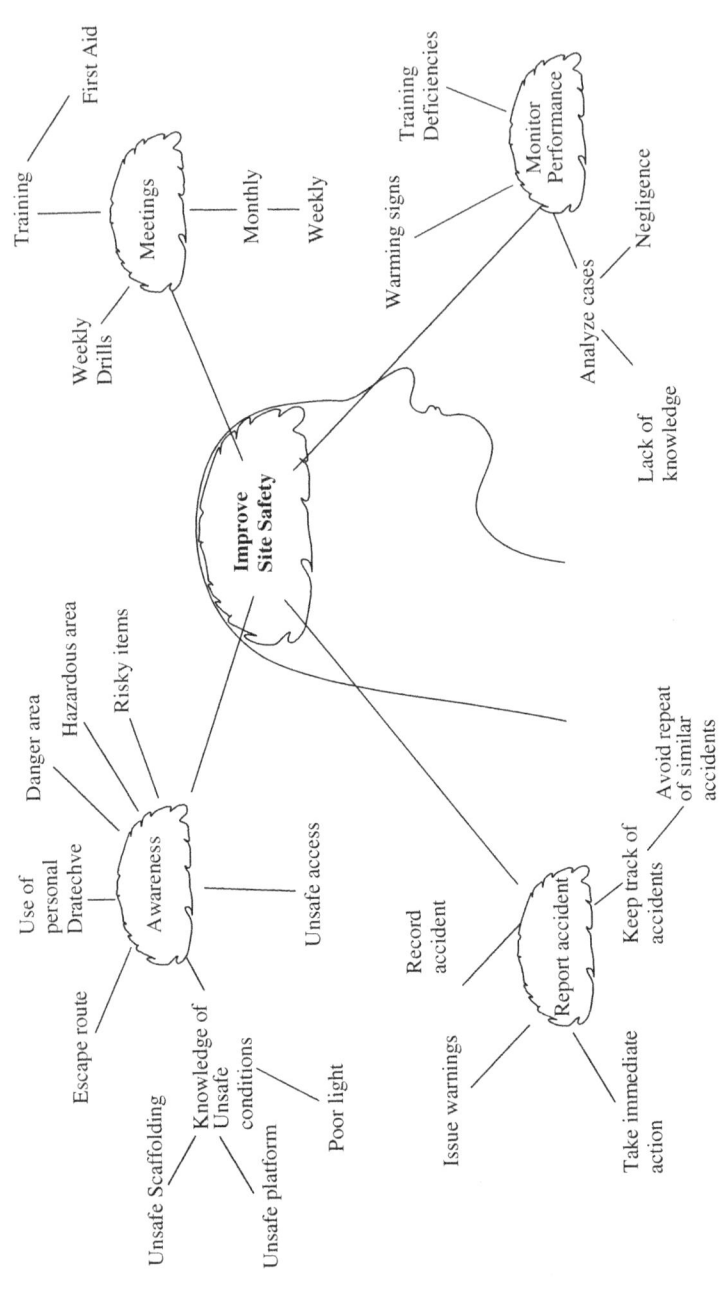

FIGURE 3.34 Mind mapping. Source: Abdul Razzak Rumane. (2013). *Quality Tools for Managing Construction Projects*, CRC Press, Florida. Reprinted with permission of Taylor & Francis Group.

Quality Tools for Sustainability

The NGT session begins with an explanation of the procedure and a statement of the question(s), preferably written by the facilitator.

3.2.5.6 Six Sigma-DMADV
Six Sigma will be discussed in detail under Section 3.2.9 of this chapter.

3.2.5.7 Triz
Triz is discussed in detail under Section 3.2.10 of this chapter.

3.2.6 Lean Tools

Table 3.12 presents Lean Tools.

TABLE 3.12
Lean Tools

Sr. No.	Name of Quality Tool	Usage
Tool 1	Cellular Design	A self-contained unit dedicated to perform all the operational requirements to accomplish sequential processing
Tool 2	Concurrent Engineering	It is used for reduction of product cycle time. It is a systematic approach to creating a product design that simultaneously considers all elements of the product life cycle.
Tool 3	5S	Used to eliminate waste that results from improper organization of the work area.
Tool 4	Just-in-Time (JIT)	Used to reduce inventory levels, improve cash flow, and reduce space requirements for storage of material.
Tool 5	Kanban	Used to signal that more material needs to be ordered. It is used to eliminate waste from inventory.
Tool 6	Kaizan	Used for continually eliminating waste from the manufacturing processes by combining the collective talent of the company.
Tool 7	Mistake proofing	Used to eliminate the opportunity for error by detecting the potential source of error.
Tool 8	Outsourcing	It involves contracting out certain works, processes, and services to a specialist in the discipline area.
Tool 9	Poka-Yoke	Used to detect the abnormality or error, fix or correct the error, and take action to prevent the error.
Tool 10	Single Minute Exchange of Die (SMED)	Used to reduce set-up time for change over to the new process.
Tool 11	Value stream mapping	Used to establishing flow of material or information and eliminating waste and adding value.
Tool 12	Visual management	It addresses both visual display and control. It focuses on waste elimination/prevention.
Tool 13	Waste reduction	It focusses on reducing waste.

Source: Abdul Razzak Ruman (2013). Quality Tools for Managing Construction Projects. Reprinted with permission of Taylor & Francis Group.

3.2.6.1 Cellular Design

A self-contained unit is dedicated to performing all the operational requirements to accomplish sequential processing. With cellular design, individual cells can be fabricated and assembled to give the same performance, while saving time.

An electrical main switch board may consist of a number of cells assembled together to perform the desired operations, facilitating easy maneuvering and assembly at the workplace to achieve proper functioning.

Figure 3.35 presents cellular design for an electrical panel.

3.2.6.2 Concurrent Engineering

A product life cycle begins with needs and extends through concept design, preliminary design, detail design, production or construction, product use, phase out, and disposal. Concurrent engineering is defined as a systematic approach to creating

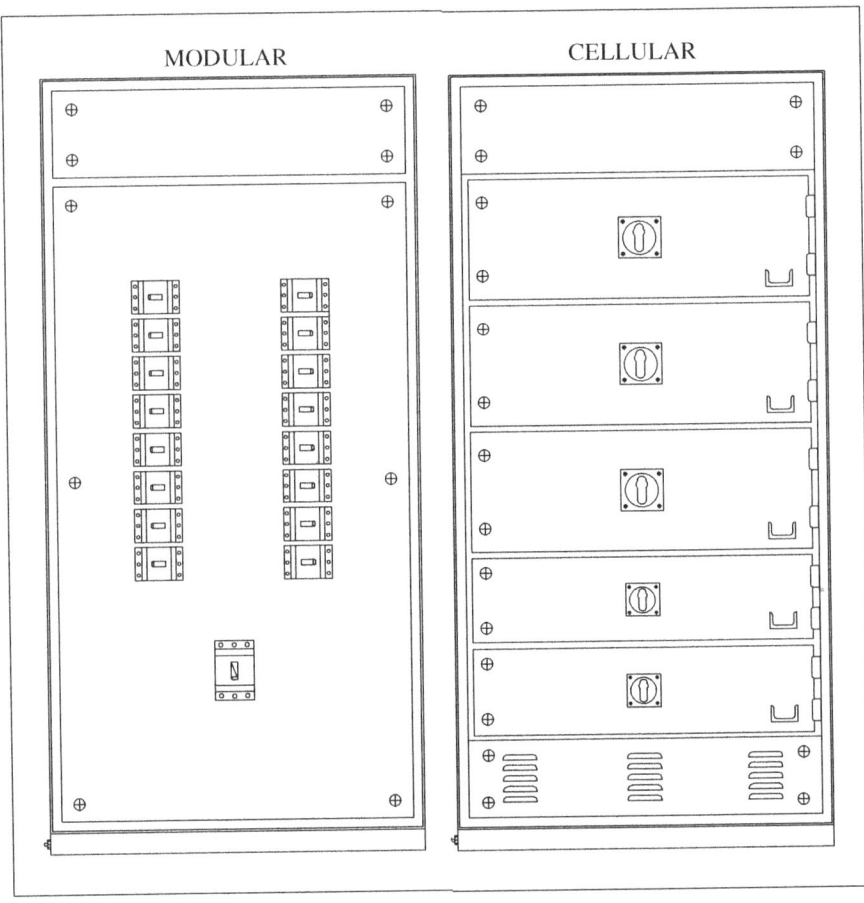

FIGURE 3.35 Cellular main switch board.

Quality Tools for Sustainability

a product design that simultaneously considers all the elements of the product life cycle, thus reducing the duration of the product life cycle. It is used to expedite the development and launch of a new product. In construction projects, construction can start while the design is under development.

Figure 3.36 presents concurrent engineering for a construction project life cycle.

3.2.6.3 5-S

5-S is a systematic approach for the improvement of quality and safety by organizing a workplace. It is a methodology which advocates;

- What should be kept.
- Where it should be kept.
- How it should be kept.

5-S is a Japanese concept of housekeeping, with reference to five Japanese words starting with the letter "S." Table 3.13 presents 5S for construction projects

3.2.6.4 Just-in-Time (JIT)

"Just-in-time" is used to reduce inventory levels, improve cash flow, and reduce storage space requirements for material. For example,

1. Concrete blocks can be received at site just before the start of block work and can be stacked near the work area, where masonry work is in progress.
2. Chiller can be received at the site and placed directly on the chiller foundation without storing in the storage yard.

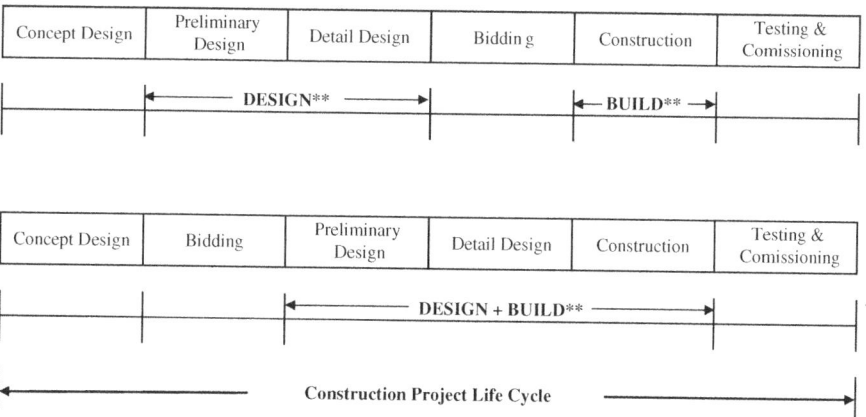

FIGURE 3.36 Concurrent engineering for construction life cycle. Source: Abdul Razzak Rumane. (2013). *Quality Tools for Managing Construction Projects*, CRC Press, Florida. Reprinted with permission of Taylor & Francis Group.

TABLE 3.13
5S for Construction Projects

Sr. No.	5S	Related Action
1	Sort	• Determine what is to be kept in the open and what to be kept under cover • Allocate an area for each type of construction equipment and machinery • Allocate an area for electrical tools • Allocate an area for hand tools • Allocate an area for construction material/equipment to be used/installed in the project • Allocate an area for hazardous, inflammable material • Allocate an area for chemicals, paints • Allocate an area for spare part for maintenance
2	Set in Order	• Keep/arrange equipment in such a way that their maneuvering/movement shall be easy • Vehicles are to be parked in the yard in such a way that frequently used vehicles are parked near the gate • Frequently used equipment/machinery are to be located near the workplace • Set boundaries for different types of equipment and machinery • Identify and arrange tools for easy access • Identify and store material/equipment as per relevant division/section of contract documents • Identify and store material in accordance with their usage as per construction schedule • Identify items which need special conditions • Mark/tag the items/material • Display the route map and location • Put the material in sequence as per their use • Frequently used consumables to be kept near the workplace • Label on the drawer with a list of contents • Keep shuttering material in one place • Determine inventory level of consumable items

3	Sweeping	• Clean site on daily basis by removing:
		  • cut pieces of reinforced bars
		  • cut pieces of plywood
		  • left out concrete
		  • cut pieces of pipes
		  • cut pieces of cables and wires
		  • used welding rods
		  • clean equipment and vehicles
		  • check electrical tools after return by the technician
		  • attend to breakdown report

4	Standardize	• Standardize the store by allocating separate areas for material used by different divisions/sections • Standardize the area for long lead items • Determine the regular schedule for cleaning the workplace • Make available standard tool kit/box for a group of technicians • Make everyone informed of their responsibilities and the area where the things are to be placed and are available • Standardize the store for consumable items • Inform suppliers/vendors in advance of the place for delivery of material
5	Sustain	• Follow the system till the end of the project

Source: Abdul Razzak Rumane (2010). Quality Management in Construction Projects. Reprinted with permission of Taylor & Francis Group.

Quality Tools for Sustainability

3.2.6.5 Kanban

It is used to signal that more material needs to be ordered. It is used to eliminate waste from inventory and inventory control, thus avoiding the need for extra storage for a large inventory. In construction projects, electrical wires for circuiting can be ordered to be received at site when the wire-pulling work is in progress. Similarly, concrete blocks and false ceiling tiles can be ordered and received as and when required.

3.2.6.6 Kaizan

It is used for continual improvements through small changes to eliminate waste from the manufacturing process by combining the collective talent of every employee of the company.

3.2.6.7 Mistake Proofing

Mistake proofing is used to eliminate the opportunity for error by detecting the potential source of error. Mistakes are generally categorized as follows:

1. Information
2. Mismanagement
3. Omission
4. Selection

Table 3.14 presents a mistake-proofing chart for eliminating design errors.

3.2.6.8 Outsourcing

Outsourcing involves contracting out certain works, processes, and services to specialists in a particular discipline area. For example, in construction projects, the following is a list of some of the works which are outsourced (sub-contracted):

1. Structural concrete
2. Waterproofing work
3. HVAC work
4. Fire suppression work
5. Water supply piping
6. Electrical work
7. Security system
8. Low voltage works
9. Landscape work

3.2.6.9 Poka-Yoke

Poke-Yoke is a quality management concept developed by Shigeo Shino to prevent human errors occurring in the production line. The main objective of Poka-Yoke is to achieve zero defects.

TABLE 3.14
Mistake Proofing for Eliminating Design Errors

Serial Number	Items	Points to be Considered to Avoid Mistakes
1	Information	1. Terms of Reference (TOR)
		2. Client's preferred requirements matrix
		3. Data collection
		4. Regulatory requirements
		5. Codes and standards
		6. Historical data
		7. Organizational requirements
2	Mismanagement	1. Compare production with actual requirements
		2. Inter-disciplinary coordination
		3. Application of different codes and standards
		4. Drawing size of different trades/ specialist consultants
3	Omission	1. Review and check the design with TOR
		2. Review and check the design with the client requirements
		3. Review and check the design with regulatory requirements
		4. Review and check the design with codes and standards
		5. Check for all required documents
4	Selection	1. Qualified team members
		2. Available material
		3. Installation methods

Source: Abdul Razzak Rumane (2013). Quality Tools for Managing Construction Projects. Reprinted with permission of Taylor & Francis Group.

3.2.6.10 Single Minute Exchange of Die (SMED):

SMED is used to reduce the set-up time for the changeover to a new process.

For example, a spare circuit breaker of similar rating can be used as an immediate replacement for a damaged circuit breaker in the electrical distribution board to avoid breakdown of electrical supply for a prolonged duration. Subsequently, a new circuit breaker can be fixed in place of the spare breaker.

3.2.6.11 Value Stream Mapping

Value stream mapping is used to establish a flow of material or information, elimination of waste, and added value. Value stream mapping is used to identify areas for improvement.

Figure 3.37 presents a value stream mapping diagram for an emergency power system.

3.2.6.12 Visual Management

Visual management addresses both visual display and control. It exposes waste elimination/prevention. Visual displays present information, whereas visual control focuses on a need to act.

Quality Tools for Sustainability

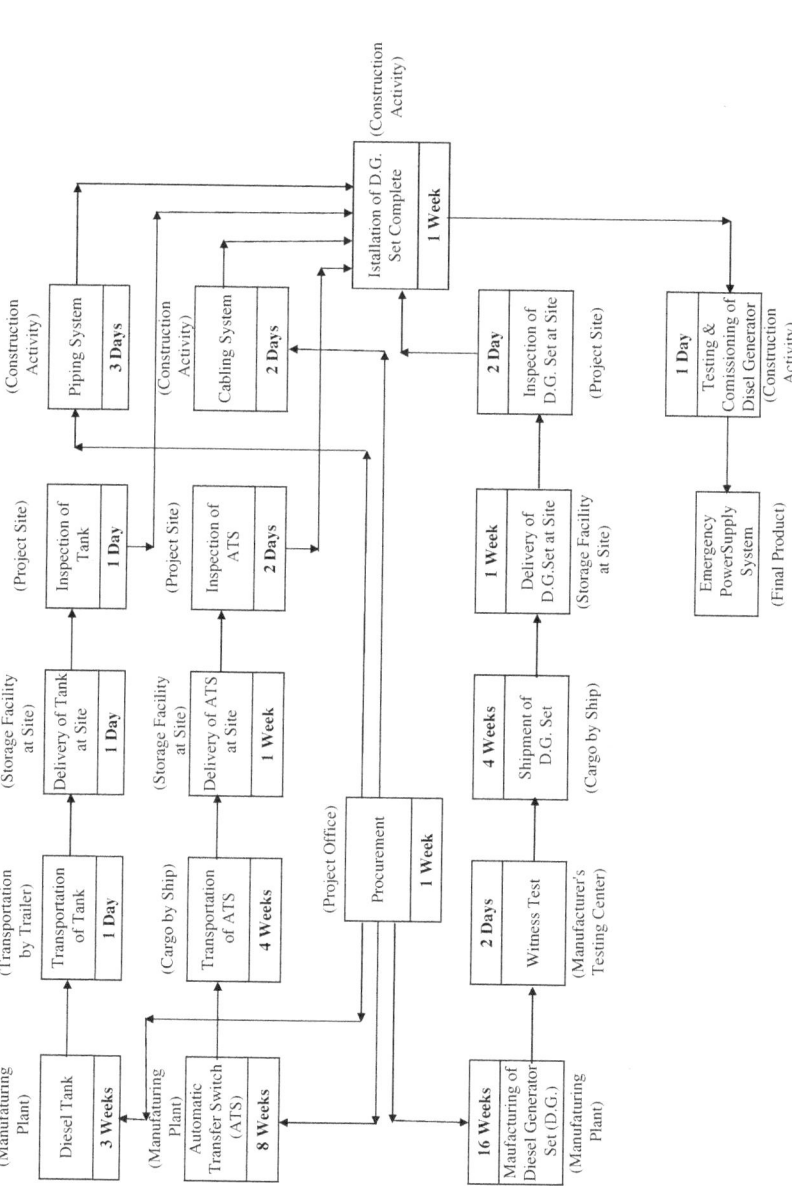

FIGURE 3.37 Value stream mapping for emergency power system. Source: Abdul Razzak Rumane. (2013). *Quality Tools for Managing Construction Projects*, CRC Press, Florida. Reprinted with permission of Taylor & Francis Group.

3.2.6.13 Waste Reduction

This strategy focuses on reducing waste. The following are the general types of waste:

1. Defective parts
2. Delays, waiting (wasted time)
3. Excess inventory
4. Mis-used resources
5. Overproduction
6. Processing
7. Transportation
8. Untapped resources
9. Wasted motion.

3.2.7 Cost of Quality

3.2.7.1 Introduction

Quality has an impact on the costs of products and services. The cost of poor quality is the annual monetary loss of products and processes that do not achieve their quality objective. The main components of the cost of low quality are:

1. Cost of conformance.
2. Cost of non-conformance.

Table 3.15 illustrates elements of the cost of quality

3.2.7.2 Categories of Costs

Costs of quality are the costs associated with providing poor-quality products or services. These costs are incurred as a result of low quality, costs that would not be incurred if things were done correctly from the beginning and at every time thereafter in order to achieve the quality objective. There are four categories of low cost:

   i. Internal failure costs: The costs associated with defects found before the customer receives the product or service. It also consists of the cost of failure to meet customer satisfaction and needs and the cost of inefficient processes.
  ii. External failure costs: The cost associated with defects found after the customer receives the product or service. It also includes lost opportunities for sales revenue.
 iii. Appraisal costs: The costs incurred to determine the degree of conformance to quality requirements.
  iv. Prevention costs: The costs incurred to keep failure and appraisal costs to a minimum.

Quality Tools for Sustainability

TABLE 3.15
Elements of the Cost of Quality

Cost of Compliance	Cost of Non-compliance
• Quality planning	• Scrap
• Process control planning	• Rework
• Quality training	• Corrective action
• Quality audit	• Additional material/inventory cost
• Design review	• Expedition
• Product design validation	• Customer complains
• Work procedure	• Product recalls
• Method statement	• Warranty
• Process validation	• Maintenance service
• Field testing	• Field Repairs
• Third party inspection	• Rectification of returned material
• Receiving inspection	• Re inspection or re-test
• Prevention action	• Downgrading
• In-process inspection	• Loss of business
• Outside endorsement	
• Calibration of equipment	
• Laboratory acceptance testing	

Source: Abdul Razzak Rumane (2013). Quality Tools for Managing Construction Projects. Reprinted with permission of Taylor & Francis Group.

These cost categories allow the use of quality cost data for a variety of purposes. Quality costs can be used for measurement of progress, for analyzing the problem, or for budgeting. By analyzing the relative size of the cost categories, the company can determine if its resources are properly allocated.

3.2.7.3 Quality Cost in Construction

The quality of construction is defined as:

1. Scope of work
2. Time
3. Budget

Cost of quality refers to the total cost incurred during the entire life cycle of the construction project in preventing non-conformance to owner requirements (the defined scope). There are certain hidden costs which may not affect the overall cost of the project directly, although it may cost consultant/designer to complete the design within the schedule stipulated to meet the owner's requirements and for conformance to all the regulatory codes/standards, and for the contractor to construct the project within the stipulated schedule, meeting all the contract requirements. Rejection/non-approval of executed/installed works by the supervisor due to non-compliance with specifications will cause the contractor loss in terms of:

- Material
- Manpower
- Time

The contractor shall have to rework or rectify the work which will need additional resources and will need extra time to do the work as specified. This may disturb the contractor's work schedule and affect execution of other activities. The contractor has to emphasize upon a "Zero Defect" policy, particularly for concrete works. To avoid rejection of works, the contractor has to take the following measures:

1. Execution of works as per approved shop drawings, using approved materials.
2. Following the approved method of statement or manufacturer's recommended method of installation.
3. Conduct the continuous inspection during the construction/installation process.
4. Employ a properly trained workforce.
5. Maintain good workmanship.
6. Identify and correct deficiencies before submitting the checklist for inspection and approval of work.
7. Coordinate requirements of other trades, for example, if any opening is required in the concrete beam for the crossing of a services pipe.

Timely completion of the project is one of the objectives to be achieved. To avoid delay in completion of the schedule, proper planning and scheduling of construction activities is necessary. Since construction projects involve many participants, it is essential that the requirements of all the participants are fully coordinated. This will ensure execution of activities as planned, resulting in timely completion of the project.

Normally, the construction budget is fixed at the inception of a project; therefore, it is required to avoid variations during the construction process as it may take time to get approval of any additional budget, resulting in time extension to the project. Quality costs related to construction projects can be summarized as follows:

Internal Failure Costs:

- Rework
- Rectification
- Rejection of checklist
- Corrective action

Quality Tools for Sustainability

External Failure Costs:

- Breakdown of installed system
- Repairs
- Maintenance
- Warranty

Appraisal Costs:

- Design review/preparation of shop drawings
- Preparation of composite/coordination drawings
- On-site material inspection/test
- Off-site material inspection/test
- Pre-checklist inspection

Prevention Costs:

- Preventive action
- Training
- Work procedures
- Method statement
- Calibration of instruments/equipment

(Source: Abdul Razzak Rumane (2017), Quality Management in Construction Projects, Reprinted with permission from Taylor & Francis Group).

3.2.8 Quality Function Deployment (QFD)

Quality function deployment (QFD) is a technique to translate customer requirements into technical requirements. It was developed in Japan by Dr Yoji Akao in the 1960s to transfer the concepts of quality control from the manufacturing process into the new product development process. QFD is referred to as the "Voice of the Customer" which helps in identifying and developing customer requirements through each stage of product or service development. It is a development process which utilizes a comprehensive matrix involving project team members.

QFD involves constructing one or more matrices containing information related to others. The assembly of several matrices showing the correlation with one another is called "the house of quality." The "house of quality" matrix is the most recognized form of QFD. QFD is being applied in virtually every industry and business from aerospace, communication, software, and transportation to manufacturing, services industry, and the construction industry. The house of quality is made up of following major components:

1. WHATS
2. HOWS
3. Correlation matrix (roof)-technical requirements
4. Interrelationship matrix
5. Target value
6. Competitive evaluation

WHATS is the first step in developing the house of quality. It is a structured set of needs/requirements ranked in terms of priority, with the levels of importance being specified quantitatively. It is generated by using questions such as:

- What are the types of finishes needed for the building?
- What type of air conditioning system is required for the building?
- What type of communication system is required for the building?
- What type of flooring material is required?
- Does the building need any security system?

HOWS is the second step, in which project team members translate the requirements (WHATS) into technical design characteristics (Specifications), and are listed across the columns

The correlation matrix identifies technical interactions or physical relationships among the technical specifications.

The interrelationship matrix illustrates the team member's perception of interrelationship between the owner's requirements and the technical specifications.

The bottom part allows for technical comparison between possible alternatives, target values for each technical design characteristic, and performance measurement.

The right side of the house of quality is used for planning purposes. It illustrates customer perceptions observed in a market survey.

QFD technique can be used to translate an owner's need/requirements into the development of a set of technical requirements during conceptual design.

Figure 3.38 presents the house of quality for hospital building.

3.2.9 SIX SIGMA

3.2.9.1 Introduction

Six Sigma is, basically, a process quality goal. It is a process quality technique that focuses on reducing variation in the process and preventing deficiencies in the product. In a process that has achieved Six Sigma capability, the variation is small compared with the specification limits.

Sigma is a Greek letter – σ – standing for standard deviation. Standard deviation is a statistical way to describe how much variation exists in a set of data, a group of items, or a process. Standard deviation is the most useful measure of dispersion.

Quality Tools for Sustainability

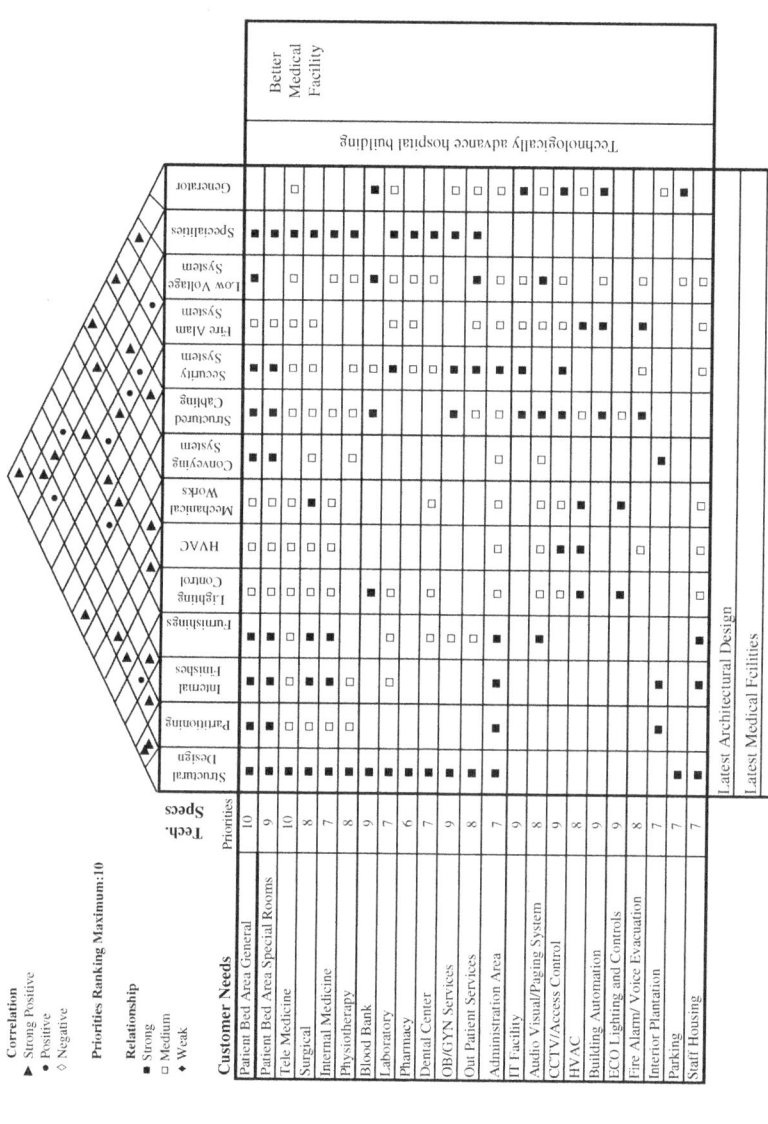

FIGURE 3.38 House of quality for hospital building project. Source: Abdul Razzak Rumane. (2013). *Quality Tools for Managing Construction Projects*, CRC Press, Florida. Reprinted with permission of Taylor & Francis Group.

Six Sigma means that, for a process to be capable at the Six Sigma level, the specification limits should be at least 6 σ from the average point. So, the total spread between the upper specification (control) limit and the lower specification (control) limit should be 12 σ. With Motorola's Six Sigma program, no more than 3.4 defects per million fall outside the specification limits, with a process shift of not more than 1.5 σ from the average or mean). Six Sigma started as a defect reduction effort in manufacturing and was then applied to other business processes for the same purpose.

Six Sigma is a measurement of "goodness," using a universal measurement scale. Sigma provides a relative way to measure improvement. Universal means that sigma can measure anything from coffee mug defects to missed chances to close a sales deal. It simply measures how many times a customer's requirements were not met (a defect), given one million opportunities. Sigma is measured in defects per million opportunities (DPMO). For example, a level of sigma can indicate how many defective coffee mugs were produced when one million were manufactured. Levels of sigma are associated with improved levels of goodness. To reach a level of Three Sigma, you can only have 66,811 defects, given a million opportunities. A level of Five Sigma only allows 233 defects. Minimizing variation is a key focus of Six Sigma. Variation leads to defects, and defects lead to unhappy customers. To keep the customers satisfied, loyal, and coming back, you have to eliminate the sources of variation. Whenever a product is created or a service performed, it needs to be done the same way every time, no matter who is involved. Only then will you truly satisfy the customer. Figure 3.39 presents the Six Sigma roadmap.

3.2.9.2 Six Sigma Methodology

Six Sigma is an overall business improvement methodology that focuses an organization on:

- Understanding and managing customer requirements.
- Aligning a key business process to achieve these requirements.
- Utilizing rigorous data analysis to minimize variation in these processes.
- Driving rapid and sustainable improvement in business process by reducing defects, cycle time, impact to the environment, and other undesirable variations.
- Timely execution.

As a management system, Six Sigma is a high-performance system for executing the business strategy. It uses concept of fact and data to drive better solutions. Six Sigma is a top-down solution to help organizations:

- Align their business strategy to critical improvement efforts.
- Mobilize teams to attack high-impact projects.
- Accelerate improved business results.
- Govern efforts to ensure that improvements are sustained.

Quality Tools for Sustainability

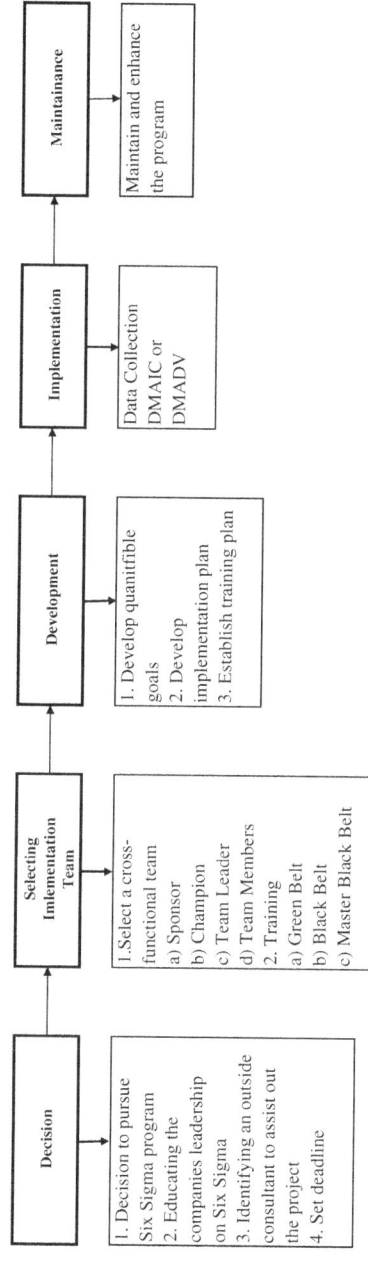

FIGURE 3.39 Six sigma roadmap. Source: Abdul Razzak Rumane (2017). *Quality Management in Construction Projects*, Second Edition. CRC Press, Florida. Reprinted with permission from Taylor & Francis Group.

Six Sigma methodology also focuses on:

- Leadership principles
- Integrated approach to improvement
- Engaged teams
- Analytic tools
- Hard-coded improvements

3.2.9.2.1 Leadership Principles

Six Sigma methodology has four leadership principles:

1. Align
2. Mobilize
3. Accelerate
4. Govern

A brief description of these leadership principles is as follows:

(1) ALIGN: Leadership should ensure that all improvement projects are in line with the organization's strategic goals.

Alignment begins with the leadership team developing a scorecard. This vital tool, the cornerstone of the Six Sigma business improvement campaign, translates strategy into tactical operating terms. The scorecard also defines the metrics which an organization can use to determine success. Just as a scoreboard at a sporting event tells you who is winning, the scorecard tells the leadership how well the company is meeting its goals.

(2) MOBILIZE: Leadership should enable teams to take action by providing clear direction, feasible scope, a definition of success, and rigorous reviews.

Mobilizing sets clear boundaries, lets people go to work, and trains them as required.

The key to mobilizing is focus – a lack of focused action was one of the downfalls of previous business improvement efforts. True focus means the project is correctly aligned with the organization's scorecard. Mobilized teams have a valid reason for engaging in improvement efforts – they can see the benefit for the customer. The project has strategic importance and they know it. They know exactly what must be done and the criteria they can use to determine success.

(3) ACCELERATE: Leadership should drive a project to rapid results through tight clock management, training as needed, and shorter deadlines

More than 70% of all improvement initiatives fail to achieve the desired results in time to make a difference. For projects to make an impact, they must achieve

Quality Tools for Sustainability

results quickly, and that is what acceleration is all about. The Accelerate Leadership Principle involves three main components:

1. Action learning
2. Clock management
3. Effective planning

Accelerate employs an "action learning" methodology to quickly bridge from "learning" to "doing." Action learning mixes traditional training with direct application. Training is received while working on a real-world project, allowing plenty of opportunity to apply new knowledge. The instructor is not simply a trainer, but a coach as well, helping work with a real-world project. Action learning accelerates improvement over traditional learning methods. It helps in receiving training, and also completing a worthwhile project at the same time. In addition to the four- to six-month time frame, Accelerate requires teams to set deadlines that are reinforced through rigorous reviews.

(4) GOVERN: Leadership must visibly sponsor projects and conduct regular and rigorous reviews to make critical mid-course corrections.

The fourth Leadership Principle is Govern. Once leadership selects an improvement opportunity, their work is not done. They must remain ultimately responsible for the success of that project. Govern requires leaders to drive for results.

While governing a Six Sigma project, you need:

- A regular communications plan and a clear review process.
- Actively sponsor teams and their projects.
- To encourage proactive dialogue and knowledge sharing on the team and throughout the organization.

3.2.9.2.2 Six Sigma Team

Teamwork is absolutely vital for complex Six Sigma projects. For teams to be effective, they must be engaged – involved, focused, and committed to meeting their goals. Engaged teams must have leadership support. There are four types of teams:

1) Black Belts
2) Green Belts
3) Breakthrough
4) Blitz

A brief description of each of these teams is as follows:

(1) **Black Belt**

Black Belt teams are led by a Black Belt and may have Green Belts and functional experts assigned to complex, high-impact process improvement projects or to design new products, services, or complex processes. Black Belts are internal Six Sigma practitioners, skilled in the application of rigorous statistical methodologies, and they are crucial to the success of Six Sigma. Their additional training and experience provide them with the skills they need to tackle difficult problems. Black Belts have many responsibilities:

- Function as a Team Leader on Black Belt projects.
- Integrate their functional discipline with statistical, project, and interpersonal skills.
- Serve as internal consultants.
- Tackle complex, high-impact improvement opportunities.
- Mentor and train Green Belts.

(2) **Green Belt**

Led by a Green Belt and comprising non-experts, Green Belt teams tackle less complex, high-impact process improvement projects. Green Belt teams are often coached by Black Belts or Master Black Belts.

Green Belts are also essential to the success of Six Sigma. They perform many of the same functions as Black Belts, but their work requires less complex analysis. Green Belts are trained in basic problem-solving skills and the statistical tools needed to work effectively as members of process improvement teams. Green Belt responsibilities include:

- Acting as Team Leader on business improvements requiring less complex analysis
- Adding their unique skills and experiences to the team
- Working with the team to come up with inventive solutions
- Performing basic statistical analysis
- Conferring with a Black Belt as questions arise

(3) **Breakthrough**

For creating simple processes, sophisticated statistical tools may not be needed. Breakthrough teams are typically used to define low-complexity, new processes.

(4) **Blitz**

Blitz teams are put in place to quickly execute improvements produced by other projects. These teams can also implement Digitization for Efficiency using a new analytic tool set.

For typical Six Sigma projects, four critical roles exist:

1. Sponsor
2. Champion

Quality Tools for Sustainability

3. Team leader
4. Team member

A Sponsor typically:

- Remains ultimately accountable for a project's impact.
- Provides project resources.
- Reviews monthly and quarterly achievements, obstacles, and key actions.
- Supports the project Champion by removing barriers as necessary.

A Champion typically:

- Reviews weekly achievements, obstacles, and key actions.
- Meets with the team weekly to discuss progress.
- Reacts to changes in critical performance measures as needed.
- Supports the Team Leader, removing barriers as necessary.
- Helps ensure project alignment.

A Team Leader typically:

- Leads improvement projects through an assigned, disciplined methodology.
- Works with the Champion to develop the Team Charter, review project progress, obtain necessary resources, and remove obstacles.
- Identifies and develops key milestones, timelines, and metrics for improvement projects.
- Establishes weekly, monthly, and quarterly review plans to monitor team progress.
- Supports the work of team members, as necessary.

Team members typically:

- Assist the Team Leader.
- Follow a disciplined methodology.
- Ensure that the Team Charter and timeline are being met.
- Accept and execute assignments.
- Add their views, opinions, and ideas.

3.2.9.3 Analytic Tool Sets

Following are the analytic tools used in Six Sigma projects.

3.2.9.3.1 FORD GLOBAL 8D TOOL

What problem needs solving? →
Who should help solve problem? →
How do we quantify symptoms? →
How do we contain them? →

What is the root cause? →
What is the permanent corrective action? →
How do we implement? →
How can we prevent this in future? →
Who should we reward? →

FORD GLOBAL 8D TOOL is primarily used to bring performance back to a previous level.

3.2.9.3.2 DMADV TOOL SET PHASES

Define → What is Important?
Measure → What is needed?
Analyze → How will we fulfill?
Design → How do we build it?
Verify → How do we know it will work?

The DMADV tool is used primarily for the invention and innovation of modified or new products, services, or processes. Using this toolset, Black Belts optimize performance before production begins. DMADV is proactive, solving problems before they begin. This tool is also called DFSS (Design for Six Sigma).

Table 3.16 lists the fundamental objectives of DMADV.

3.2.9.3.2.1 DMADV PROCESS

DEFINE PHASE: What is important?
(Define the project goals and customer deliverables)

TABLE 3.16
Fundamental Objectives of Six Sigma-DMADV Tool

DMADV	Phase	Fundamental Objective
1	**Define** – What is important?	Define the project goals and customer deliverables (internal and external)
2	**Measure** – What is needed?	Measure and determine customer needs and specifications
3	**Analyze** – How we fulfill?	Analyze process options and prioritize according to the capabilities to satisfy customer requirements
4	**Design** – How we build it?	Design detailed process(es) capable of satisfying customer requirements
5	**Verify** – How do we know it will work?	Verify the design performance capability

Source: Abdul Razzak Rumane. (2013). Quality Tools for Managing Construction Projects. Reprinted with permission of Taylor & Francis Group

Key deliverables of this phase are:

- Establish the goal
- Identify the benefits
- Select the project team
- Develop the project plan
- Project charter

MEASURE PHASE: What is needed?
(Measure and determine customer needs and specifications)
The key deliverable in this phase is:

- Identify the specification requirements

ANALYZE PHASE: How do we fulfill?
(Analyze the process options and prioritize based on the capability to satisfy customer requirements)
Key deliverables in this phase are:

- Design generation (data collection)
- Design analysis
- Risk analysis
- Design model (prioritization of the data under major variables)

DESIGN PHASE: How do we build it?
(Design the detailed process(es) capable of satisfying customer requirements)
Key deliverables in this phase are:

- Constructing a detailed design
- Converting CTQs (Critical to Quality) into CTPs (Critical to Process elements)
- Estimating the capabilities of the CTPs in the design
- Preparing a verification plan

VERIFY PHASE: How do we know it works?
(Verify the design performance capability)
The key deliverable in this phase is:
- Designing a control and transition plan

3.2.9.3.3 DMAIC TOOL

Define → What is important?
Measure → How are we doing?
Analyze → What is wrong?
Improve → What needs to be done?
Control → How do we guarantee performance?

The DMAIC tool refers to a data-driven quality strategy and is used primarily for improvement of an existing product, service, or process.

Table 3.17 lists the fundamental objectives of DMAIC.

3.2.9.3.3.1 THE DMAIC PROCESS The majority of the time, Black and Green Belts approach their projects with the DMAIC analytic tool set, driving process performance to never-before-seen levels.

DMAIC has following fundamental objectives:

1. Define Phase: Define the project and the customer deliverables.
2. Measure Phase: Measure the process performance and determine the current performance.
3. Analyze: Collect, analyze, and determine the root cause(s) of variation and process performance.
4. Improve: Improve the process by diminishing defects with an alternative remedy.
5. Control: Control the improved process performance.

The DMAIC process contains five distinct steps that provide a disciplined approach to improving existing processes and products through the effective integration of project management, problem solving, and statistical tools. Each step has fundamental objectives and a set of key deliverables, so the team member will always know what is expected of him/her and his/her team.

DMAIC stands for the following:

- **D**efine opportunities
- **M**easure performance

TABLE 3.17
Fundamental Objectives of Six Sigma DMAIC Tool

DMAIC	Phase	Fundamental Objective
1	**Define**—What is important?	Define the project goals and customer deliverables (internal and external)
2	**Measure**—How are we doing?	Measure the process to determine current performance
3	**Analyze** – What is wrong?	Analyze and determine the root cause(es) of the defects
4	**Improve** – What needs to be done?	Improve the process by permanently removing the defects
5	**Control** – How do we guarantee performance?	Control the improved process's performance to ensure sustainable results

Source: Abdul Razzak Rumane (2013). Quality Tools for Managing Construction Projects. Reprinted with permission of Taylor & Francis Group.

Quality Tools for Sustainability

- **A**nalyze opportunity
- **I**mprove performance
- **C**ontrol performance

DEFINE OPPORTUNITIES (What is important?)

The Objective of this phase is:

To identify and/or validate the improvement opportunities that will achieve the organization's goals and provide the largest payoff, develop the business process, define critical customer requirements, and prepare to function as an effective project team.

Key deliverables in this phase include:

- Team charter
- Action plan
- Process map
- Quick win opportunities
- Critical customer requirements
- Prepared team

MEASURE PERFORMANCE (How are we doing?)

The Objectives of this phase are:

- To identify critical measures that are necessary to evaluate the success or failure, meet critical customer requirements, and begin developing a methodology to effectively collect data to measure process performance.
- To understand the elements of the Six Sigma calculation and establish baseline sigma for the processes the team is analyzing.

Key deliverables in this phase include:

- Input, process, and output indicators
- Operational definitions
- Data collection format and plans
- Baseline performance
- Productive team atmosphere

ANALYZE OPPORTUNITY (What is wrong?)

The Objectives of this phase are:

- To stratify and analyze the opportunity to identify a specific problem and define an easily understood problem statement.
- To identify and validate the root causes and thus the problem the team is focused on.
- To determine true sources of variation and potential failure modes that lead to customer dissatisfaction.

Key deliverables in this phase include:

- Data analysis
- Validated root causes
- Sources of variation
- Failure modes and effects analysis (FMEA)
- Problem statement
- Potential solutions

IMPROVE PERFORMANCE (What needs to be done?)
The Objectives of this phase are:

- To identify, evaluate, and select the appropriate improvement solutions.
- To develop a change management approach to assist the organization in adapting to the changes introduced through solution implementation.

Key deliverables in this phase include:

- Solutions
- Process maps and documentation
- Pilot results
- Implementation milestones
- Improvement impacts and benefits
- Storyboard
- Change plans

CONTROL PERFORMANCE (How do we guarantee performance?)
The Objectives of this phase are:

- To understand the importance of planning and executing the plan, and determining the approach to be taken to ensure achievement of the targeted results.
- To understand how to disseminate lessons learned, identify replication and standardization opportunities/processes, and develop related plans.

Key deliverables in this phase include:

- Process control systems
- Standards and procedures
- Training
- Team evaluation
- Change implementation plans
- Potential problem analysis
- Solution results
- Success stories
- Trained associates

Quality Tools for Sustainability

- Replication opportunities
- Standardization opportunities

Six Sigma methodology is not so commonly used in construction projects, although the DMAIC tool can be applied at various stages in construction projects. These stages are:

1. Detailed Design Stage – To enhance the coordination method in order to reduce repetitive work.
2. Construction Stage – Preparation of the builders' workshop drawings and composite drawings, as they need lots of coordination among different trades.
3. Construction Stage – Preparation of the contractor's construction schedule.
4. Execution of Works.

3.2.9.3.4   DMADDD TOOL

Define	→	Where must we be leaner?
Measure	→	What's our baseline?
Analyze	→	Where can we free capacity and improve yields?
Design	→	How should we implement?
Digitize	→	How do we execute?
Draw Down	→	How do we eliminate parallel paths?

The DMADDD TOOL is primarily used to drive the cost out of a process by incorporating digitization improvements. These improvements can drive efficiency by identifying non-value-added tasks and using simple web-enabled tools to automate certain tasks and improve efficiency. In doing so, employees can be freed up to work on more value-added tasks.

Table 3.18 lists the fundamental objectives of DMADDD.

3.2.9.4 Impact of the Six Sigma Strategy

The Six Sigma strategy affects five fundamental areas of business:

1. Process improvement
2. Product and service improvement
3. Customer satisfaction
4. Design methodology
5. Supplier improvement

(Source: Abdul Razzak Rumane (20170), Quality Management in Construction Projects, Reprinted with permission from Taylor & Francis Group).

3.2.10 TRIZ

TRIZ is short for *teirija rezhenijia izobretalenksh zadach* (theory of inventive problem-solving), developed by the Russian scientist Genrish Altshuller. TRIZ provides systematic methods and tools for analysis and innovative problem-solving to support the decision-making process.

TABLE 3.18
Fundamental Objectives of the Six Sigma DMADDD Tool

DMADDD	Phase	Fundamental Objective
1	**Define** – Where must we be learners?	Identify potential improvements
2	**Measure** – What's our baseline?	Analog touch points
3	**Analyze** – Where can we free capacity and improve yields?	Task elimination and consolidated ops. Value-added/non-value-added tasks Free capacity and yield
4	**Design** – How should we implement?	Future state vision Define specific projects Define drawdown timing Define commercialization plans
5	**Digitize** – How do we execute?	Execute the project
6	**Drawdown** – How do we eliminate parallel paths?	Commercialize the new process Eliminate the parallel path

Source: Abdul Razzak Rumane (2013). Quality Tools for Managing Construction Projects. Reprinted with permission of Taylor & Francis Group.

Continuous and effective quality improvement is critical for an organization's growth, sustainability, and competitiveness. The cost of quality is associated with both chronic and sporadic problems. Engineers are required to identify, analyze the causes, and solve these problems by applying various quality improvement tools. Any of these quality tools taken individually does not allow a quality practitioner to carry out a whole problem-solving cycle. These tools are useful for solving a particular type of problem and need a combination of various tools and methods to find problem solutions. TRIZ is an approach which starts at a point where fresh thinking is needed to develop a new process or to redesign a process. It focuses on a method for developing ideas to improve a process, get something done, design a new approach, or redesign an existing approach. TRIZ offers a more systematic, although still universal, approach to problem-solving. TRIZ has advantages over other problem-solving approaches in terms of time efficiency and is also a low-cost quality improvement solution. The pillar of TRIZ is the realization that contradictions can be methodically resolved through the application of innovative solutions. Altshuller defined an inventive problem as one containing a contradiction. He defined contradiction as a situation where an attempt to improve one feature of a system detracts from another feature.

3.2.10.1 TRIZ Methodology

Traditional processes for increasing creativity have a major flaw in that their usefulness decreases as the complexity of the problem increases. At times, a trial-and-error method is used in every process and the number of trials increases with the complexity of the inventive problem. In 1946, Altshuller determined to improve the inventive process by developing the "Science" of creativity, which led to the creation of TRIZ. TRIZ was developed by Altshuller as a result of analysis of many thousands of patents. He reviewed over 200,000 patents looking for problems and how they are solved. He selected 40,000 as being representative of inventive solutions, while the rest were direct improvements easily recognized within the system. Altshuller recognized a pattern where some fundamental problems were solved with solutions that were repeatedly used from one patent to another, although the patent subject, applications, and timings varied significantly. Altshuller categorized these patterns into five levels of inventiveness. Table 3.19 summarizes the level of inventiveness of Altshuller's findings.

He noted that, with each succeeding level, the source of the solution required broader knowledge and more solutions to consider before an ideal solution could be found.

TRIZ is a creative thinking process which provides a highly structured approach for generating innovative ideas and solutions for problem solving. It provides tools and methods for use in problem formulation, system analysis, failure analysis, and the pattern of system evolution. TRIZ is in contrast to techniques such as brainstorming and aims to create an algorithmic approach to the invention of new systems and the refinement of old systems. Using TRIZ requires some training and a good deal of practice.

The TRIZ Body of Knowledge contains 40 creative principles drawn from an analysis of how complex problems have been solved.

- The laws of systems solution.
- The algorithm of inventive problem-solving.
- Substance-field analysis.
- Seventy-six standard solutions.

3.2.10.2 Application of TRIZ

Engineers can use TRIZ for solving the following problem types in construction projects:

- Non-availability of a specified material.
- Regulatory changes to use certain types of material.
- Failure of a dewatering system.
- Casting of a lower grade of concrete to that of a specified higher grade.
- Collapse of a trench during excavation.

TABLE 3.19
Level of Inventiveness

Level	Degree of Inventiveness	% of Solutions	Source of Solution
1	Obvious solution	32%	Personal skill
2	Minor improvement	45%	Knowledge within existing systems
3	Major improvement	18%	Knowledge within the industry
4	New concept	4%	Knowledge outside industry and are found in science, not in technology
5	Discovery	1%	Outside the confines of scientific knowledge

Source: Abdul Razzak Rumane (2010). Quality Management in Construction Projects. Reprinted with permission of Taylor & Francis Group.

- Collapse of formwork.
- Collapse of a roof slab while casting is in progress.
- Chiller failure during peak hours in the summer.
- Modifying a method statement.
- A quality auditor can use TRIZ to develop corrective actions to the audit findings during auditing.

3.2.10.3 TRIZ Process

Altshuller recommended four steps to invent new solutions to a problem:

Step 1 – Identify the problem
Step 2 – Formulate the problem
Step 3 – Search for a precisely well-solved problem
Step 4 – Generate multiple ideas and adapt a solution

The above referenced methods are primarily used for low-level problems. To solve more difficult problems, more precise tools are used. These are as follows:

1. ARIZ (Algorithm for Inventive Problem Solving)
2. Separation Principles
3. Substance-Field Analysis
4. Anticipator Failure Determination
5. Direct Product Evaluation

The quality function deployment (QFD) matrix is also used to identify new functions and performance levels to achieve truly existing levels of quality by eliminating technical bottlenecks at the conceptual stage. QFD may be used to feed data into TRIZ, especially using the "rooftop" to help develop contradictions.

The different schools for TRIZ and individual practitioners have continued to improve and add to the methodology.

3.3 TOOLS FOR PROJECT DEVELOPMENT

Project development is a process spanning from the project inception through to the completion of the project with close out and finalizing project records after project construction and handing over of the project. The project development process is initiated in response to an identified need. It covers a range of time-framed activities extending from identification of a project need to a finished set of contract documents and construction.

Project development process has four major elements/stages:

1. Study
2. Design
3. Bidding and tendering
4. Construction

As the project develops, more information and specifications are developed.

Table 3.20 describes the project development stages.

TABLE 3.20
Project Development Stages

Sr. No.	Stages	Elements	Description
1	Study	Problem statement/needs identification	Project needs, goals, and objectives
		Needs assessment	Identification of needs
			Prioritization of needs
			Leveling of needs
			Deciding what needs to be addressed
		Need analysis	Perform the project need analysis/study to outline the scope of issues to be considered in the planning phase
		Needs statement	Develop the project need statement
		Feasibility study	Technical studies, economics assessment, financial assessment, scheduling, market demand, risk, environmental and social assessment
		Establish project goals and objectives	Scope, time, cost, quality
		Identify alternatives	Identify alternatives based on a predetermined set of performance measures
		Preliminary schedule	Estimate the duration for completion of the project/facility
		Preliminary financial implications	Preliminary budget estimates of total project cost (life cycle cost) on the basis of any known research and development requirements. This will help arrange the finances (funding agency)
		Preliminary resources	Estimate resources
		Project risk	Risks, constraints
		Authority clearance	Identify issues, sustainability, impacts, and potential approvals (environmental, authorities, permits) required for subsequent design and authority approval processes.
		Select preferred alternative	Select preferred alternative considering technological and economical feasibility
		Identify project delivery system	Establish how the participants, owner, designer (A/E), and contractor will be involved to construct the project/facility. (Design/Build/Bid, Design/Build, Guaranteed Maximum Price, CM type, PM type, BOT, Turnkey, etc.)

(Continued)

TABLE 3.20 CONTINUED
Project Development Stages

Sr. No.	Stages	Elements	Description
		Identify type of contracting system/pricing	Select contract pricing system such as firm fixed price or lump sum, unit price, cost reimbursement (cost plus), re-imbursement, target price, time and material, guaranteed maximum price
		Develop the project charter	Terms of Reference (TOR)
		Identify project team	Select Designer (A/E) firm if design/bid/build type of contract system is selected. Select other team members based on project delivery system requirements
		Project launch	Proceed with concept design
2	Design	Develop concept design	Report, drawings, models, presentation
		Regulatory approvals	Obtain regulatory approvals
		Project planning	Prepare project plan
		Schematic design	Preliminary design, value engineering
		Design development	Detailed design
		Construction documents	Construction contract documents
3	Bidding and Tendering	Bid documents	Tendering documents
		Selection of Contractor	Advertisement for bidders
			Pre-qualification of contractors
			Issuing tender documents/ Request for proposal
			Receipt of bid documents
			Tabulation of proposals
			Analysis of proposal
			Selection of contractor
			Most competitive bidder (low-bid, qualification-based)
		Award Contract	Signing of contract
			Bonds and Insurance
			Notice to proceed
4	Construction	Construction	Contractor to execute contracted works
		Monitoring and Control	Monitor and control scope, schedule, cost, quality, risk, procurement of the project
		Commissioning and Handover	Testing, commissioning and handover
		Project close out	Close the project

Source: Abdul Razzak Rumane (2013). Quality Tools for Managing Construction Projects. Reprinted with permission of Taylor & Francis Group.

Quality Tools for Sustainability

3.3.1 Tools for the Study Stage

Table 3.21 lists the major elements of the study stage and related tasks performed during this stage.

Figure 3.40 presents a logic flow diagram in the study stage (conceptual design stage)

Table 3.22 presents the points to be considered for a Needs Assessment for the development of the project.

Table 3.23 presents the points to be considered for a Feasibility Study for the development of the project.

Table 3.24 describes the contents of the Terms of Reference (TOR).

Some of the tools discussed earlier can be applied during the study stage. Table 3.25 is an example of how 5W2H (under the Innovation and Creative tool) can be used to analyze the project needs.

3.3.2 Tools for the Design Stage

Table 3.26 presents major elements of the design stage and related tasks performed during this stage.

Table 3.27 presents points to be considered for Data Collection to facilitate development of concept design.

Table 3.28 presents points to be considered for the Development of Concept Design.

Figure 3.41 presents the preliminary Project Plan.

Table 3.29 presents points to be considered for Development of Schematic Design.

Figure 3.42 presents major activities in the Detailed Design Phase.

Some of the tools discussed earlier can be applied during the design stage. These are as follows:

- Innovative and Creative Tool
  - Six Sigma-DMADV
- Process Analysis Tool
  - Cost of Quality
- Process Improvement Tool
  - PDCA Cycle
- Lean Tool
- Mistake Proofing

3.3.3 Tools for the Bidding and Tendering Stage

Table 3.30 describes the major activities during the Bidding and Tendering stage and related tasks performed during this stage.

TABLE 3.21
Major Elements of the Study Stage

Serial Number	Elements	Description	Related Tasks
1	Problem Statement/Need Identification	Project needs, goals, and objectives.	Strategic objectives, policies, and priorities
2	Need Assessment	Identification of needs Prioritization of needs Leveling of needs Deciding what needs to be addressed	Ensure that the owner's business case has been properly considered
3	Need Analysis	Perform project needs analysis/study to outline the scope of issues to be considered in the planning phase	Perform Need Analysis
4	Need Statement	Develop project need statement	Develop Need Statement
5	Feasibility Study	Technical studies, economics assessment, financial assessment, scheduling, market demand, risk, environmental and social assessment	Perform Feasibility Study. Statement
6	Establish Project Goals and Objectives	Scope, time, cost, quality	Project initiation documents developed on SMART concept
7	Identify alternatives	Identify alternatives based on a predetermined set of performance measures	Select conceptual alternatives
8	Preliminary schedule	Estimate the duration for completion of project/facility	Establish the project schedule
9	Preliminary financial implications	Preliminary budget estimates of total project cost (life cycle cost) on the basis of any known research and development requirements This will help arrange the finances (funding agency)	Determine the project budget
10	Preliminary resources	Estimate resources	Confirm availability of resources, manpower, material, equipment
11	Project risk	Identify project risk, constraints	Establish risk response, mitigation plan

(Continued)

TABLE 3.21 CONTINUED
Major Elements of the Study Stage

Serial Number	Elements	Description	Related Tasks
12	Authorities' clearance	Identify issues, sustainability, impacts, and potential approvals (environmental, authorities, permits) required for subsequent design and authority approval processes	Establish requirements for statutory approvals and other regulatory authorities
13	Select preferred alternative	Assess technological and economical feasibility and compare to the preferred option/alternative to prepare business case	Discuss relative merits of various alternative schemes and evaluate the performance measures to meet the owner's needs/ requirements. Consider social, economical, environmental impact, safety, reliability, and functional capability
14	Identify project delivery system	Establish how the participants, owner, designer (A/E), and contractor will be involved in constructing the project/facility (Design/Build/Bid, Design/Build, Guaranteed Maximum Price, CM type, PM type, BOT, Turnkey, etc.)	Select suitable project delivery system as per strategic decision and suitability of appropriate system
15	Identify type of contracting system	Select contract pricing system such as firm fixed price or lump sum, unit price, cost reimbursement (cost plus), re-imbursement, target price, time and material, guaranteed maximum price.	Select contracting/pricing most appropriate for the benefit of owner
16	Identify project team	Select Designer (A/E) firm if the design/bid/build type of contract system is selected. Select other team members based on project delivery system requirements	Select project team considering the selected type of project delivery system and procurement process of the organization
17	Project launch	Project charter	Prepare Terms of Reference (TOR)

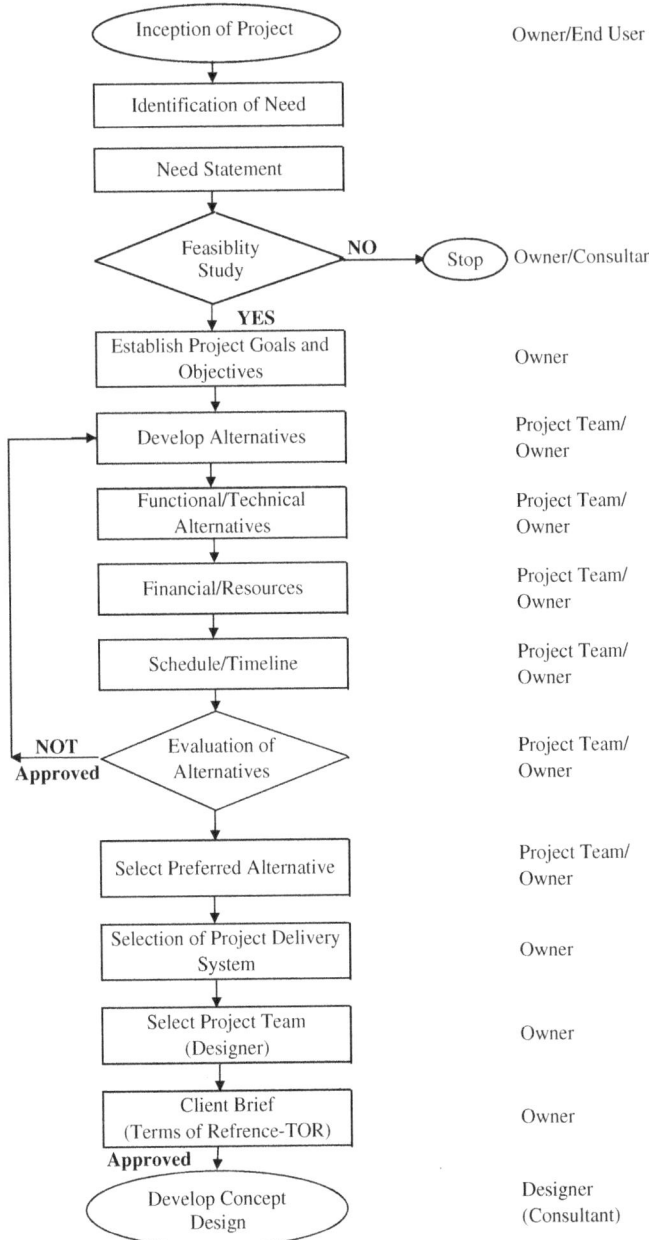

FIGURE 3.40 Logic flow diagram of activities in the study stage.

TABLE 3.22
Need Assessment for Development of Project

Serial Number	Points to be Considered
1	Ensure that the need is properly defined
2	Confirm that the need outcome will benefit the owner/end-user
3	Gather and analyze the owner's/end-user's requirements
4	Set priorities and establish criteria for solutions to meet the targeted demand or requirements
5	Identify and measure areas for improvement to achieve the objectives of the project

TABLE 3.23
Major Considerations for Feasibility Study for Development of a Project

Serial Number	Points to be Considered
1	Technical suitability of facility for intended use by the owner/end-user
2	Economical feasibility to ascertain value of benefit that results from the project exceeding the cost that results from the project
3	Financial payback period
4	Market demand
5	Environmental impact
6	Social and cultural assessment
7	Legal and regulatory impacts
8	Political aspects
9	Resources availability
10	Scheduling of the project
11	Operational
12	Risk analysis

TABLE 3.24
Typical Contents of Terms of Reference (TOR) Documents

Serial Number	Topics
1	Project Objectives
	1.1 Background
	1.2 Project information
	1.3 General requirements
	1.4 Special considerations
2	Project requirements
	2.1 Scope of work
	2.2 Work program
	2.2.1 Study phase
	2.2.2 Design phase
	2.2.3 Tender stage
	2.2.4 Construction phase
	2.3 Reports and presentations
	2.4 Schedule of requirements
	2.5 Drawings
	2.6 Energy conservation considerations
	2.7 Cost estimates
	2.8 Time program
	2.9 Interior finishes
	2.10 Asthetics
	2.11 Mechanical
	2.12 HVAC
	2.13 Lighting
	2.14 Engineering systems
3	Opportunities and constraints
	3.1 Site location
	3.2 Site conditions
	3.3 Land size and access
	3,2 Climate
	3.3 Time
	3.4 Budget
4	Performance target
	4.1 Financial performance
	4.1.1 Performance bond
	4.1.2 Insurance
	4.1.3 Delay penalty
	4.2 Energy performance target
	4.2.1 Energy conservation
	4.3 Work program schedule
5	Environmental considerations
6	Design approach
	6.1 Procurement strategies
	6.2 Design parameters
	6.2.1 Architectural design
	6.2.2 Structural design

(*Continued*)

TABLE 3.24 CONTINUED
Typical Contents of Terms of Reference (TOR) Documents

	6.2.3 Mechanical design
	6.2.4 HVAC design
	6.2.5 Electrical design
	6.2.6 Information and Communication Technology
	6.2.7 Conveying system
	6.2.8 Landscape
	6.2.9 External works
	6.2.10 Parking
	6.3 Sustainable architecture
	6.4 Engineering systems
	6.5 Value engineering study
	6.6 Design review by client
	6.7 Selection of products/systems
7	Specifications and contract documents
8	Project control guidelines
9	Submittals
	9.1 Reports
	9.2 Drawings
	9.3 Specifications
	9.4 Models
	9.5 Sample boards
	9.6 Mock up
10	Presentation
11	Project team members
	8.1 Number of project personnel
	8.2 Staff qualification
	8.3 Selection of specialists
12	Visits

Source: Abdul Razzak Rumane (2013). Quality Tools for Managing Construction Projects. Reprinted with permission of Taylor & Francis Group.

TABLE 3.25
5W2H Analysis for Project Need

Serial Number	Why/What/Who etc.	Related Analyzing Question
1	Why	Why develop a new project?
2	What	What advantage will it have over other similar projects?
3	Who	Who will be the customer for this project?
4	Where	Where can we find the market for the project?
5	When	When will the project be ready?
6	How many	How many such projects are in the market?
7	How much	How much market share will we have from this project

TABLE 3.26
Major Elements of the Design Stage

Serial Number	Elements	Description	Related Task
1	Develop concept design	Report, drawings, models, presentation	Data collection Design development points
2	Regulatory approvals	Obtain regulatory approvals	Submission of drawings and related documents to authorities and obtain their approvals
3	Project planning	Prepare project plan	Develop project plan considering
4	Schematic design	Preliminary design, value engineering	Develop schematic design. Preliminary (outline) specifications, authority approvals, value engineering
5	Design development	Detailed design	Develop detailed design
6	Construction/Project documents	Construction/Project contract documents	Preparation of particular specifications, contract drawing.

TABLE 3.27
Major Items for Data Collection During the Concept Design Phase

Serial Number	Items to be Considered
1	Certificate of title a. Site legalization b. Historical records
2	Topographical Survey a. Location plan b. Site visits c. Site coordinates d. Photographs
3	Geotechnical investigations
4	Field and laboratory test of soil and soil profile
5	Existing structures in/under the project site
6	Existing utilities/services passing through the project site
7	Existing roads, structure surrounding the project site
8	Shoring and underpinning requirements with respect to adjacent area/structure
9	Requirements to protect neighboring area/facility
10	Environmental studies
11	Daylighting requirements
12	Wind load, seismic load, dead load, and live load
13	Site access/traffic studies
14	Applicable codes, standards, and regulatory requirements
15	Usage and space program
16	Design protocol
17	Scope of work/client requirements

TABLE 3.28
Development of the Concept Design

Serial Number	Points to be Considered
1	Project goals
2	Usage
3	Incorporate requirements from collected data
4	Technical and functional capability
5	Esthetics
6	Constructability
7	Sustainability (environmental, social, and economic)
8	Health and Safety
9	Reliability
10	Environmental compatibility
11	Sustainability
12	Fire protection measures
13	Supportability during maintenance/Maintainability
14	Cost-effective over the entire life cycle (Economy)
15	LEED (Equivalent) compliance
16	Reports, drawings, models,

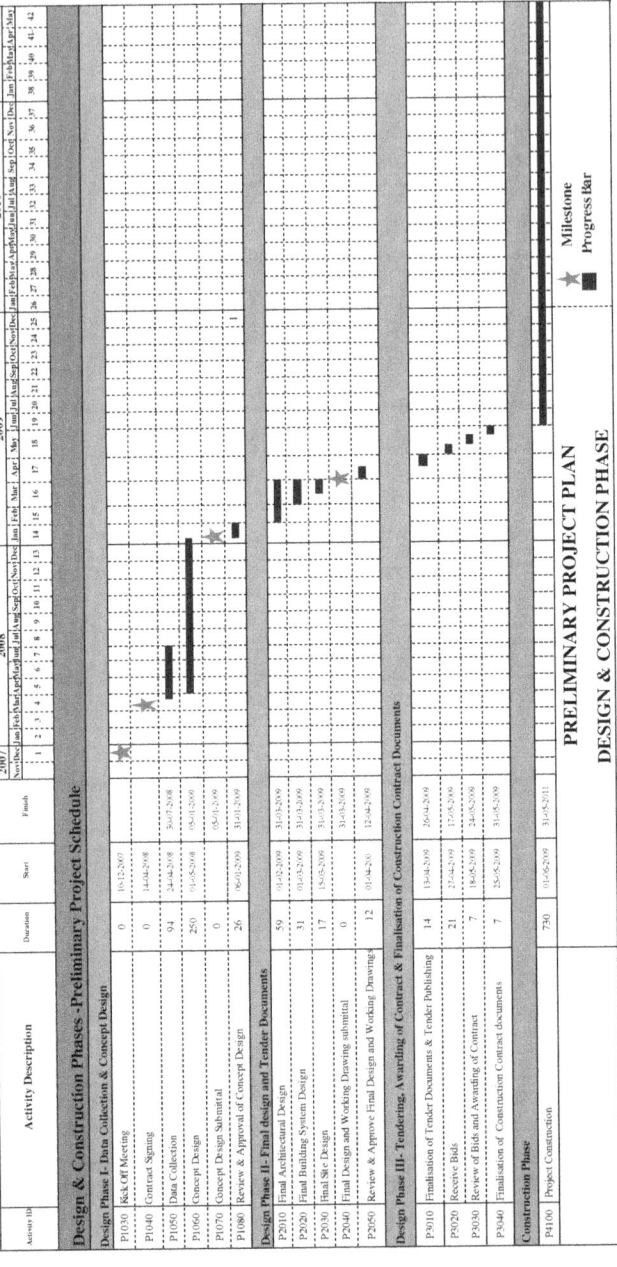

FIGURE 3.41 Preliminary schedule for construction project.

TABLE 3.29
Development of the Schematic Design for a Construction Project

Serial Number	Points to be Considered
1	Concept design deliverables
2	Calculations to support the design
3	System schematics for an electromechanical system
4	Coordination with other members of the project team
5	Authorities' requirements
6	Availability of resources
7	Constructability
8	Health and Safety
9	Reliability
10	Energy conservation issues
11	Environmental issues
12	Selection of systems and products which support functional goals of the entire facility
13	Sustainability
14	Requirements of all stakeholders
15	Optimized life cycle cost (value engineering)

3.3.4 Tools for the Construction Stage

Table 3.31 presents the major activities during the Construction stage and related tasks performed during this stage.

Table 3.32 presents major activities to be performed by the contractor during the construction phase.

Figure 3.45 presents the Project Monitoring and Controlling Process cycle.

Some of the tools discussed earlier can be applied during the Construction stage. These are as follows:

- Management and Planning Tool
  - Networking Arrow Diagram
- Process Analysis Tool
  - Root Cause Analysis
- Process Improvement Tool
  - Six DMAIC
  - PDCA Cycle
- Classic Quality Tool
  - Flowchart for concrete casting

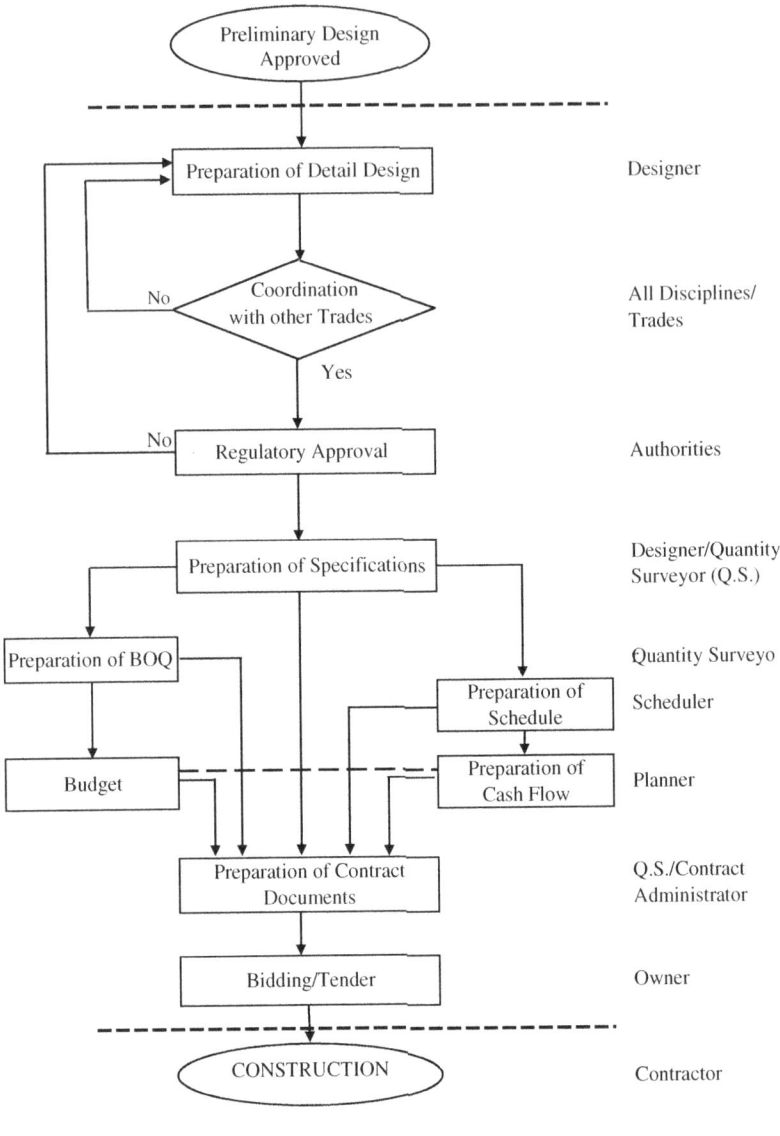

FIGURE 3.42 Major activities in the detailed design phase. Functional relationship. Source: Abdul Razzak Rumane. (2017). *Quality Management in Construction Projects*. Second Edition, CRC Press, Florida. Reprinted with permission of Taylor & Francis Group.

Quality Tools for Sustainability

Internal Cost	External Cost
• Redesign/Redraw to meet fully coordinated design • Rewrite specifications/documents to meet requirements of all other trades	• Incorporate design review comments by Client/Project Manager • Incorporate specifications/documents review comments by Client/Project Manager • Incorporate comments by Regulatory Authority (ies) • Resolve RFI (Request for Information) during Construction
Appraisal Cost	**Prevention Cost**
• Review of Design Drawings • Review of Specifications • Review of Contract Documents to ensure meeting Owner's Needs, Quality Standards, Constructability, and Functionality • Review for Regulatory requirements, Codes	• Conduct technical meetings for proper coordination • Follow quality system • Meeting Submission Schedule • Training of project team members • Update of software used for design

FIGURE 3.43 Cost of quality during the design stage.

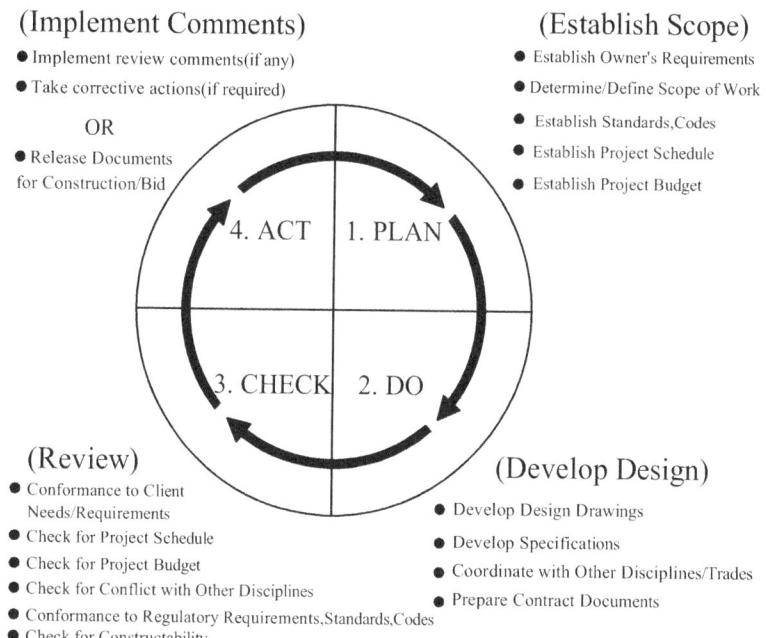

FIGURE 3.44 PDCA cycle for construction projects(design phases). Source: Abdul Razzak Rumane. (2010). *Quality Management in Construction Projects*, CRC Press, Florida. Reprinted with permission from Taylor & Francis Group.

TABLE 3.30
Major Elements of the Bidding and Tendering Stage

Serial Number	Elements	Description	Related Task
1	Bidding	Bid documents	Prepare Tender documents
2	Selection of Contractor	Tender announcement	Advertisement for bidders
			Pre-qualification of contractors
3	Issue tender documents	Request for proposal	Issuing tender documents/Request for proposal
4	Receipt of proposal	Receipt of bid documents	
5	Tender analysis	Tabulation of proposals	Analysis of quotation
			Low-bid
			Qualification-based
6	Selection of Contractor		Most competitive bidder (low-bid or qualification-based as per the company's methodology)
7	Award of contract	Signing of contract	Performance bid
			Insurance
8	Notice to proceed		Contractor to commission

TABLE 3.31
Major Elements of the Construction Stage

Serial Number	Elements	Description	Related Task
1	Construction	Contractor to execute contracted works	Mobilization, managing resources, work execution, QA/QC, works approval (Please refer Table 5.13)
2	Monitoring and control	Monitor and control scope, schedule, cost, quality, risk, procurement of the project	Scheduling, monitoring, and control (Please refer to Figure 5.63)
3	Commissioning and handover	Testing, commissioning, and handover	Project start-up procedures
4	Project closeout	Close the project	Project closeout documents and finalization of claims and payments.

TABLE 3.32
Major Activities by Contractor During the Construction Phase

Serial Number	Activities to be performed by Contractor
1	Mobilization
2	Staff approval
3	Selection of sub contractor(s)
4	Selection of material
5	Selection of resources
6	Preparation of shop drawings
7	Execution of works
8	Project monitoring and control
9	Auditing/Installation of executed works
10	Quality management
11	Approval of works
12	Health, Safety, and Environmental compliance

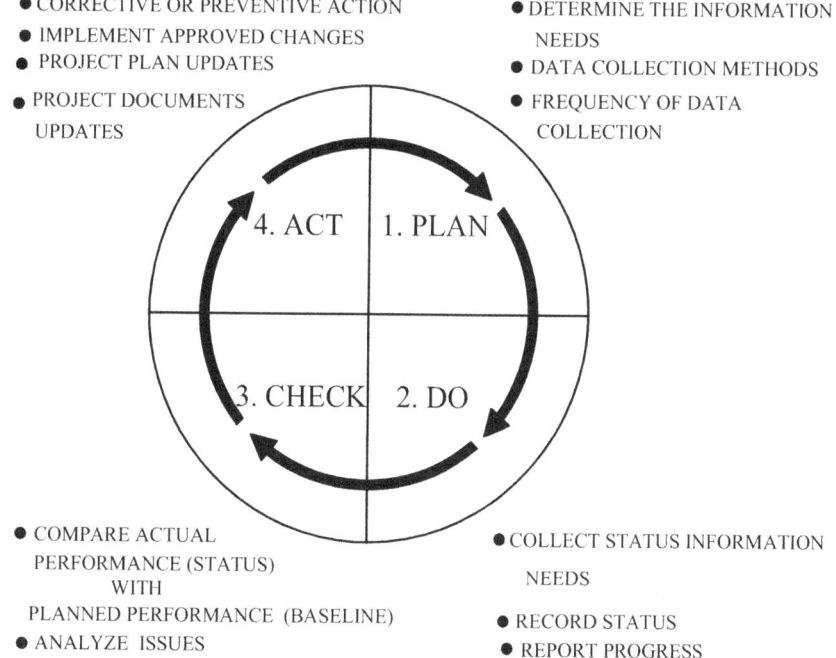

FIGURE 3.45 Project monitoring and controlling process cycle. Source: Abdul Razzak Rumane. (2013). *Quality Tools for Managing Construction Projects*, CRC Press, Florida. Reprinted with permission of Taylor & Francis Group.

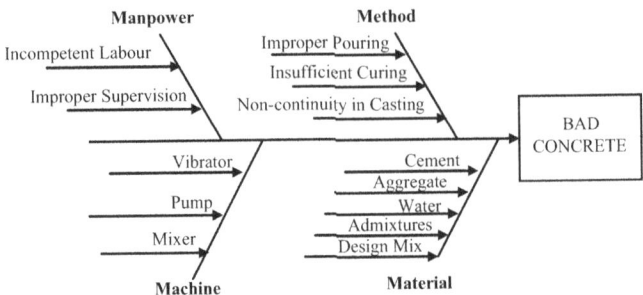

FIGURE 3.46 Root cause analysis for bad concrete. Source: Abdul Razzak Rumane. (2017). *Quality Management in Construction Projects*, CRC Press, Florida. Reprinted with permission of Taylor & Francis Group.

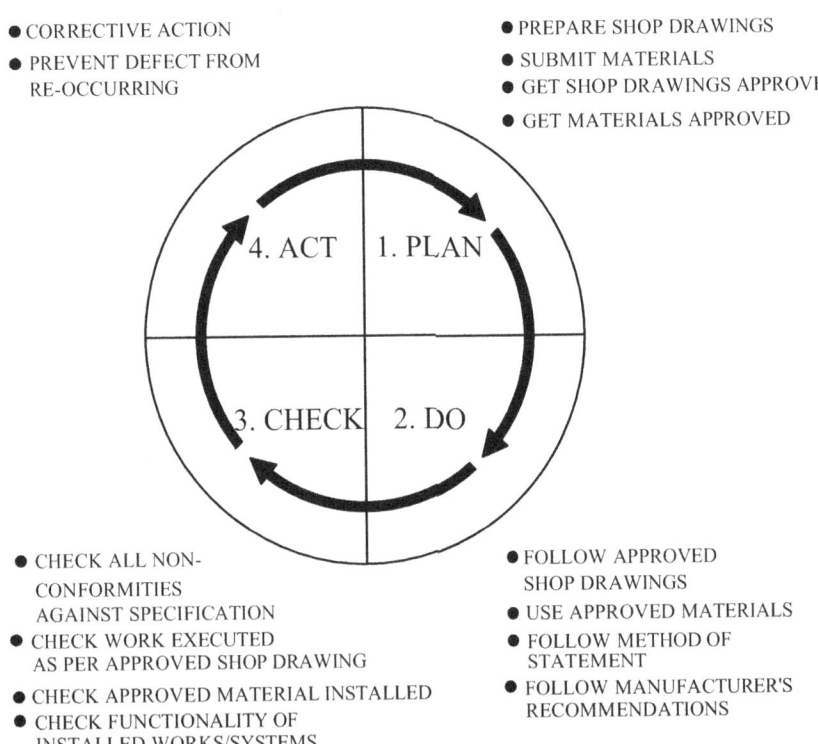

FIGURE 3.47 PDCA cycle (Deming Wheel) for execution of works. Source: Abdul Razzak Rumane. (2013). *Quality Tools for Managing Construction Projects*, CRC Press, Florida. Reprinted with permission of Taylor & Francis Group.

Quality Tools for Sustainability

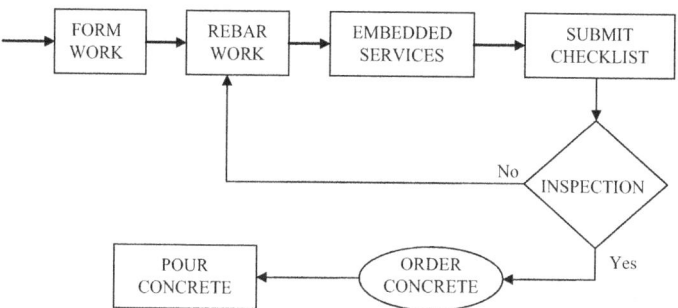

FIGURE 3.48 Flowchart for concrete casting. Source: Abdul Razzak Rumane. (2017). *Quality Management in Construction Projects*, CRC Press, Florida. Reprinted with permission of Taylor & Francis Group.

4 Management of Quality for Sustainability

4.1 INTRODUCTION

Sustainability economical, environmental, and social challenges and requirements to develop a sustainable project have become more complex. Sustainable development is a systematic concept relating to economical, environmental, and social elements.

Some of the characteristics of project sustainability are:

1. Adaptability (affordability)
2. Implementation ability
3. Audit ability
4. Scalability
5. Extensibility
6. Maintainability
7. Manageability

Maintaining the quality and the continuous improvement of this quality are critical components of sustainability in the projects.

The following major elements are to be considered while developing a sustainable project.

1. **Economical**
    - Cost management
    - Economical dimension
    - Economical performance
    - Economical growth
    - Financial benefits and good practices
    - Financial performance
    - Improving productivity
    - Quality management
2. **Environmental**
    - Adherence to environmental policies
    - Biodiversity
    - Eco-efficiency
    - Energy resources
    - Environmental dimension
    - Environmental impact
    - Global warming
    - Natural resources

Management of Quality for Sustainability

- Renewable natural resources
- Water resources
- Waste treatment
- Social dimension

3. **Social**
   - Concept of social justice
   - Engagement of stakeholders
   - Human rights
   - Improving health, comfort, and well-being
   - Labor practices and working conditions
   - Optimizing social benefits
   - Regulatory requirements
   - Relationships with the local community
   - Social reports

Project sustainability is now a common approach related to the management of projects, programs, institutions, organizations, people, and other entities, requiring effective and efficient production, marketing, distribution, and the delivery of products and services.

Table 4.1 lists the benefits of sustainability in the design and construction of projects.

Generally, for projects to be sustained, certain metrics and standards need to be set, from project identification through feasibility studies, formulation, design,

TABLE 4.1
Benefits of Sustainability in the Design and Construction of the Project

1. Economical
1.1	Affordable project
1.2	Energy efficiency
1.3	Optimized lifecycle cost
1.4	Optimal space program
1.5	Reduced operation and maintenance cost

2. Environmental
2.1	Environment-preferred material
2.2	Green Building concept
2.3	Improved air and water quality
2.4	Material conservation
2.5	Noise control
2.6	Pollution control
2.7	Reduce global warming
2.8	Waste management

3. Social
3.1	Enhanced local employment and business opportunities
3.2	Fair labor practices
3.3	Health and safety
3.4	Support esthetic qualities
3.5	Utilization of local resources

appraisal, funding, implementation, monitoring, and evaluation. It is a proven truism that most projects are failing because of the lack of an appropriate sustainability plan. It is therefore very necessary for a comprehensive analysis to be undertaken of the social, economic, legal, cultural, educational, and political environments for project implementation. The project philosophy, mission, vision, values, goals, and objectives should be fully articulated and stated in the plan. The involvement of stakeholders and advocates is of paramount importance since it facilitates some logistical preparation. Beneficial assessment, legal and regulatory framework studies, marketing and competition analysis, partnership development and institutional analysis give room for effective and efficient implementation

In order to conveniently manage and control the project, the project is divided into a number of stages/phases based on the principles of a systems engineering approach. A systems engineering approach to construction projects helps to understand the entire process of project management and to manage and control its activities at different levels of various stages/phases to ensure timely completion of the project with economical use of resources to achieve sustainability in the project.

Each stage/phase is further subdivided into work breakdown structure (WBS) principles to reach a level of complexity where each element/ activity/process can be treated as a single unit which can be conveniently managed.

Systems engineering starts from the complexity of the large-scale problem as a whole and moves toward the structural analysis and partitioning process until the questions of interest are answered. This process of decomposition is called a work breakdown structure (WBS). The WBS is a hierarchical representation of system levels. Being a family tree, the WBS consists of a number of levels, starting with the complete system at level 1 at the top and progressing downward through as many levels as necessary to obtain elements that can be conveniently managed.

The benefits of systems engineering applications are:

- Reduction in the cost of system design and development, production/construction, system operation and support, system retirement, and material disposal
- Reduction in system acquisition time
- Greater visibility and reduction in the risks associated with the design decision-making process

It is difficult to generalize the project life cycle to a system life cycle, considering that there are innumerable processes that make up the construction process and the technologies and processes as applied to systems engineering can also be applied to construction projects. However, the number of stages/phases will depend on the complexity of the project.

In this book, the author has considered four stages for the development of the project. These are:

1. Study stage
2. Design stage
3. Bidding and tendering stage
4. Construction stage

Each stage can further be subdivided into the WBS principle to reach a level of complexity where each element/activity can be treated as a single unit to be conveniently managed. WBS represents a systematic and logical breakdown of the project phase into its components (activities). It is constructed by dividing the project into major elements with each of these being divided into sub-elements. This is done until a breakdown is done in terms of manageable units of work for which responsibility can be defined. WBS involves envisioning the project as a hierarchy of goals, objectives, activities, sub-activities, and work packages. The hierarchical breakdown of activities continues until the entire project is displayed as a network of separately identified and non-overlapping activities. Each activity will be single purposed, of a specific time duration, and manageable; its time and cost estimates will be easily derived, deliverables clearly understood, and responsibility for its completion clearly assigned. The work breakdown structure helps in:

- Effective planning by dividing the work into manageable elements, which can be planned, budgeted, and controlled
- Assignment of responsibility for work elements to project personnel and outside agencies
- Development of control and information systems

WBS facilitates the planning, budgeting, scheduling, and control activities for the project manager and their team. By application of WBS elements, the construction phases are further divided into various activities. Division of these phases will improve the control and planning of the construction project at every stage before a new phase starts. From the perspective of risk management, it will be easy to identify risks in each activity of the project lifecycle.

4.2 PROCESS GROUP

PMBOK® has described five management groups required for the successful completion of any project. These process groups are as follows:

1. Initiating process group
2. Planning process group
3. Executing process group
4. Monitoring and controlling process group
5. Closing process group

These process groups are independent of application areas or industry focus. These groups consist of 47 Project Management Processes and are further grouped into 13 separate Knowledge Areas. These knowledge areas are listed below:

1. Integration Management
2. Stakeholder Management
3. Scope Management
4. Schedule Management
5. Cost Management
6. Quality Management
7. Resource Management

8. Communication Management
9. Risk Management
10. Contract Management
11. Health, Safety, and Environment Management (HSE)
12. Financial Management
13. Claim Management

In order to achieve sustainability in projects, related major activities in these process groups need to be considered. The major activities evolved, based on the Management Processes, are described under five Project Management Process Groups. These activities help to manage and control the project and can also help identify risks involved in each of the activities and treat the risks to achieve project objectives.

4.2.1 Initiating Process Group

Table 4.2 illustrates Major Construction Activities relating to the Project Initiating Process Group.

4.2.2 Planning Process Group

Table 4.3 illustrates Major Construction Activities relating to the Project Planning Process Group.

TABLE 4.2
Major Project Activities Relating to the Initiating Process Group

Serial Number	Management Processes	Activities	Elements
1	Integration Management	1.1 Develop Project Charter	1.1.1 Project Inception
			1.1.2 Problem Statement/ Need Identification
			1.1.3 Need Analysis
			1.1.4 Need Statement
			1.1.5 Need Feasibility
			1.1.6 Project Goals and Objectives
			1.1.7 Project Deliverables
			1.1.8 Design Deliverables
		1.2 Develop Preliminary Scope Statement	1.2.1 Project Terms of Reference (TOR)
			1.2.2 Contract Documents
2	Stakeholder Management	2.1 Identify Stakeholders	2.1 Project Delivery System
			2.2 Project Life Cycle
			2.3 Project Team Members
			2.4 Other Parties

Source: Abdul Razzak Rumane (2016). Handbook of Construction Management: Scope, Schedule, and Cost Control, Second Edition. Reprinted with permission of Taylor & Francis Group.

TABLE 4.3
Major Project Activities Relating to the Planning Process Group

Serial Number	Management Processes	Activities	Elements
1	Integration Management	1.1 Project Baseline Plan	1.1.1 Preliminary Plans
2	Stakeholder Management	2.1 Responsibilities Matrix	2.1.1 Owner, Designer, Contractor, Other Stakeholders
		2.2 Stakeholders Requirement (Work Progress)	2.2.1 Design Progress
			2.2.2 Construction Progress
			2.2.3 Testing, Commissioning, and Handover
		2.3 Change Reporting	2.3.1 Updated Schedule
			2.3.2 Variation Report
			2.3.3 Cost Variation
		2.4 Project Updates	
		2.5 Status Reports	2.5.1 Status Logs
			2.5.2 Performance Reports
			2.5.3 Issue Log
		2.6 Meetings	2.6.1 Kickoff Meeting
			2.6.2 Progress Meetings
			2.6.3 Coordination Meetings
			2.6.4 Other Meetings
		2.7 Payments	2.7.1 Payment Status
3	Scope Management	3.1 Establish Scope Baseline Plan	
		3.2 Collect Requirements	3.2.1 Need Statement
			3.2.2 Project Goals and Objectives
			3.2.3 Project Terms of Reference (TOR)
			3.2.4 Owner's Preferred Requirements
		3.3 Project Scope Documents	3.3.1 Design Development
			• Concept Design
			• Schematic Design
			• Detail Design
			3.3.2 Final Design
			3.3.3 Bill of Quantities
			3.3.4 Project Specifications
			3.3.5 Construction Documents
			3.3.6 Project Deliverables
		3.4 Organizational Breakdown Structure	3.4.1 Project Delivery System
			3.4.2 Organizing
			3.4.3 Staffing
			3.4.4 Project Design

(Continued)

TABLE 4.3 CONTINUED
Major Project Activities Relating to the Planning Process Group

Serial Number	Management Processes	Activities	Elements
		3.5 Work Breakdown Structures	3.5.1 Project Life Cycle
			3.5.2 Work Packages
4	Schedule Management	4.1 Bill of Quantities	4.1.1 Quantities Takeoff
			4.1.2 Sequencing of Activities
			4.1.3 Estimate Activity Resources
			4.1.4 Estimate Duration of Activity
		4.2 Identify Project Assumption	4.2.1 Dependencies
			4.2.2 Risks and Constraints
			4.2.3 Milestone
		4.3 Develop Baseline Schedule	
		4.4 Develop Schedule	4.4.1 Pre-Design Stage
			4.4.2 Design Development
			• Concept Design
			• Schematic Design
			• Detail Design
			4.4.3 Contract Documents
			4.4.4 Bidding, Tendering, and Contract Award
			4.4.5 Construction Phase
			4.4.6 Testing, Commissioning, and Handover
		4.5 Construction Schedule	4.5.1 Contractor's Construction Schedule
5	Cost Management	5.1 Estimate Cost	5.1.1 Conceptual Estimate
			5.1.2 Preliminary Estimate
			5.1.3 Detail Estimate
			5.1.4 Definitive Estimate
		5.2 Estimate Budget	5.2.1 Prepare Budget
		5.3 Determine Project Cost Baseline	5.3.1 S-Curve
			5.3.2 Cost Loading
			5.3.3 Resource Loading
		5.4 Estimate Cost	5.4.1 Estimate Project Resources Cost
			5.4.2 Estimate Project Material Cost
			5.4.2 Estimate Project Equipment Cost
			5.4.3 Bill of Quantities
			5.4.4 BOQ Price Analysis
		5.5 Contracted Project Value	5.5.1 Progress Payments
		5.6 Change Order Procedure	5.6.1 Change Order
			5.6.2 Cost Variation

(Continued)

TABLE 4.3 CONTINUED
Major Project Activities Relating to the Planning Process Group

Serial Number	Management Processes	Activities	Elements
6	Quality Management	6.1 Project Quality Management Plan	6.1.1 Quality Codes and Standards to be Compiled
			6.1.2 Design Criteria
			6.1.3 Design Procedure
			6.1.4 Quality Matrix (Design Stage)
			6.1.5 Well-defined Specifications
			6.1.6 Detailed Construction Drawings
			6.1.7 Quality Matrix (Construction Phase)
			6.1.8 Construction Process
			6.1.9 Detailed Work Procedures
			6.1.10 Quality Matrix (Inspection, Testing during Execution)
			6.1.11 Defect Prevention/Rework
			6.1.12 Quality Matrix (Testing and Handing-Over Start-Up)
			6.1.13 Regulatory Requirements
			6.1.14 Quality Assurance/ Quality Control Procedures
			6.1.15 Reporting Quality Assurance/ Quality Control Problems
			6.1.16 Stakeholders' Quality Requirements
7	Resource Management	7.1 Project Human Resources	7.1.1 Construction/Project Manager
			7.1.2 Designer's Team
			7.1.3 Supervision Team
		7.2 Construction Resources	7.2.1 Contractor's Core Team
			7.2.2 Construction Material
			7.2.3 Construction Equipment
			7.2.3 Construction Labor
			7.2.4 Subcontractor(s)
8	Communication Management	8.1 Communication Plan	8.1.1 Communication Matrix
		8.2 Communication Methods	8.2.1 Design Progress
			8.2.2 Work Progress
			8.2.3 Project Issues
			8.2.4 Project Variations
			8.2.5 Authorities
		8.3 Submittal Procedures	8.3.1 Submittal Procedure
			8.3.1 Progress Payments

(*Continued*)

TABLE 4.3 CONTINUED
Major Project Activities Relating to the Planning Process Group

Serial Number	Management Processes	Activities	Elements
			8.3.2 Progress Reports
			8.3.1 Minutes of Meetings
			8.3.2 Other Meetings
		8.4 Documents	8.4.1 Design Documents
			8.4.2 Contract Documents
			8.4.3 Construction Documents
			8.4.4 As-Built Documents
			8.4.5 Authority-Approved Documents/Drawings
		8.5 Logs	8.5.1 Issue Log
			8.5.2 Correspondence with Stakeholders
			8.5.3 Correspondence with Team Members
			8.5.4 Regulatory Authorities
9	Risk Management	9.1 Risk Identification	9.1.1 During Inception
			9.1.2 During Design
			9.1.3 During Bidding
			9.1.4 During Construction
			9.1.5 During Testing and Commissioning
			9.1.6 During Handing-Over
		9.2 Managing Risk	9.2.1 Risk Register
			9.2.2 Risk Analysis
			9.2.3 Risk Response
10	Contract Management	10.1 Project Delivery System	10.1.1 Selection of CM
			10.1.2 Selection of Designer
		10.2 Bidding and Tendering	10.2.1 Prequalification of Contractors
			10.2.2 Issue Tender Documents
			10.2.3 Acceptance of Tender
11	Health, Safety, and Environment	11.1 Environmental Compatibility	
		11.2 Safety Management Plan	11.2.1 Safety Consideration in Design
			11.2.2 HSE Plan for Construction Site Safety
			11.2.3 Emergency Evacuation Plan
		11.3 Waste Management Plan	

(Continued)

TABLE 4.3 CONTINUED
Major Project Activities Relating to the Planning Process Group

Serial Number	Management Processes	Activities	Elements
12	Financial Management	12.1 Financial Planning	12.1.1 Payments to Designer (Consultant), Construction/Project Manager, Contractor
			12.1.2 Material Procurement
			12.1.3 Equipment Procurement
			12.1.4 Project Staff Salaries
			12.1.5 Bonds, Insurance, Guarantees
			12.1.6 Cash Flow
13	Claim Management	13.1 Claim Identification	13.1.1 Design Errors
			13.1.2 Additional Works
			13.1.3 Delays in Payment
		13.2 Claim Quantification	13.2.1 Change Order Procedures
			• Cost
			• Time

Source: Abdul Razzak Rumane (2016). Handbook of Construction Management: Scope, Schedule, and Cost Control, Second Edition. Reprinted with permission of Taylor & Francis Group.

4.2.3 EXECUTING PROCESS GROUP

Table 4.4 illustrates Major Construction Activities relating to the Project Executing Process Group.

4.2.4 MONITORING AND CONTROLLING PROCESS GROUP

Table 4.5 illustrates Major Construction Activities during the Project Monitoring & Controlling Process Group.

4.2.5 CLOSING PROCESS GROUP

Table 4.6 illustrates Major Construction Activities relating to the Project Closing Process Group.

In addition to major activities, elements at various stages derived from the WBS concept, the sustainability elements under the sustainability process (economical, environmental, and social) are to be considered while developing the project.

Construction projects are mainly capital investment projects. They are customized and non- repetitive in nature. Construction projects have become more complex and technical, and the relationships and the contractual grouping of those who are

TABLE 4.4
Major Project Activities Relating to the Project Executing Group

Serial Number	Management Processes	Activities	Elements
1	Integration Management	1.1 Design Development	1.1.1 Concept Design
			1.1.2 Schematic Design
			1.1.3 Detail Design
		1.2 Construction	1.2.1 Notice to Proceed
			1.2.2 Mobilization
			1.2.3 Submittals
			1.2.4 Management Plans
			1.2.5 Execution
			1.2.6 Corrective Actions
			1.2.7 Project Deliverables
		1.3 Implement Changes	1.3.1 Approved Changes
			1.3.1 Preventive Actions
			1.3.2 Defect Repairs
			1.3.3 Rework
			1.3.4 Update Scope
			1.3.5 Update Plans
			1.3.6 Update Contract Documents
2	Stakeholder Management	2.1 Project Status/ Performance Report	2.1.1 Updated Plans
		2.2 Payments	2.2.1 Progress Payments
		2.3 Change Requests	2.3.1 Site Work Instruction
			2.3.2 Change Orders
			2.3.3 Schedule
			2.3.2 Materials
		2.4 Conflict Resolution	
		2.5 Issue Log	
3	Scope Management		
4	Schedule Management		
5	Cost Management		
6	Quality Management	6.1 Quality Assurance	6.1.1 Design Compliance to TOR
			6.1.2 Design Coordination with all Disciplines
			6.1.3 Material Approval
			6.1.4 Shop Drawing Approval
			6.1.5 Method Approval
			6.1.6 Method Statement
			6.1.7 Mock-up
			6.1.8 Quality Audit
			6.1.9 Functional and Technical Compatibility
7	Resource Management	7.1 Project Staff	7.1.1 Project/ Construction Manager Staff
			7.1.2 Supervision Staff

(*Continued*)

TABLE 4.4 CONTINUED
Major Project Activities Relating to the Project Executing Group

Serial Number	Management Processes	Activities	Elements
		7.2 Project Manpower	7.2.1 Core Staff
			7.2.2 Site Staff
			7.2.3 Workforce
		7.3 Team Management	7.3.1 Team Behavior
			7.3.2 Conflict Resolution
			7.3.3 Demobilization Project Workforce
		7.4 Construction Resources	7.4.1 Material
			7.4.2 Equipment
			7.4.3 Subcontractor(s)
8	Communication Management	8.1 Submittals	8.1.1 Shop Drawings
			8.1.2 Material
			8.1.3 Change Orders
			8.1.4 Payments
		8.2 Documentation	8.2.1 Status Log
			8.2.3 Issue Log
			8.2.4 Minutes of Meetings
			8.2.5 Contract Documents
			8.2.6 Specifications
			8.2.7 Payments
		8.3 Correspondence	8.3.1 Stakeholders
			8.3.2 Regulatory Authorities
			8.3.3 Correspondence among Team Members
9	Risk Management	9.1 Manage Risk	9.1.1 Risk Register
			9.1.2 Risk Response
10	Contract Management	10.1 Contract Documents	10.1.1 Notice to Proceed
		10.2 Selection of Subcontractor(s)	
		10.3 Selection of Materials, Systems and Equipment	
		10.4 Execution of Works	
11	Health, Safety, and Environment	11.1 HSE Management Plan	11.1.1 Site Safety
			11.1.2 Preventive and Mitigation Measures
			11.1.2 Temporary Fire Fighting
			11.1.3 Environmental protection
			11.1.4 Waste Management
			11.1.5 Safety Hazards
12	Financial Management		
13	Claim Management		

Source: Abdul Razzak Rumane (2016). Handbook of Construction Management: Scope, Schedule, and Cost Control, Second Edition. Reprinted with permission of Taylor & Francis Group.

TABLE 4.5
Major Project Activities Relating to Monitoring and Controlling Processes Group

Serial Number	Management Processes	Activities	Elements
1	Integration Management	1.1 Project Performance	1.1.1 Design Performance
			1.1.2 Construction Performance
			1.1.3 Project Status
			1.1.4 Management Plans
			1.1.5 Forecasted Schedule
			1.1.6 Forecasted Cost
			1.1.7 Issues
		1.2 Change Management System	1.2.1 Design Changes
			1.2.2 Design Errors
			1.2.3 Change Requests
			1.2.4 Scope Change
			1.2.5 Variation Orders
			1.2.6 Site Work Instruction
			1.2.7 Alternate Material
			1.2.8 Specifications/Methods
		1.3 Change Analysis	1.3.1 Review, Evaluate Changes
			1.3.2 Approve, Delay, Reject Changes
			1.3.3 Corrective Actions
			1.3.4 Preventive Actions
		1.4 Compliance with Contract Documents	
2	Stakeholder Management	2.1 Project Performance	2.1.1 Progress Reports
			2.1.2 Updates
			2.1.3 Safety Report
			2.1.4 Risk Report
		2.2 Project Updates	2.2.1 Contract Documents
		2.3 Payments	10.3.1 Payment Certificate
		2.4 Change Requests	2.4.1 Site Work Instruction
			2.4.2 Change Orders
		2.5 Issue Log	2.5.1 Anticipated Problems
		2.6 Minutes of Meetings	2.6.1 Progress Meetings
			2.6.2 Other Meetings
3	Scope Management (Contract Documents)	3.1 Validate Scope	3.1.1 Conformance to TOR
			3.1.2 Review of Design Documents
			3.1.3 Conformance to Contract Documents
			3.1.4 Approval of Changes

(*Continued*)

TABLE 4.5 CONTINUED
Major Project Activities Relating to Monitoring and Controlling Processes Group

Serial Number	Management Processes	Activities	Elements
			3.1.5 Authorities' Approval of Deliverables
			3.1.6 Stakeholders' Approval of Deliverables
			3.1.7 Quality Audit
		3.2 Scope Change Control	3.2.1 Variation Orders
			3.2.1 Change Orders
		3.3 Performance Measures	
4	Schedule Management	4.1 Schedule Monitoring	4.1.1 Project Status
		4.2 Schedule Control	4.2.1 Progress Curve
		4.3 Schedule Changes	4.3.1 Approved Changes
		4.4 Progress Monitoring	4.4.1 Planned v. Actual
		4.5 Submittals Monitoring	4.5.1 Subcontractors
			4.5.2 Material
			4.5.3 Shop Drawings
5	Cost Management	5.1 Cost Control	5.1.1 Work Performance
			5.1.2 S-Curve
			5.1.3 Forecasted Cost
		5.2 Change Orders	
		5.3 Progress Payment	
		5.4 Variation Orders	
6	Quality Management	6.1 Control Quality	6.1.1 Quality Metrics
			6.1.2 Quality Checklist
			6.1.3 Material Inspection
			6.1.4 Work Inspection
			6.1.5 Rework
			6.1.6 Testing
			6.1.7 Regulatory Compliance
7	Resource Management	7.1 Conflict Resolution	
		7.2 Performance Analysis	
		7.3 Material Management	
8	Communication Management	8.1 Meetings	8.1.1 Progress Meetings
			8.1.2 Coordination Meetings
			8.1.3 Safety Meetings
			8.1.4 Quality Meetings
		8.2 Submittal Control	8.2.1 Drawings
			8.2.2 Material
		8.3 Documents Control	8.3.1 Correspondence

(Continued)

TABLE 4.5 CONTINUED
Major Project Activities Relating to Monitoring and Controlling Processes Group

Serial Number	Management Processes	Activities	Elements
9	Risk Management	9.1 Monitor and Control Risk	9.1.1 Scope Change Risk 9.1.2 Schedule Change Risk 9.1.3 Cost Change Risk 9.1.4 Mitigate Risk 9.1.5 Risk Audit
10	Contract Management	10.1 Inspection 10.2 Checklists 10.3 Handling of Claims, Disputes	
11	Health, Safety, and Environment	11.1 Prevention Measures	11.1.1 Accidents Avoidance/Mitigation 11.1.2 Fire Fighting System 11.1.3 Loss Prevention Measures
		11.2 Application of Codes and Standards	
12	Financial Management	12.1 Financial Control	12.1.1 Payments to Project Team Members 12.1.2 Payments to Contractor(s)/Sub Contractor(s) 12.1.3 Material Purchases 12.1.4 Variation Order Payment 12.1.5 Insurance and Bonds
		12.2 Cash Flow	
13	Claim Management	13.1 Claim Prevention	13.1.1 Proper Design Review 13.1.2 Unambiguous Contract Documents Language 13.1.3 Practical Schedule 13.1.4 Qualified Contractor(s) 13.1.5 Competent Project Team Members 13.1.6 RFI Review Procedure 13.1.7 Negotiations 13.1.8 Appropriate Project Delivery System

Source: Abdul Razzak Rumane (2016). Handbook of Construction Management: Scope, Schedule, and Cost Control, Second Edition. Reprinted with permission of Taylor & Francis Group.

Management of Quality for Sustainability

TABLE 4.6
Major Project Activities Relating to the Closing Process Group

Serial Number	Management Processes	Activities	Elements
1	Integration Management	1.1 Close Project or Phase	1.1.1 Testing and Commissioning
			1.1.2 Authorities' Approvals
			1.1.3 Punch List/Snag List
			1.1.4 Handover of Project/Facility
			1.1.5 As Built Drawings
			1.1.6 Technical Manuals
			1.1.7 Spare Parts
			1.1.8 Lesson Learned
2	Resource Management	2.1 Close Project Team	2.1.1 Demobilization
			2.1.2 New Assignment
		2.2 Material and Equipment	2.2.1 Excess Material Removal/Disposal
			2.2.2 Equipment Removal
3	Contract Management	3.1 Close Contract	3.1.1 Project Acceptance/Takeover
			3.1.2 Issuance of Substantial Completion Certificate
			3.1.3 Occupancy
4	Financial Management	4.1 Financial Administration and Records	4.1.1 Payments to all Contractors, Subcontractors, and Other Team Members
			4.1.2 Bank Guarantees/Warranties
5	Claim Management	5.1 Claim Resolution	5.1.1 Settlement of Claims

Source: Abdul Razzak Rumane (2016). Handbook of Construction Management: Scope, Schedule, and Cost Control, Second Edition. Reprinted with permission of Taylor & Francis Group.

involved are also more complex and contractually varied. The products used in construction projects are expensive, complex, immovable, and long-lived.

Construction projects are constantly increasing in technological complexity. In addition, the requirements of construction clients are on the increase and, as a result, construction products (buildings, products in case of oil and gas projects) must meet varied performance standards and satisfy the customer. Therefore, to ensure the adequacy of the client brief (the project charter), which addresses the numerous complex client/user needs, it has become the responsibility of the designer to evaluate the requirements in terms of activities and their relationships, and follow health, safety, and environmental regulations while designing any building.

Quality management is an organization-wide approach to understanding customer needs and delivering solutions to fulfill these needs and satisfy the customer. Quality management involves managing and implementation of quality system to achieve customer satisfaction at the lowest overall cost to the organization while continuing to improve the process. Quality system is a framework for quality management. It embraces the organization structure, policies, procedures, and processes needed to implement the quality management system.

Quality management system in construction projects mainly consists of:

- Quality management planning
- Quality assurance
- Quality control

Quality tools and techniques provide an efficacious framework to deliver and maintain sustainability in the project.

Construction projects involve a cross-section of many different participants.

The following sections discuss the major stages for the development of sustainability in the projects. Traditionally, the projects involve three main groups of participants. These are:

1. Project owner
2. Designer/Consultant
3. Contractor/Builder/Constructor

In order to achieve sustainability in projects, the goals and objectives of the project are to be based on sustainability requirements, taking into consideration the construction project quality.

The following sections discuss the major stages for the development of sustainability in the projects.

4.3 SUSTAINABILITY IN THE STUDY STAGE

The study stage is the first stage/phase of the construction project lifecycle. It is the project inception stage, sometimes known as the inception phase. It is a relatively short period at the start of a project and is about understanding the project goals and objectives and getting enough information to confirm that the project should proceed or to convince that it should not proceed but abort. Successful completion of inception stage activities and approval of the project by the stakeholder/sponsor mark the initiation of the project.

The primary objectives of this stage are:

- Establish a justification or business case for the project
- Conduct a feasibility study to assess whether the project is technically feasible and economically viable
- Obtain the stakeholders' approval to proceed with the project or to stop the project

Management of Quality for Sustainability

- Establish the project goals and objectives
- Establish the scope and boundary conditions
- Estimate the overall schedule for and cost of the project
- Conduct cost/benefit analysis
- Identify the project quality requirements
- Identify the potential risks associated with the project
- Outline the key requirements that will drive design trade-offs and identify which requirements are critical
- Identify and evaluate the alternatives/options
- Select the preferred alternative
- Develop the design basis
- Establish the project delivery and contracting system
- Develop the project charter

Sustainability focuses on:

- Economical factors
- Environmental issues
- Social requirements

These elements have to be considered at this stage in order to develop the sustainability of the project.

Figure 4.1 illustrates the flowchart for the study stage.

Figure 4.2 illustrates the major activities relating to the study stage process based on the project management process group methodology.

The project development process begins with the project initiation and ends with the project closeout and finalization of project records, settling claims (if any) after completion of the construction of the project. The project initiation starts with the identification of a need and a business case.

4.3.1 Identification of Need

Project development is initiated with the identification of the need to develop a new facility or renovate/refurbish an existing facility. The need for the project is created by the owner. The owner of the facility could be an individual, a public/private sector company, or a governmental agency. The need for the project created by the owner is linked to the available financial resources to develop the facility/project. The owner's needs must be well defined, describing the minimum requirements of quality, performance, project completion date, and approved budget for the project.

The owner's needs are quite simple and are based on the following:

- To have best facility/project for the money, i.e., to have the maximum profit or services at a reasonable cost
- To achieve on-time completion, i.e., to meet the owner's/user's schedule
- To achieve completion within budget, i.e., to meet the investment plan for the facility.

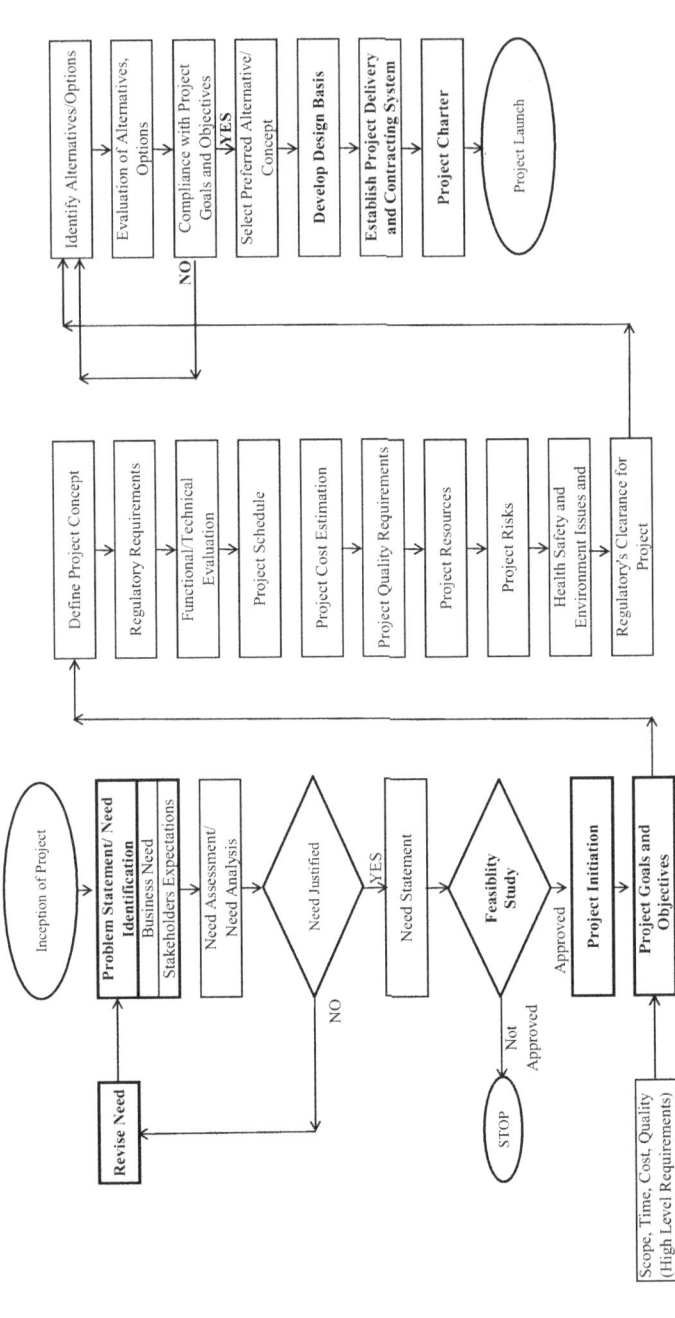

FIGURE 4.1 Flowchart for study stage. Source: Abdul Razzak Rumane. (2021). *Quality Management in Oil and Gas Projects*. Reprinted with permission of Taylor & Francis Group

Management of Quality for Sustainability

Study Stage Phase →

Management Processes	Initiating Process	Planning Process	Execution Process	Monitoring & Controlling Process	Closing Process
Integration Management	Problem Statement/Need Identification	Preliminary Project Management Plan		Monitor Study Stage Process Progress	Project Charter
	Need Analysis			Manage Owner Need, Changes	
	Need Statement			Compliance with SMART concept	
	Feasibility Study			Review of Feasibility Study	
	Project Goals and Objectives			Compliance with Project Goals and Objectives	
		Identification of Alternatives	Development of Alternatives	Evaluation and Approval of Alternative	
	Project Terms of Reference		Development of TOR	Reiew of TOR	
Stakeholder Management	Project Delivery System	Responsibility Matrix	Study Stage Performance Report	Study Stage Progress/Status	
	Project Team Members	Establish Stakeholders Requirements		TOR Progress/Status	
	Regulatory Authorities			Regulatory Clearance	
Scope Management				Compliance to Owner's Need	
				Stakeholder's Approval of TOR	
				Authorities Approval	
Schedule Management		Preliminary Schedule		Study Stage Progress Submission Shedule	
				Approval of Project Schedule	
Cost Management		Conceptual Estimate		Control Project Cost	
Quality Management		Quality Codes and Standards, Regulatory Requirements		Conformance to Technical and Functional Capability	
		Project Quality Requirements			
Resouce Management		Assign Study Stage Team Members		Performance of Team Members	Assign New Project
		Roles and Responsibilities of Team Members			
Communication Management		Study Stage Process Progress Information		Study Stage Status Information	
Risk Management		Project Risk		Control Project Risk	
Contract Management		Project Delivery System		Administer Project Delivery System Requirements	
		Selection of Consultant for Feasibility Study			
HSE Management		HSE Issues and Impact	HSE Consideration is Design (TOR)	Check for Regulatory Requirements	
Financial Management		Consultant Payment		Cost Effectiveness over the Project Life Cycle	Consultant Payment
Claim Management		Management of Owner's Need Changes		Control Changes/Claims	Settle Claims by Consultant

FIGURE 4.2 Major activities relating to study stage process groups. Note: These activities may not be strictly sequential; however, the breakdown allows implementation of project management function to be more effective and easily manageable at different stages of the project phase.

The need should be based on real (or perceived) requirements or deficiencies.

Table 4.7 illustrate the 5W2H quality tool for Identification of Need.

4.3.1.1 Need Analysis

The identified need is then assessed and analyzed to develop the need statement. Need assessment is a systematic process for determining and addressing needs or

TABLE 4.7
5W2H Quality Tool for the Identification of Need

Serial Number	Why?	Related Analyzing Question
1	Why?	Why is there a new project?
2	What?	What advantage will it have over other similar projects?
3	Who?	Who will be the customer(s) for the new project?
4	Where?	Where from the material/equipment to develop the project will be acquired? Will the project comply with the sustainability requirements of materials to be used?
5	When?	When will the project be ready and will it meet sustainability requirements?
6	How many?	How many competitors are there in the market for this project?
7	How much?	How much market share will there be for this project?

"gaps" between current conditions and the desired conditions' "want." Need analysis is the process of identifying and evaluating needs. The need statement is written based on the need analysis and is used to perform a feasibility study to develop project goals and objectives, and subsequently to prepare project scope documents.

Business need assessment is essential to ensure that the owner's business case has been properly considered before the initial Project Brief (Need Statement) is developed.

Table 4.8 lists illustrate the major points to be considered for need analysis.

4.3.1.2 Need Statement

It is essential to get a clear definition of the identified need or the problem to be solved by the new project. The owner's needs must be well defined, indicating the minimum requirements of quality and performance, an approved main budget, and the required completion date. Sometimes, the project budget is fixed and therefore the quality of the building system, materials, and finishes of the project need to be balanced with the budget. A business case typically addresses the business need for the project and the value the project brings to the business (project value proposition). A value proposition is a promise of the value to be delivered by the project. The following questions address the value proposition:

a. How the project solves the current problems or improves the current situation?
b. What specific benefits the project will deliver?
c. Why the project is the ideal solution for the problem?

The Need Statement is developed based on the assessment and analysis of the Identification of Need

TABLE 4.8
Major Points to be Considered for Need Analysis

Serial Number	Points to be Considered
1	Is the project in line with the organization's strategy/strategic plan and mandated by management in support of a specific objective?
2	Is the project a part of the mission statement of the organization?
3	Is the project a part of the vision statement of the organization?
4	Is the need mandated by regulatory body?
5	Is the need to meet government regulations?
6	Is the need to fill a deficiency/gap of such types of project(s) on the market?
7	Is the need created to meet market demand?
8	Is the need to meet the research and development requirements?
9	Is the need for technical advances?
10	Is the need generated to construct a facility/project which is innovative in nature?
11	Is the need to develop infrastructure?
12	Is the need aiming to improve the existing facility?
13	Is the need a part of mandatory investment?
14	Is the need created to resolve a specific problem?
15	Is the need to serve the community and fulfill social responsibilities?
16	Does the need have any time frame to implement?
17	Does the need have an effect on the environment?
18	Does the need have a major risk?
19	Does the need have financial constraints?
20	Will the need be beneficial to the organization's objectives?
21	Is the need within the capability of the owner/client, either alone or in cooperation with other organizations?
22	Can the need be managed and implemented?
23	Is the need realistic and genuine?
24	Is the need measurable?
25	Is the need in compliance with the government's health and safety regulations?
26	Does the need meet the sustainability development requirements as per government policies and programs and regulatory requirements?
27	Is the need in compliance with environmental protection agency requirements (sustainability requirements)?
28	Is the need in compliance with using local resources (sustainability requirements)?
29	Will the need have an impact on global warming/climate change?

The Need Statement is written based on the need analysis and is used to perform a feasibility study to develop project goals and objectives and subsequently to prepare project scope documents.

Table 4.9 illustrates points to be considered for the Need Statement.

4.3.2 Feasibility Study

The feasibility study takes its starting point from the output of the project identification need. The Need Statement is the input to perform a feasibility study. The main purpose of the feasibility study is to evaluate and review the technical/financial viability and conformity with sustainability requirements of the project need to give sufficient information to the client and decide whether to proceed with the project or to abort it. A feasibility study is undertaken to analyze the ability to complete a project successfully, taking into account various factors such as economic, technological, scheduling, etc. A feasibility study looks into the positive and negative effects of a project before investing the company resources, viz., time and money. Depending on the circumstances, the feasibility study may be short or lengthy, simple or complex.

The feasibility study assists decision-makers (investors/owners/clients) in determining whether or not to implement the project. Since the feasibility study stage is a very crucial stage, in which all kinds of professionals and specialists are required

TABLE 4.9
Need Statement

Serial Number	Points to be Considered
1	Project purpose and need a) Project description
2	What is the purpose of the project? a) Project justification
3	Why is the project needed now?
4	How did the need of the project determine? a) Supporting data
5	Is it important to have the needed project?
6	Is such a facility/project required?
7	What are the factors contributing to the need?
8	What is the impact of the need?
9	Will the need improve the existing situation and be beneficial?
10	What are the hurdles?
11	What is the timeline for the project?
12	What are the funding sources for the project?
13	What are the benefits of the project?
14	What is the social impact of the project?
15	What is the environmental impact of this project?
16	What is the social impact of this project?

to bring many kinds of knowledge and experience into a broad-ranging evaluation of feasibility, it is necessary to engage a firm having expertise in the related fields. The feasibility study establishes the broad objectives for the project and so exerts an influence throughout subsequent stages. The successful completion of the feasibility study marks the first of several transition milestones and is therefore most important in determining whether or not to implement a particular project or program. The feasibility study decides the possible design approaches that can be pursued to meet the need.

In any case, it is the principal requirement in project development as it gives the owner/client an early assessment of the viability of the project and the degree of risk involved.

A feasibility study can be categorized into the following functions:

1. Legal
2. Marketing
3. Technical and engineering
4. Financial
5. Sustainability
   - Economical
   - Environmental
   - Social
6. Risk
7. Scheduling of project

The project feasibility study is usually performed by the owner through their own team or by engaging a specialist agency or individual.

The following are the contents of a feasibility study report:

1. Purpose of the feasibility study
2. Project history (project background information)
3. Description of the proposed project.
   a. Project location
   b. Plot area
   c. Interface with the adjacent/neighboring area
   d. Expected project deliverables
   e. Key performance indicators
   f. Constraints
   g. Assumptions
4. Business case
   a. Project need
   b. Stakeholders
   c. Project benefits
   d. Estimated time
   e. Financial benefits
   f. Estimated cost

g. Justification
5. Feasibility study details
    a. Technical
    b. Time scale
    c. Financial
    d. Political, legal
    e. Ecological
    f. Sustainability requirements
        i. Economical considerations
        ii. Environmental impact considerations
        iii. Social impacts
6. Risk
7. Final recommendation

4.3.3 Project Goals and Objectives

After the completion and approval of the feasibility study, it is possible to establish project goals and objectives. The outcome of the feasibility study helps selection of a defined project, which meets the stated project objectives, together with a broad plan of implementation. If the feasibility study shows that the objectives of the owner are best met through the ideas generated, then the project is moved to the next stage to deliver the intended objectives of the project, passing through different stages of the project life cycle.

Project goals and objectives are prepared taking into consideration the final recommendations/outcome of the feasibility study. Clear goals and objectives provide the project team with appropriate boundaries to make decisions about the project and ensure the project/facility will satisfy the owner's/end-user's requirements in fulfilling the owner's needs. Establishing properly defined goals and objectives are the most fundamental elements of project planning. Therefore, the project goals and objectives must be:

1. Specific (is the goal specific?)
2. Measurable (is the goal measurable?)
3. Agreed upon/Achievable (is the goal achievable?)
4. Realistic (is the goal realistic or result-oriented?)
5. Time (cost) limited (does the goal have a time element?)

The project objectives definition usually includes the following information:

1. Project scope and project deliverables
2. Preliminary project schedule
3. Preliminary project budget
4. Specific quality criteria that the deliverables must meet
5. Type of contract to be employed
6. Design requirements

Management of Quality for Sustainability

7. Regulatory requirements
8. Logistic requirements
9. Environmental considerations
10. Sustainability requirements
11. Potential project risks

4.3.4 Regulatory Clearance

Regulatory clearance needs to be obtained, if applicable, to proceed with the project development activities.

4.3.5 Identification of Alternatives/Options

Once the owner/client defines the project objectives, a project team (in-house or outside agency) is selected to start the identification of alternatives. Normally, the owner assigns a specialist consultant (in certain cases, it may be the designer of the project) to identify conceptual alternatives, evaluate conceptual alternatives, and select preferred conceptual alternatives in consultation with the owner and project manager-consultant (PM), if one is commissioned as applicable and is included in the Terms of Reference/Project Charter.

The project goals and objectives serve as a guide for the development of alternatives. The team develops several alternative schemes and solutions. Each alternative is based on a predetermined set of performance measures to meet the owner's requirements. In the case of construction projects, it is mainly the extensive review of development options which is discussed between the owner and the team members. The team provides engineering advice to the owner to enable them to assess the feasibility and relative merits of various alternative schemes to meet the owner's requirements.

Quality tools, such as Brainstorming, the Delphi technique, and 5W2H, can be used to identify alternatives.

4.3.5.1 Analyze Alternatives

Qualitative and quantitative comparison, evaluation, and analysis of the identified alternatives are carried out by considering the advantages and disadvantages of each item systematically. Social, economic (time and cost), sustainability, performance of equipment, environmental impacts, functional capability, and safety and reliability should be considered during the development of alternatives. Each alternative is compared by considering the advantages and disadvantages of each element systematically to meet a predetermined set of performance measures and owner's requirements. Evaluation of the alternatives requires cooperative efforts between the owner and other team members involved in performing evaluation and analysis, regulatory authorities, and other stakeholders who have involvement, interest, or impact on the processes of the construction project. The team makes a brief presentation to the owner, and the project is selected based on the preferred conceptual alternatives.

The following elements are considered to evaluate and analyze each of the identified alternatives:

1. Suitability to the purpose and objectives
2. Sustainability (environmental, social, economical)
3. Cost efficiency
4. Life cycle costing
5. Material handling and optimization
6. Environmental impact
7. Environment-preferred materials and products
8. Physical properties, thermal comfort, insulation, fire resistance
9. Utilization of space
10. Accessibility
11. Ventilation
12. Air quality
13. Power consumption and energy-saving measures – use of renewable energy, alternative energy
14. Day lighting
15. High-performance lighting
16. Green building concept
17. Aesthetic
18. Availability of resources
19. Safety and security
20. Statutory/regulatory requirements
21. Quality requirements
22. Codes and standards
23. Any other critical issues

4.3.5.2 Select Preferred Alternative

Based on the analysis of identified alternatives, the preferred alternative that satisfies the project goals and objectives is selected.

With the approval of the preferred alternative by the owner, the project proceeds towards the next stage of the project development process.

Figure 4.3 illustrates the Evaluation and Analysis Method to select the preferred alternative.

4.3.6 Development of the Design Basis

The selected preferred alternative is the base for development of the concept design. The project charter (TOR) defines the activities to be performed by the designer and also the deliverables that have to be prepared for further proceedings.

4.3.7 Establish Project Delivery and Contracting System

After establishment of the project objectives, the owner/client develops the project procurement strategy by selecting a particular type of project delivery system. The

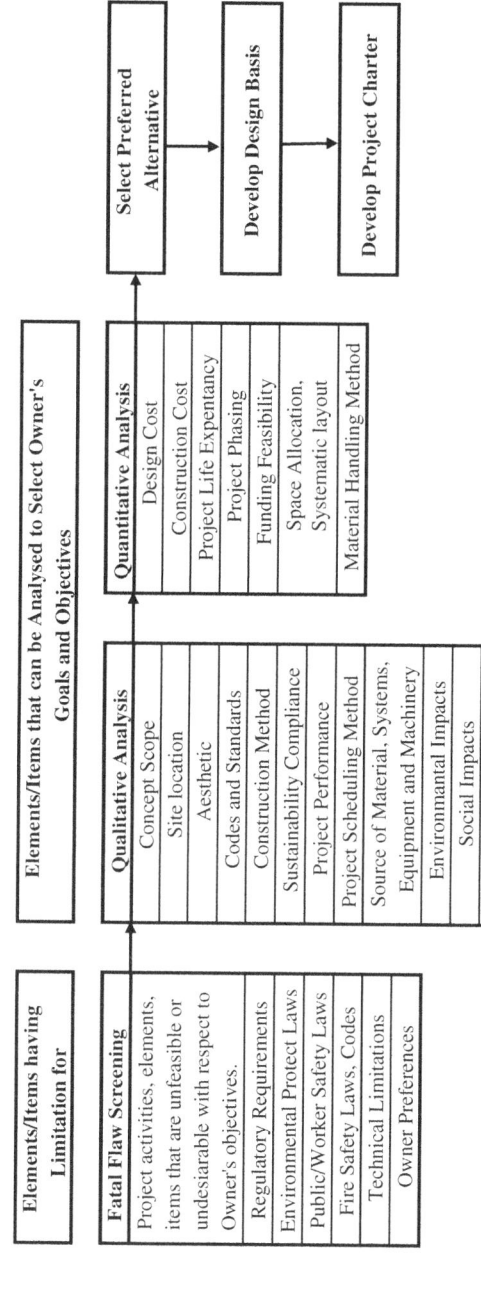

FIGURE 4.3 Evaluation and analysis of alternatives method to select preferred alternative.

type of project delivery system varies from project to project, taking into consideration the objectives of the project.

A project delivery system is defined as the organizational arrangement among various participants comprising the owner, designer, contractor, and many other professionals involved in the design and construction of the project/facility to translate/transform the owner's needs/goals/objectives into a finished facility/project to satisfy the owner's/end-user's requirements. The project delivery method is a system to achieve satisfactory completion of a construction project from its inception to occupancy. The project delivery system establishes responsibility for how the project is delivered to the owner.

The project delivery system:

- Establishes scope and responsibility for how the project is delivered to the owner
- Includes the project design and construction
- Defines responsibility/obligations that each of the participants is expected to perform, such as scheduling, cost control, quality management, safety management, and risk management during the various phases of the construction project life cycle
- Is the approach by which the project is delivered to the owner, but is separate and distinct from the contractual arrangements for financial compensation
- Establishes the procedures, actions, and sequence of events to be carried out.

The project delivery system defines the responsibilities/obligations that each of the participants is expected to perform during the study stage, design stage, bidding and tendering stage, and construction stage. The project delivery system is the approach by which the project is delivered to the owner but is separate and distinct from the contractual arrangements for financial compensation.

There are different types of project delivery systems followed in construction projects, although each of the project delivery systems is a variation on the project delivery systems.

Table 4.10 illustrates the different categories of project delivery systems.

4.3.8 Project Charter

The Project Charter defines the objectives, scope, deliverables, and overall approach for the new project. In a construction project, the project charter is the document prepared by the owner/client, or by the project manager on behalf of the owner, describing the project objectives and requirements to deliver the project. The Project Charter is also known as the:

- Client Brief
- Definitive Project Brief
- Terms of Reference

TABLE 4.10
Categories of Project Delivery Systems

Serial Number	Category	Classification	Sub-Classification
1	Traditional System (Separated & Cooperative)	Design-Bid-Build	Design-Bid-Build
		Variant of Traditional System	Sequential Method
			Accelerated Method
2	Integrated System	Design-Build	Design-Build
		Design-Build	Joint Venture (Architect and Contractor)
		Variant of Design – Build System	Package Deal
		Variant of Design-Build System	Turnkey Method (EPC)-(Engineering, Procurement, Construction)
		Variant of Design-Build System (Turnkey)	Build-Operate-Transfer (BOT)
			Build-Own-Operate-Transfer (BOOT)
			Build-Transfer-Operate (BTO)
			Design-Build-Operate-Maintain (DBOM)
		Variant of Design-Build System (Funding Option)	Lease-Develop-Operate (LDO)
			Wraparound (Public-Private Partnership)
		Variant of Design – Build System	Build-Own-Operate (BOT)
			Buy-Build-Operate (BBO)
3	Management Oriented System	Management Contracting	Project Manager (Program Management)
		Construction Management	Agency Construction Manager
			Construction Manager-At-Risk
4	Integrated Project Delivery System	Integrated Form of Contract	

Source: Abdul Razzak Rumane (2013). Quality Tools for Managing Construction Projects. Reprinted with permission of Taylor & Francis Group.

The Terms of Reference (TOR) are developed based on the Need Statement, which is further analyzed to assess the feasibility of the project. The outcome of the feasibility report is used to prepare the Project Brief (Project Goals and Objectives). Alternatives are identified, and the preferred alternative is selected after evaluation/analysis of the alternatives. Thereafter, the Project Delivery System is selected based on the complexity of the project and the strategy to be followed by the owner.

The Project Charter is prepared by the owner/client, or by a project management consultant (PMC) on behalf of the owner, describing the objectives and requirements to develop the project. A Client Brief (TOR) defines the objectives of the project and agrees on a Project Brief to guide the next stage of the project. The Terms of Reference (TOR) are developed to proceed with the design and construction of the project.

A Terms of Reference is a written document stating what will be done by the designer (consultant) to develop the project/facility. It is issued to the designer (consultant) by the owner to develop the project design and construction documents. A well-prepared accurate and comprehensive client brief (TOR) is essential to achieve a qualitative and competitive project. The TOR generally requires a designer (consultant) to perform the following:

- Development of the Concept Design
- Preparation of the Preliminary/Schematic Design
- Preparation of the Detail Design
- Preparation of the construction documents
- Preparation of the project schedule and budget, and obtaining of the authorities' approvals

The TOR gives the project team (designer) a clear understanding for the development of the project. Further, the TOR is used throughout the project as a reference to ensure that the established objectives are achieved. The client brief or TOR describes information such as:

- Project objectives which have triggered the project
- Proposed location of the project
- Project/facility to be developed
- Project assumptions and constraints
- Project function and size
- Performance characteristics of the project
- Design approach
- Estimated timescale
- Estimated cost
- Codes, standards and specifications
- Project quality
- Project control guidelines
- Sustainability requirements
- Regulatory requirements
- Initial list of defined risks
- Description of approval requirements
- Drawings
- Report
- Models (if applicable)
- Presentation

For the development of construction projects, TOR generally details the services to be performed by the designer (consultant) which include, but are not limited to, the following
    requirements for a building construction project, normally mentioned in TOR, to be prepared by the designer during the conceptual phase for submission to the owner:

1. Site Plan
   A) Civil
   B) Services
   C) Landscaping, external works
   D) Irrigation
2. Architectural Design
3. Building and Engineering Systems
   A) Structural
   B) Conveyance system (elevator, escalator)
   C) Mechanical (HVAC)
   D) Public health
   E) Fire suppression systems
   F) Electrical
   G) Low-voltage systems
   H) Others
4. Schedules
5. Cost Estimates

Figure 4.4 illustrates a flowchart for the development of Terms of Reference.

4.4 SUSTAINABILITY IN THE DESIGN STAGE

During the design stage, the construction project documents (project scope, design, and contract documents) are developed based on the requirements described in the Terms of Reference (TOR) prepared by the owner/client or construction manager/project manager on behalf of the owner/client. The TOR gives the designer (consultant) a clear understanding for the development of the project. The designer (consultant) utilizes the TOR to develop contract documents to suit the project delivery system. The contract documents for the Design-Bid-Build type of project delivery system mainly consist of:

1. Scope of work
2. Design drawings
3. Technical specifications
4. Conditions of contract
5. Bill of quantities
6. Project schedule
7. Project cost
8. Construction documents/bidding and tendering documents

In the case of the Design-Build type of project delivery system, bidding and tendering documents are prepared, taking into consideration performance specifications for the project and select Design-Build contractor.

Figure 4.5 illustrates the project scope development process (the design stage)

154 Quality Management

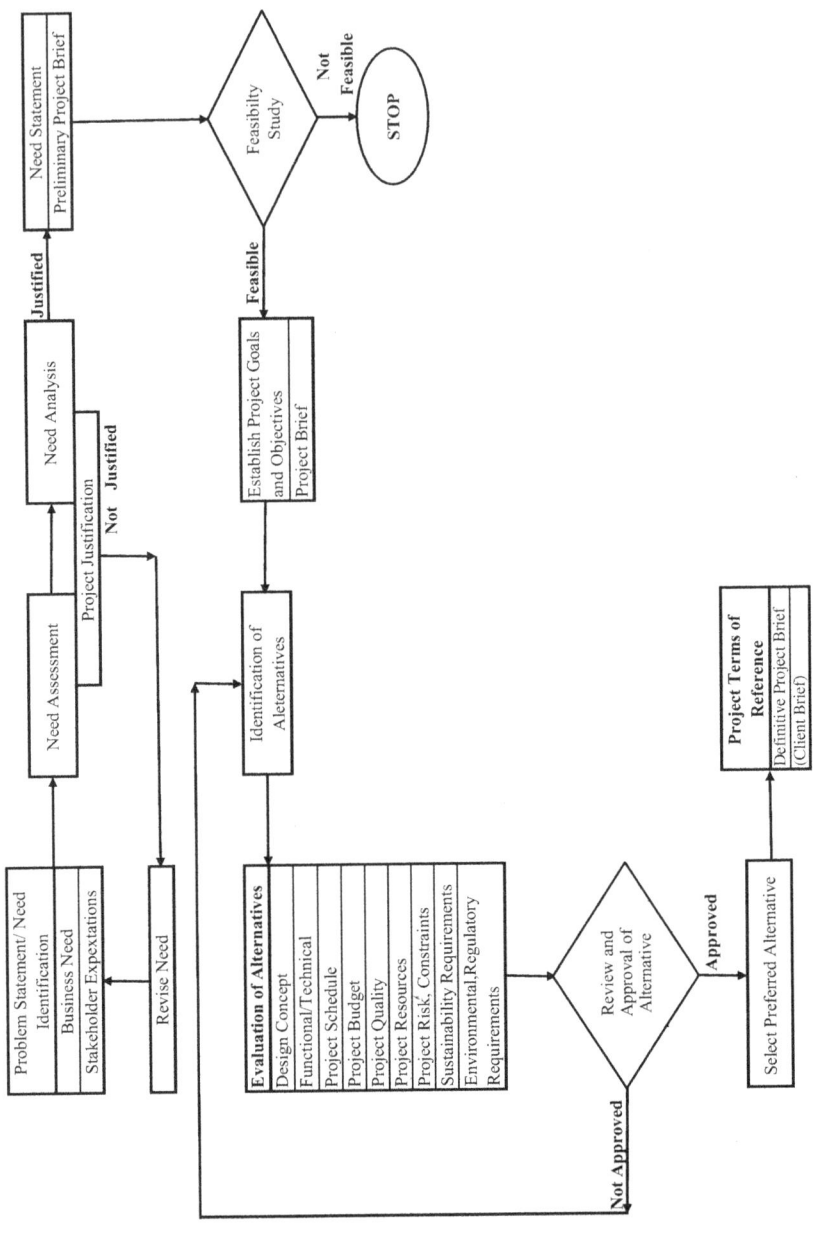

FIGURE 4.4 Flowchart for development of terms of reference.

Management of Quality for Sustainability

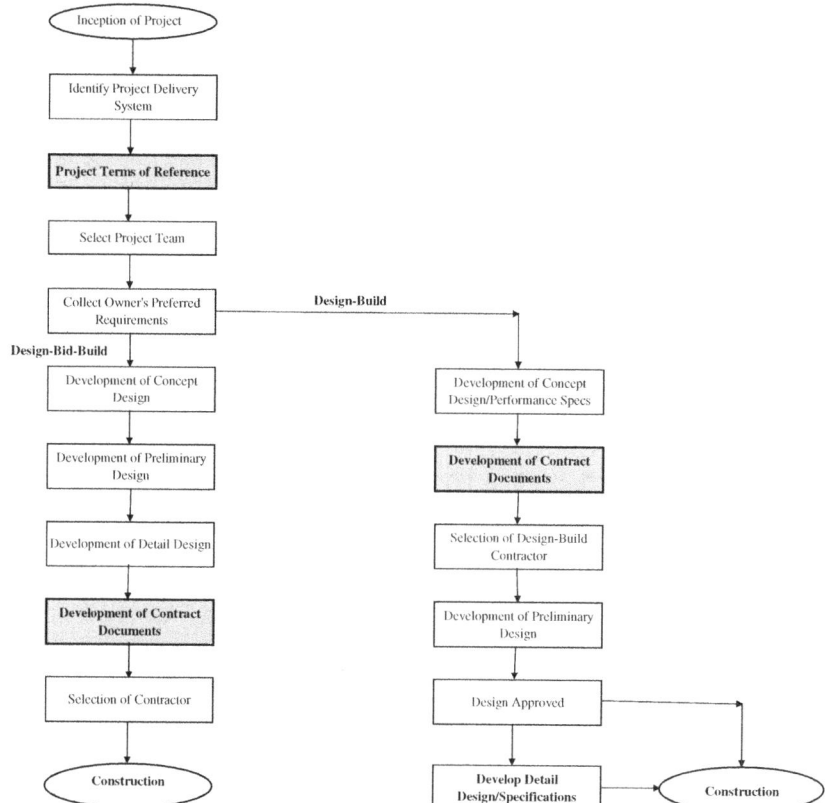

FIGURE 4.5 Development of project scope documents.

The design stage consists of developing the following:

1. Concept design
2. Preliminary/schematic design
3. Detail design
4. Construction documents

In order to develop scope documents/design documents for the sustainability of the project, the designer has to consider all the related activities to develop design documents, and also take into consideration all the elements for sustainability of the project as discussed in Chapter 1 (Please refer to Figure 1.2)

4.4.1 THE CONCEPT DESIGN

In order to develop the concept design, it is necessary to identify the concept design deliverables. The TOR provides guidelines to develop the concept design deliverables. Figure 4.6 illustrates a logic flowchart for the development of the concept design phase.

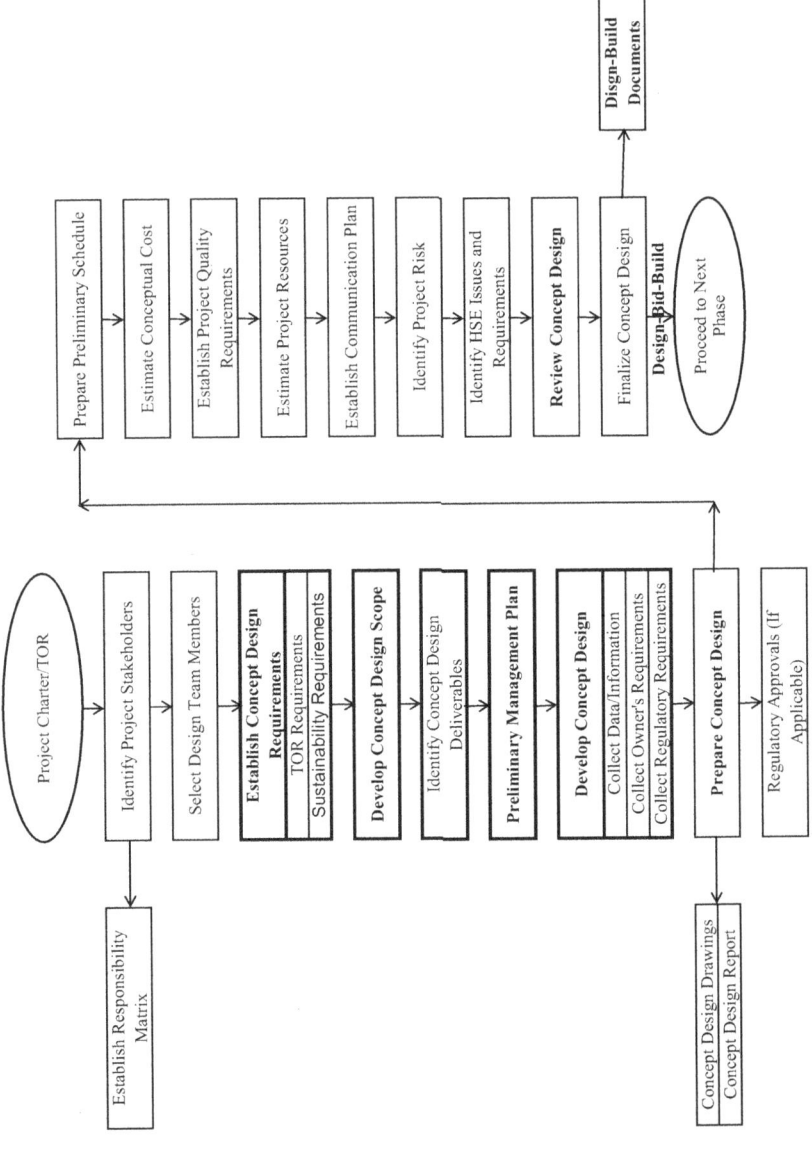

FIGURE 4.6 Logic flowchart for concept design phase.

Management of Quality for Sustainability

Figure 4.7 illustrates the major activities relating to the concept design phase based on the project management process groups methodology.

4.4.1.1 Identify Stakeholders

The following stakeholders have direct involvement in the project during the conceptual design:

- Owner
- Consultant

Concept Design Phase

Management Processes	Project Management Process Groups				
	Initiating Process	Planning Process	Execution Process	Monitoring & Controlling Process	Closing Process
Integration Management	Project Goals and Objectives	Preliminary Project Management Plan	Data Collection/Information	Monitor Concept Design Progress	Concept Design Deliverables
	Project Terms of Reference (TOR)		Development of Concept Design	Manage Owner Need, Changes	
			Implement Changes	Review of Concept Design	
Stakeholder Management	Project Stakeholders	Responsibility Matrix	Concept Drawings, Reports, Models		
	Project Team Members	Establish Stakeholders Requirements	Design Performance Report	Design Progress	
	Regulatory Authorities			Design Status	
Scope Management		Concept Design Scope		Regulatory Approval	
		Data Collection		Aesthetics, Constructability	
		Sustainability Requirements		Sustainability, Economy, Environmental Compatibility	
		Owner's Requirements		Authorities Approval	
		Design Deliverables		Compliance to Owner's Need	
		Authorities' Approval		Stakeholder's Approval of Concept Design	
Schedule Management		Preliminary Schedule		Concept Design Submission Shedule	
				Approval of Project Schedule	
Cost Management		Conceptual Estimate		Control Project Cost	
Quality Management		Quality Codes and Standards, Regulatory Requirements	Design Compliance to Codes, Standards and Regulatory Requirements	Conformance to Technical and Functional Capability, Energy Efficiency,	
Resouce Management		Assign Project Design Team	Manage Team Members from Different Disciplines	Performance of Team Members	Assign New Phase/Project
		Roles and Responsibilities of Team Members			
Communication Management		Design Progress Information	Liaison and Coordination with All Parties	Design Status Information	
			Coordination Meetings		
Risk Management		Management of Design Risk		Control Design Risk	
		Project Delivery System	Design to Comply Contract Type/Pricing	Administer Project Delivery System Requirements	Demobilize Team Members
Contract Management		Contracting System			
		Selection of Consultant for Feasibility Study			
		Selection of Design Team (Consultant)		Check for Contracting System	
HSE Management		SafetyConsideration in Design	HSE Consideration is Design	Check for Regulatory Requirements	
Financial Management		Designer/Consultant Payment		Cost Effectiveness over the Project Life Cycle	Designer/Consultant Payment
Claim Management		Management of Owner's Need Changes		Control Changes/Claims	Settle Claims by Designer

FIGURE 4.7 Major activities relating to concept design process groups. Note: These activities may not be strictly sequential; however the breakdown allows implementation of project management function to be more effective and easily manageable at different stages of the project phase.

- Designer
- Regulatory authorities
- Project/construction manager (if the owner decided to engage one during this phase)

Table 4.11 illustrates the responsibilities of the various participants during the development of the concept design

4.4.1.2 Select the Design Team

The design team is selected as per the procurement strategy. Refer Table 4.10 for categories of project delivery system.

For the Design-Bid-Build type of contract system, the first thing the owner has to do is to select design professionals/consultants. Generally, the owner selects a designer/consultant on the basis of qualifications (Qualifications Based System, QBS) and prefers to use one they have used before and with whom they have had a satisfactory result.

On selection of the designer (A/E), an agreement is made between the owner and designer (A/E). The following are typical contents of the contract between the owner and the designer (A/E).

1. Project definition
2. Scope of work

TABLE 4.11
Responsibilities of Various Participants During the Concept Design Phase

	Responsibilities		
Phase	Owner/Project Manager	Designer	Regulatory
Concept Design	• Approval of team members • Approval of time schedule • Approval of budget • Approval of Concept Design	• Identification of sustainability requirements • Data collection • Preparation of schedule • Estimation of project cost • Project quality requirements (concept design quality, project quality) • Estimation of project resources • Identification of project risks • Identification of HSE Issues and requirements • Regulatory approvals • Development of Concept Design • Analyze/review design	• Approval of project submittals

3. Project schedule
4. Design deliverables
5. Owner responsibilities
6. Variation order
7. Appointment of subconsultant
8. Selection of team members
9. Duties (responsibilities)
10. Compliance with authorities' requirements
11. Fees for the services
12. Penalty for delay
13. Liability toward design errors
14. Insurance
15. Arbitration/dispute resolutions
16. Taxes
17. Suspension of contract
18. Termination
19. Glossary

Once the designer has signed the contract with the project owner (client) to design the project and offer other services, the designer (consultant) assigns a project manager to execute the contract and is responsible for managing the development of design and contract documents, as per the project charter (TOR), to meet the client's needs and objectives. The project manager coordinates with other departments and recruits design team members to develop the project design. A project team leader, along with the respective design engineer(s), quality engineer, and AutoCAD technician from each trade, is assigned to work on the project. The project team is briefed by the project manager about the project objectives and the roles and responsibilities and authorities of each team member. A quality manager also joins the project design team to ensure compliance with the organization's quality management system. Figure 4.8 illustrates the project design organization structure.

4.4.1.3 Establish Concept Design Requirements

In order to establish the concept design phase requirements, the designer has to review all the documents which are part of the contract that the designer has signed with the owner of the project. It basically includes the Project Charter (TOR). The TOR generally requires the designer to perform the following:

- Development of the project design
- Development of the bidding and tendering documents

4.4.1.3.1 Identify the Sustainability Requirements

The designer has to identify the sustainability requirements. The following are the basic elements that have to be taken care of by the designer to develop the concept design:

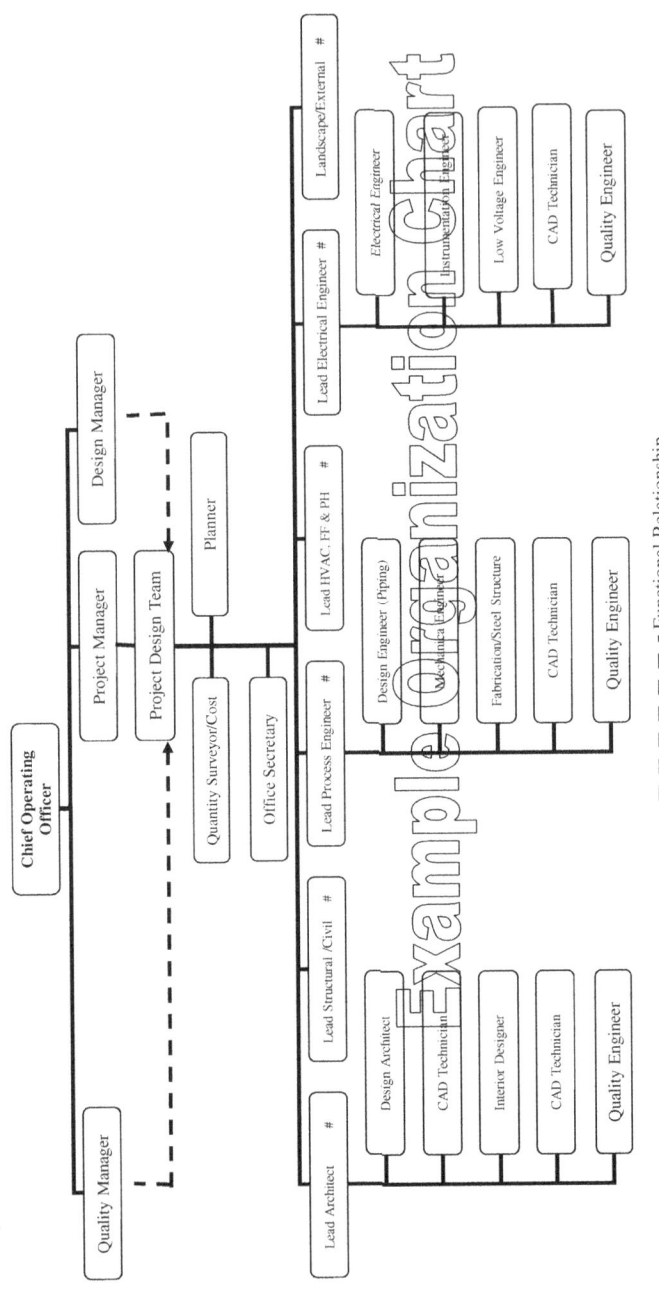

FIGURE 4.8 Project design team organization chart.

- The owner's financial capability
- Quality management
- Efficient resources
- Recycling systems
- Energy efficiency
- Energy-efficient equipment, systems
- Water conservation
- Waste management
- Pollution control
- Renewable energy
- Green building
- Sensor-controlled lighting system
- Landscape and plantation
- Local resources
- Regulatory requirements
- Security practices
- Health and safety

4.4.1.4 Develop the Concept Design Scope

The concept design scope is developed, considering TOR and sustainability requirements.

4.4.1.4.1 *Identify Concept Design Deliverables*

The following are the concept design deliverables for a building (non-process type) project:

1. Concept design report (narrative/descriptive report)
    a) Space program
    b) Building exterior
    c) Building interior
    d) Structural system
    e) MEP systems (energy-efficient systems)
    f) Fire safety
    g) Conveyance system
    h) Information and Communication Technology (ICT)
    i) Landscape
    j) Traffic plan (if applicable)
    k) HSE issues
2. Drawings
    a) Overall site plan
    b) Floor plan
    c) Elevations
    d) Sketches
    e) Sections (indicative to illustrate overall concept)

3. Data collection, studies reports
4. Existing site conditions
5. Concept schedule of materials and finishes
6. Lighting/daylight studies/sensor-controlled lighting
7. LEED requirements
8. Facility management requirements
9. Preliminary project schedule
10. Preliminary cost estimate
11. Regulatory approvals
12. Models
13. Evaluation criteria of the alternatives (if part of the TOR)

4.4.1.5 Develop a Preliminary Management Plan

The designer prepares a preliminary project management plan. It should document the key management and oversight tasks and will be updated throughout the project. The plan should include definitions of the owner's goals and objectives, technical requirements, schedules, budget, resources, quality plan and standards, risk plan, and financial plan.

4.4.1.6 Develop the Concept Design

The project charter (TOR) is the basis for the development of the concept design. While developing the concept design, the designer must consider the following:

1. Project goals
2. Usage
3. Technical and functional capability
4. Aesthetics
5. Constructability
6. Sustainability (economical, environmental, and social requirements)
7. Applicable codes and standards
8. Reliability
9. Project risks
10. Health and Safety
11. Environmental compatibility
12. Fire protection measures
13. Provision for facility management requirements
14. Supportability during maintenance/maintainability
15. Energy conservation
16. Waste management
17. Scalability to next stage
18. Cost-effectiveness over the entire life cycle (economic)

The design has a significant impact on the product and on customer satisfaction. The designer can use quality tools, such as benchmarking, for the development of product

Management of Quality for Sustainability

and quality function deployment (QFD) to translate the owner's needs into technical specifications.

The benchmarking tool identifies and measures the factors critical to the success of the product compared with similar products from other businesses. It is used to determine the performance level of the product with other products and achieve functional performance requirements.

QFD is a technique to translate customer needs into technical requirements when developing a product or project. It is also known as the "House of Quality" because of its shape. It was developed in Japan by Dr Yoji Akao in the 1960s to transfer the concepts of quality control from the manufacturing process into the new product development process.

Figure 4.9 illustrates the concept of QFD or the "House of Quality."

It is the designer's responsibility to pay greater attention to improving the environment and to achieving sustainable development. The designer has to address environmental and social issues and comply with local environmental protection codes. Numerous UN meetings (such as the first United Nations conference on Human Development, held in Stockholm in 1972; the 1992 Earth Summit in Rio de Janeiro; the 2002 Earth Summit in Johannesburg; and the 2005 World Summit) and the Brundtland Commission on Environment and Development in 1987 emphasize "sustainability," whether it be a sustainable environment, sustainable economic development, sustainable agricultural and rural development, and so on. Accordingly, the designer has to address environmental and social issues and comply with local environmental protection codes. A number of tools and rating systems have been created by LEED (USA), BREEAM (UK), and HQE (France) in order to assess and compare the environmental performance of the buildings. These initiatives have a great impact on how the buildings are designed, constructed, and maintained. Therefore, during the development of building construction projects, the following need to be considered:

1. Consolidation with the natural environment by using natural resources such as sunlight, solar energy, ventilation configuration
2. Energy conservation by energy-efficient measures to diminish energy consumption (energy-efficient construction)
3. Material-conserving design and construction
4. Material handling and optimization
5. Environmental protection to reduce environmental impact
6. Sustainable usage of renewable resources
7. Decreased water consumption
8. Use of material in harmony with the environment
9. Good air quality
10. Comfortable temperature
11. Energy conservation
12. Comfortable lighting
13. Day lighting
14. Sensor-controlled lighting

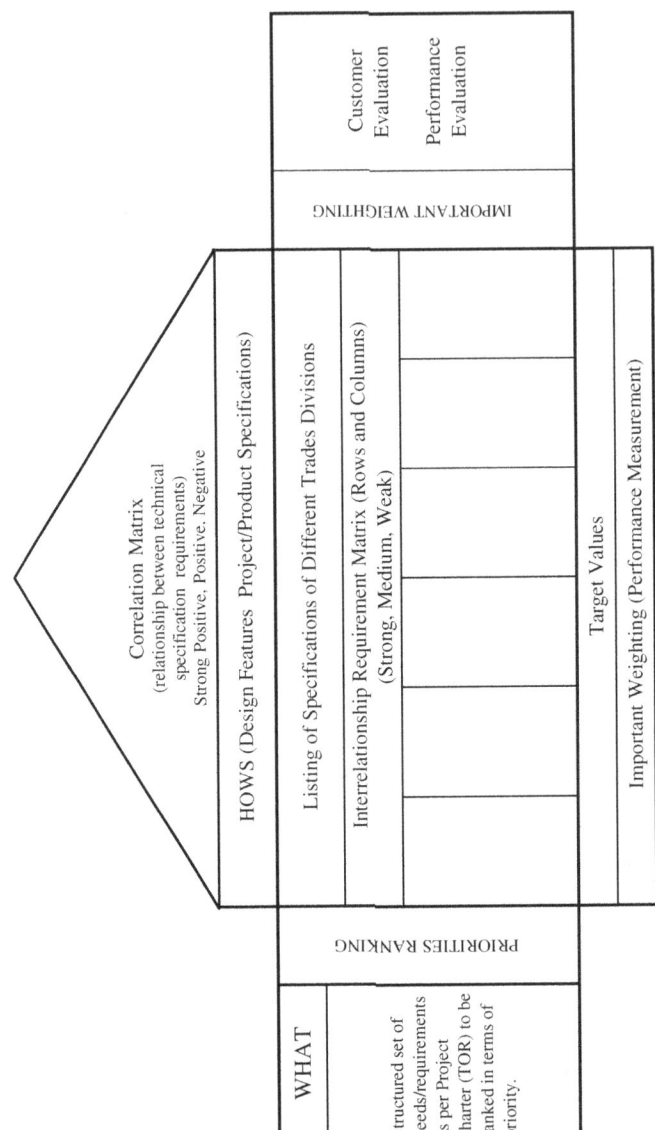

FIGURE 4.9 Concept of QFD "house of quality."

Management of Quality for Sustainability

15. Comfortable sound
16. Clean water
17. Landscape and plantation
18. Waste management
19. Pollution control
20. Aesthetic harmony between a structure and its surrounding nature and built environment
21. Integration with the social and cultural environment
22. Utilization of locally available resources

During the design stage, the designer must work jointly with the owner to develop details regarding the owner's needs, and give due consideration to each part of the requirements. The owner, on their part, should ensure that the project objectives are:

- Specific
- Measurable
- Agreed upon by all the stakeholders
- Realistic
- Possible to complete within the defined time
- Possible to complete within the budget

The designer can use Design for Six Sigma (DFSS), an organizational approach for translating customer requirements into design features and then developments of design to meet the customers' requirements. The DMADV tool can be used to develop the concept design. The phases of DMADV are:

D – Define: Requirements listed under Projects Charter (TOR) to develop the new project/product
M – Measure: Identify the needs/requirements and then translate them into the technical requirements (Critical to Quality, CTQ) and specifications
A – Analyze: Analyze the product/process options and prioritize based on their capabilities to satisfy customer requirements (TOR)
D – Design: Design the product/process to satisfy customer requirements as per TOR
V – Verify: Verify and review the design to ensure that the design meets customer needs

4.4.1.6.1 *Collect Data/Information*

The purpose of data collection is to gather all the relevant information under the existing conditions, both on the project site and the surrounding area that will impact the planning and design of the project. Data related to the following major elements needs to be collected by the designer:

1. Certificate of title
    a. Site legalization

b. Historical records
2. Topographical survey
   a. Location plan
   b. Site visits
   c. Site coordinates
   d. Photographs
3. Geotechnical investigations
4. Field and laboratory tests of soil and soil profile
5. Existing structures in/under the project site
6. Existing utilities/services passing through the project site
7. Existing roads, structure surrounding the project site
8. Shoring and underpinning requirements with respect to adjacent area/structure
9. Requirements to protect neighboring area/facility
10. Environmental studies
11. Daylighting requirements
12. Wind load, seismic load, dead load, and live load
13. Site access/traffic studies
14. Applicable codes, standards, and regulatory requirements
15. Usage and space program
16. Design protocol
17. Scope of work/client requirements

4.4.1.6.2 Collect the Owner's Requirements

Based on the scope of work/requirements mentioned in the Terms of Reference
(TOR), a detailed list of requirements is prepared by the designer (consultant). The project design is developed, taking into consideration the owner's requirements and all other design criteria. The designer can prepare checklists of various trades by listing the items to obtain the owner's preferences and requirements.

Table 4.12 is a sample check list to collect the owner's preferred requirements.

4.4.1.6.3 Collect the Regulatory Requirements

The designer has to collect the regulatory requirements to be taken into consideration while preparing the concept design.

The designer should also collect the sustainability requirements that are essential to comply with the authorities' requirements.

4.4.1.7 Prepare the Concept Design

The designer can use techniques such as quality function deployment (QFD) to translate the owner's needs into technical specifications. Figure 4.10 illustrates the House of Quality for a college building project, based on certain specific requirements by the customer.

While preparing the concept design, the designer has to consider various available options/systems to ensure project economy, performance, and operation. The concept design should be suitable for further developments.

TABLE 4.12
Checklist for Owner's Requirements

Sr. No.	Description of Item/Activity	Yes	No	Notes
	Owner's Preferred Requirements (Trade Name)			
1				
2				
3				
4				
5				
6				
7				
8				
9				
10				
11				
12				
13				
14				
15				
16				
17				
18				
19				
N…				

Sample Check List

4.4.1.7.1 Concept Design Drawings

Table 4.13 lists the elements which comprise the concept design drawings of various trades (disciplines) in a building construction project.

4.4.1.7.2 Concept Design Report

Concept Design reports should be concise, covering all the related information about concept design considerations. Reports are prepared for each of the trades mentioned in the TOR. Below is a sample Table of Contents for a Concept Design report for architectural works and civil/structural works for a building construction project.

A. **Table of Contents for Architectural Works**
   1. Introduction
      1.1 Project goals and objectives
   2. Owner's schedule requirements
   3. Project directory
   4. Architecture
      4.1 Applicable codes and standards

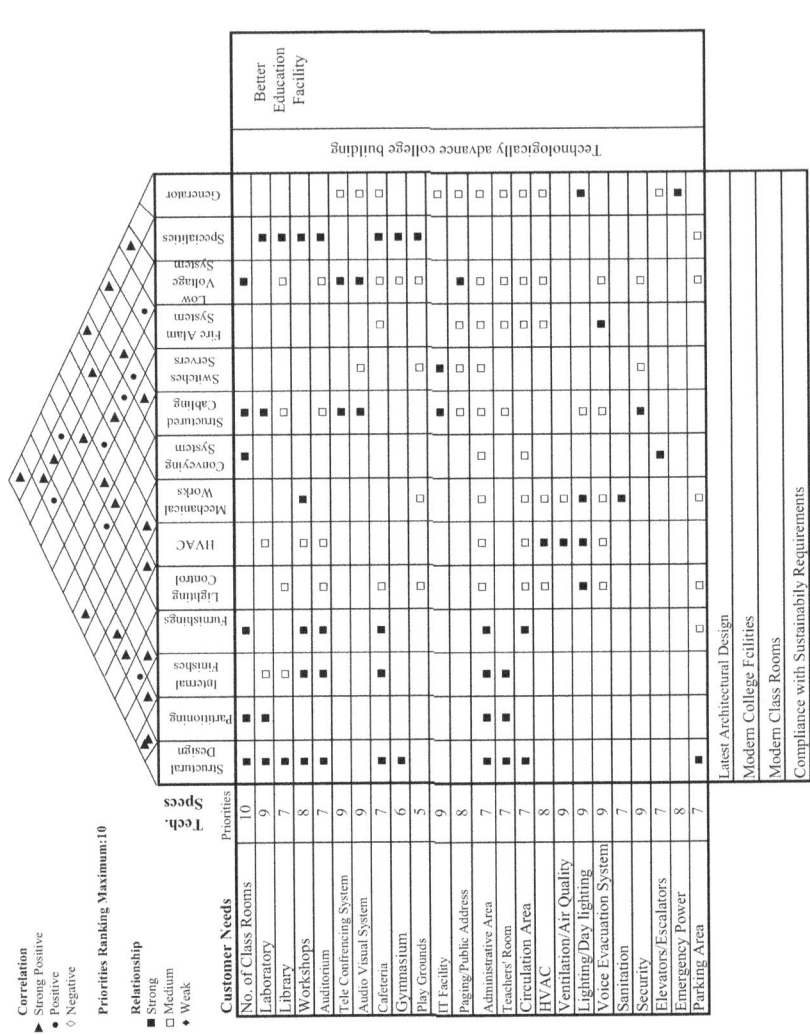

FIGURE 4.10 House of quality for college building project. Source: Abdul Razzak Rumane. (2013). *Quality Tools for Managing Construction*.

TABLE 4.13
Elements to be Included in Concept Design Drawings

Serial Number	Elements to be Included in Drawing
1	Architectural
1.1	Overall site plans
	a. Existing site plan
	b. Location of building, roads, parking, access, and landscape
	c. Project boundary limits
	d. Site utilities
	e. Water supply, drainage, stormwater lines
	f. Zoning
	g. Reference grids and axis
	h. Demolition plan, if any
1.2	Floor plans
	a. Floor plans of all floors
	b. Structural grids
	c. Vertical circulation elements
	d. Vertical shafts
	e. Partitions
	f. Doors
	g. Windows
	h. Floor elevations
	i. Designation of rooms
	j. Preliminary finish schedule
	k. Services closets
	l. Raised floor, if required
1.3	Roof plans
	a. Roof layout
	b. Roof material
	c. Roof drains and slopes
2	Structural
2.1	a. Building structure
	b. Floor grade and system
	c. Foundation system
	d. Tentative size of columns, beams
	e. Stairs
	f. Roof and general sections
3	Elevator
3.1	a. Traffic studies
	b. Elevator/escalator location
	c. Equipment room
4	Plumbing and fire suppression
4.1	a. Sprinkler layout plan
	b. Piping layout plan
	c. Water system layout plan
	d. Water storage tank location
	e. Development of preliminary system schematics
	f. Location of the mechanical room

(*Continued*)

TABLE 4.13 CONTINUED
Elements to be Included in Concept Design Drawings

Serial Number	Elements to be Included in Drawing
5	HVAC
5.1	a. Ducting layout plan
	b. Piping layout plan
	c. Development of preliminary system schematics
	d. Calculations to allow preliminary plant selections
	e. Establishment of primary building services distribution routes
	f. Establishment of preliminary plant location and space requirements
	g. Determination of heating and cooling requirements based on heat dissipation of equipment, lighting loads, type of wall, roof, glass, … etc.
	h. Estimation of HVAC electrical load
	i. Development of BMS schematics showing interface
	j. Location of plant room, chillers, cooling towers
6	Electrical
6.1	a. Lighting layout plan
	b. Power layout plan
	c. System design schematic without any sizing of cables and breakers
	d. Substation layout and location
	e. Total connected load
	f. Location of electrical rooms and closets
	g. Location of MLTPs, MSBs, SMBs, EMSB, SEMBs, DBs, EDBs, … etc.
	h. Location of starter panels, MCC panels, …, etc.
	i. Location of generator, UPS
	j. Raceway routes
	k. Riser requirements
	l. Information and Communication Technology (ICT)
	a) Information Technology (computer network)
	b) IP Telephone System (telephone network)
	c) Smart Building System
	m. Loss Prevention Systems
	a) Fire alarm system
	b) Access control security system layout
	c) Intrusion system
	n. Public address, audiovisual system layout
	o. Schematics for FA and other loss prevention systems
	p. Schematics for ICT System, and other low-voltage systems
	q. Location of LV equipment

(*Continued*)

TABLE 4.13 CONTINUED
Elements to be Included in Concept Design Drawings

Serial Number	Elements to be Included in Drawing
7	Landscape
7.1	a. Green area layout
	b. Selection of plants
	c. Irrigation system
8	External
8.1	a. Street/road layout
	b. Street lighting
	c. Bridges (if any)
	d. Security system
	e. Location of electrical panels (feeder pillars)
	f. Pedestrian walkways
	g. Existing plans
9	Narrative description

    4.2 International codes
    4.3 Local codes
    4.4 Building height
    4.5 Projections
    4.6 Fire safety requirements
    4.7 Usage of building
    4.8 Stairways
5. Project requirements
    5.1 Basement
    5.2 Ground floor
    5.3 Other floors
6. Existing site conditions
    6.1 Location maps
    6.2 Project site and surrounding area survey reports
7. Data collection
    7.1 Site analysis
    7.2 Traffic study
    7.3 Existing site views
8. Design options
9. Building structure
    9.1 Design loads
    9.2 Wind loads
    9.3 Deflection criteria
10. Design program software
11. Geographical investigation
12. Shoring
13. Dewatering
14. Drawings

B. **Table of Contents for Structural/Civil Works**
    1. Introduction
        a. Project goals and objectives
        b. Existing site conditions
    2. Codes and standards
    3. Regulatory compliance
    4. Design criteria
    5. Design systems
        a. Design system alternatives
        b. Key construction elements
        c. Environmental risk assessment
        d. Geotechnical investigation and foundation design
        e. Analysis model
        f. Material
        g. Durability
        h. Fire safety
        i. Design loads
        j. Lateral loads
        k. Structural movement criteria
        l. System suitability for structure
        m. Shoring and dewatering
        n. Retaining wall and lateral stability system
        o. Substructure
        p. Design program/software
    5. Testing systems
    6. Risk assessment

C. **Table of Contents for Mechanical Works**
    1. Introduction
    2. Codes and standards
    3. Regulatory compliance
    4. Incoming services routes and regulatory requirements
    5. Fire suppression work
        a. Design criteria
        b. Sprinkler system
        c. Smoke ventilation system
        d. Firefighting pumps
        e. Piping system
        f. Fire suppression system for the diesel generator room
        g. Fire suppression for the low-voltage equipment room
    7. Mechanical (plumbing) work
        a. Design criteria
        b. Water supply system
            i. Cold water
            ii. Hot water
        c. Water distribution network
        d. Water storage system

Management of Quality for Sustainability

    e. Plumbing fittings and fixtures
    f. Plant room (pumps, heaters) location
    g. Storm water, drainage system
    h. Sewerage system
    i. Irrigation water system
    j. Piping for mechanical works

D. **Table of Contents for HVAC Works**
   1. Introduction
   2. Codes and standards
   3. Regulatory compliance
   4. HVAC system design criteria
   5. Environmental conditions
   6. HVAC equipment selection
      a. Chiller
      b. Cooling system
      c. AHU
      d. Energy recovery equipment
      e. Direct digital control system
      f. Pumps
      g. Fans and ventilation
      h. Ducting
      i. Duct insulation
      j. Piping
      k. Risers
      l. Piping insulation
      m. Acoustic and vibration
   6. Location of plant room
   7. Building Management System (BMS)

E. **Table of Contents for Electrical Works**
   1. Introduction
   2. Codes and standards
   3. Regulatory compliance
   4. Design criteria
      a. Safety and protection
   5. Electrical substation
   6. Cables and wires
   7. Main electrical supply
      a. Main distribution system
      b. Secondary distribution system
      c. Distribution equipment (panels, boards, switchgear)
      d. Circuits
   8. Internal and external lighting
      a. Light fixtures
   9. Power receptacles
   10. Emergency power system

11. Earthing (grounding) system
12. Lightning protection system
13. Alternate energy (solar) system
14. Fire alarm and voice evacuation system
15. Audiovisual system
16. Security system (CCTV, access control)
17. Communication system

The designer also has to prepare the following reports;

1. Interior works
2. Elevator works
3. External works
4. Landscape works

The contents of the design report are illustrated in Table 4.14

TABLE 4.14
Contents of Concept Design Report (Trade Name)

Section	Topic	
1	Introduction	
	1.1	Description of project
	1.2	Project goals and objectives
2	Owner's schedule requirements	
3	Project directory	
4	Trade name (Process, Architecture, MEP, … etc.)	
	4.1	Applicable codes and standards
	4.2	International codes
	4.3	Local codes
	4.4	Regulatory requirements
	4.5	Applicable design system/details of the project trade item/element
5	Project requirement	
6	Existing site conditions	
7	Data collection	
8	Design options/strategy/criteria	
9	Design software	
10	Geographical investigation	
11	Risk assessment	
12	HSE issues	
13	Drawings	
14	Models (if applicable)	

4.4.1.8 Prepare the Preliminary Schedule

The duration of the construction project is finite. It has a definite beginning and a definite end; therefore, during the conceptual phase, the expected time schedule for the completion of the project/facility is worked out. The expected time schedule is important from both a financial viewpoint and the acquisition of the facility by the owner/end-user. It is the owner's goal and objective that the facility is completed in time for subsequent usage. During this phase, very limited information and details about project activities are available and have a very wide variance in the schedule.

In order to improve the understanding and the communication among stakeholders involved with preparing, evaluating, and using a project schedule, AACE International has published the Guidelines to classify schedules into five classes and five levels. Figure 4.11 illustrates the Schedule Classes and Schedule Levels that can be developed and/or presented.

Table 4.15 illustrates the Generic Schedule Classification Matrix

4.4.1.9 Estimate the Conceptual Cost

In construction projects, the cost estimates vary as the project design progresses. At the inception of the project, the cost estimate is based on the Rough Order of Magnitude. When the detail design is available, the cost estimates become definitive.

The cost estimated by the designer (consultant) is based on assumptions and historical data available from experience on similar projects.

Table 4.16 illustrates Cost Estimation Levels for Construction Projects.

A cost estimate during the concept design is required by the owner to know how much the capital cost of construction is. This is required by the owner to arrange the financial resources. The conceptual cost is also known as the budgetary cost. Parametric cost-estimation methodology is used to estimate the conceptual cost. The designer has to ensure that the conceptual cost does not exceed the cost estimated during the feasibility stage. The accuracy and validity of conceptual cost estimation is related to the level of information available when developing the concept design. The designer has to properly estimate the resources required for the successful completion of the project. Table 4.17 illustrates the quality check for the cost estimation.

4.4.1.10 Establish Project Quality Requirements

During this phase, the designer has to plan and establish quality criteria for the project. This includes mainly the following:

- Owner's requirements
- Quality standards and codes to be complied with
- Regulatory requirements
- Conformance to the owner's requirements
- Conformance to the requirements listed under the project charter (TOR)
- Design of the review procedure
- Drawings of the review procedure
- Document the review procedure
- Quality management during all the phases of the project life cycle

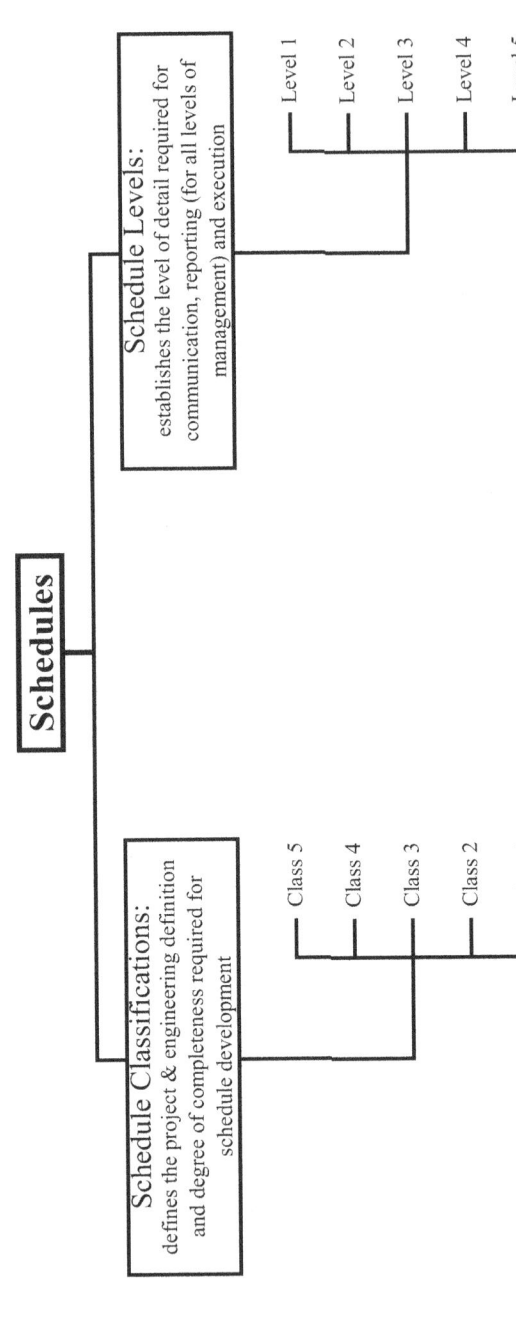

FIGURE 4.11 Schedule: Classifications versus levels. Source: Adapted from AACE International Recommended Practice No. 27R-03. Copyright @ 2010 AACE International. Reprinted with permission of AACE International.

Management of Quality for Sustainability

TABLE 4.15
Generic Schedule Classification Matrix

Schedule Class	Primary Characteristic — Degree of Project Definition (expressed as % of complete definition)	Secondary Characteristic — End Usage	Secondary Characteristic — Scheduling Methods Used
Class 5	0% to 2%	Concept screening	Top-down planning using high-level milestones and key project events
Class 4	1% to 15%	Feasibility study	Top-down planning using high-level milestones and key project events. Semi-detailed
Class 3	10% to 40%	Budget, authorization, or control	"Package" top-down planning using key events. Semi-detailed
Class 2	30% to 75%	Control or bid/tender	Bottom-up planning. Detailed
Class 1	65% to 100%	Bid/tender	Bottom-up planning. Detailed

Source: The table is from AACE International Recommended Practice No. 27R-03 Copyright @ 2010 by AACE International. Reprinted with permission of AACE International, 726 East Park Ave., #180, Fairmont, WV 26554. Email: info@aacei.org info@aacei.org. Phone: +1.304.296.8444; website: web.aacei.org. All AACE International content is copyright protected ©2010, All rights reserved.

Table 4.18 lists the contents of the Designer's Quality Control Plan.

4.4.1.10.1 Manage Concept Design Quality

During this phase, the designer has to manage the quality for the development of concept design and also plan and establish quality criteria for quality management during all the phases of the project life cycle.

Figure 4.12 illustrates the quality management procedure for the development of the concept design.

4.4.1.11 Estimate Project Resources

The designer has to estimate the resources required to complete the project. This includes the estimation of manpower required during the construction phase, testing, commissioning, and handover phase. The designer has to give preference to the available local resources to comply with the sustainability requirements.

TABLE 4.16
Cost Estimation Levels for Construction Projects

Project Stage/ Phase	Tools/ Methodology	Accuracy	Purpose
Inception	Analogous	−50% to +100%	Project Initiation (Rough Order of Magnitude)
Feasibility	Analogous	−25% to +75%	Justification to Proceed (Screening Estimate)
Concept Design	Parametric	−10% to +25%	Budgetary (Conceptual Estimate)
Preliminary Design	Elemental Parametric	−10% to +25%	Budgetary (Preliminary Estimate)
Detail Design	Elemental Parametric/ Detailed Costing	−5% to +10%	Detailed Estimate
Bidding and Tendering	Detailed Costing	−5% to +5%	Bid Estimate/ Definitive Estimate
Construction	Detailed Costing	Project Cost Baseline (Contracted Value)	Contract Cost (Control Estimate)

TABLE 4.17
Quality Check for Cost Estimate During Concept Design

Serial Number	Points to be Checked	YES/NO
1	Check for use of historical data	
2	Check if estimate factors are used to adjust historical data	
3	Check if estimate is updated with revision or update of concept	
4	Check if the updates/revisions are chronologically listed and cost estimate is updated	
5	Check whether the scope of work is descriptive/narrative, well enough for estimation purposes	
6	Is the estimate based on area schedule, overall plan for the project provided by the architect?	
7	Is the estimate based on schedule of equipment, machinery for the project provided by the process engineer?	
8	Is the estimate based on schedule of piping material provided by the piping/mechanical engineer?	
9	Check whether estimate is updated taking into consideration feedback from each trade	
10	Does the cost estimate clearly identify the quantities and associated work?	
11	Whether all the assumption are as per current market data?	
12	Does cost estimates from all the relevant trades included in the final sum?	
13	Does the estimate include all the requirements of specialist consultants?	
14	Has the total estimate been reviewed and verified?	

Source: Modified from Abdul Razzak Rumane (2013). Quality Tools for Managing Construction Projects. Reprinted with permission of Taylor & Francis Group.

TABLE 4.18
Contents of Designer's Quality Control Plan

Section			Topic
1			Introduction
2			Description of project
3			Quality control organization
4			Qualification of QC staff
5			Responsibilities of QC personnel
6			Procedure for submittals
7			Quality control procedure for Concept Design
	7.1		Establish Concept Design requirements
		7.1.1	Review of Project Charter, Terms of Requirements (TOR)
		7.1.2	Identify project stakeholders
		7.1.3	Identify project design team
		7.1.4	Establish Concept Design development scope
		7.1.5	Identify sustainability requirements
		7.1.6	Establish Concept Design deliverables
		7.1.7	Establish Concept Design procedure
	7.2		Identify quality management requirements
		7.2.1	Plan quality
			7.2.1.1 Owner's requirements
			7.2.1.2 TOR requirements
			7.2.1.3 Numbers of drawings, reports, models
			7.2.1.4 Scope of design work
			7.2.1.5 Responsibility matrix
			7.2.1.6 Codes and standards
			7.2.1.7 Regulatory requirements
			7.2.1.8 Submittal plan
			7.2.1.9 Drawings, specifications, documents
			7.2.1.10 Design review plan
		7.2.2	Quality assurance
			7.2.2.1 Information/data collection
			7.2.2.2 Site investigation
			7.2.2.3 Engineering surveys
			7.2.2.4 Preparation of drawings
			7.2.2.5 Interdisciplinary coordination
			7.2.2.6 Project risks
			7.2.2.7 Environmental issues
			7.2.2.8 Preparation of specification, documents
			7.2.2.9 Functional and technical compatibility
		7.2.3	Quality control
			7.2.3.1 Design drawings
			7.2.3.2 Quality of drawings
			7.2.3.3 Project schedule
			7.2.3.4 Project cost
			7.2.3.5 Project resources
			7.2.3.6 Specification, contract documents
8			Project-specific design requirements
9			Design software
10			Design development procedure
	10.1		Design criteria
	10.2		Preparation of drawings
	10.3		Interdisciplinary coordination
	10.4		Design review

(Continued)

TABLE 4.18 CONTINUED
Contents of Designer's Quality Control Plan

Section	Topic
11	Company's quality manual and procedure.
12	Subconsultant's work
13	Value engineering
14	Quality auditing program
15	Quality control record
16	Innovative and latest technology
17	Quality updating program

The designer also has to prepare equipment and machinery data sheets and the schedule of materials.

4.4.1.12 Establish the Communication Plan

The designer has to develop a communication plan taking into consideration the assigned responsibilities of the stakeholders and design team members. The designer has to prepare:

- Communication matrix
- Communication method for the design progress
- Design the submittal procedure
- Design the progress report
- Design the progress meetings

4.4.1.13 Identify Project Risks

The designer has to identify the risks which will affect the successful completion of the project. The following are typical risks which normally occur during the conceptual design phase:

- Lack of input from the owner about the project goals and objectives
- Project charter (TOR) is not defined clearly
- The related project data and information collected are incomplete
- The related project data and information collected are likely to be incorrect and wrongly estimated
- Regulatory requirements
- Errors in estimating the project schedule
- Errors in cost estimation
- Errors in resource estimation
- Environmental considerations
- Environmental issues are not correctly identified
- Sustainability elements are not identified and considered
- Design basis is not suitable for further development/development of the concept design

Management of Quality for Sustainability

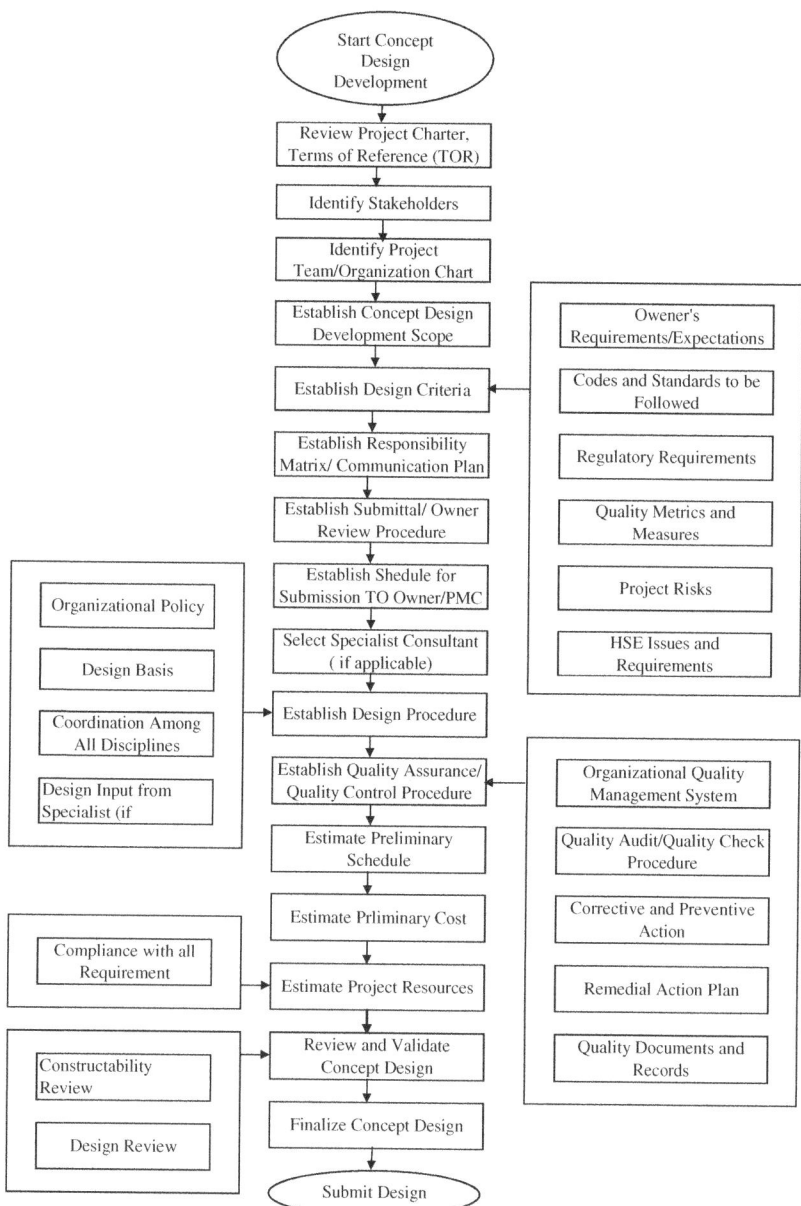

FIGURE 4.12 Quality management procedure for development of concept design.

The designer has to take into account the above-mentioned risk factors while developing the concept design.

Furthermore, the designer has to consider the following risks while planning the duration for completion of the conceptual phase:

- Impractical conceptual design preparation schedule
- Delay in data collection
- Delay in obtaining the authorities' approval
- Delay in obtaining environmental approval

The designer has to prepare a Risk Assessment Report for submission to the owner along with other documents while submitting the concept design package. The typical contents of the report are as follows:

1. Purpose of the report
2. Description of the project (goals and objectives)
3. References (relevant codes, standards, owner references)
4. Scope of the work (during all the life cycle phases of the project)
5. Hazard identification and assessment method

4.4.1.14 Identify HSE Issues and Requirements

The designer has to identify the HSE issue that could affect the environment due to construction of the project. When developing the design, the designer has to consider the following:

- Hazardous properties of the materials or products used in the project
- Safety in process and operations
- Hazardous emissions
- Pollution and its impact
- Impact on health
- Safety in design (safe design)
- Regulatory and other environmental protection agencies' requirements
- Environmental compatibility

The designer has to prepare an Environmental Impact Assessment Report for submission to the owner, along with other documents when submitting the concept design package. The typical contents of the environmental impact assessment report are as follows:

1. Purpose of the report
2. Description of the project (goals and objectives)
3. References (relevant codes, standards, owner references)
4. Scope of work (during all the life cycle phases of the project)

The designer is also required to submit the HSE Management Plan during this phase.

4.4.1.15 Review the Concept Design

Table 4.19 illustrates the major points for review of the concept design.

TABLE 4.19
Major Points for Review of Concept Design

Serial Number	Description	Yes	No	Notes
A: Concept Design				
A.1	Does the design support the owner's project goals and objectives?			
A.2	Does the design meet all the elements specified in the project charter (TOR)?			
A.3	Does the design meet all the performance requirements/parameters?			
A.4	Has constructability been taken care of?			
A.5	Is usage of space optimal?			
A.6	Has technical and functional capability been considered?			
A.7	Has the suitability of the project for the defined purpose and objectives been considered?			
A.8	Has ease of operations been considered while selecting the equipment?			
A9	Has accessibility been considered?			
A10	Does the design achieve ease of maintenance?			
A.11	Does the design meet all the specified codes and standards?			
A.12	Has the designer considered all the data and information collected into the design?			
A.13	Does the design conform to fire and egress requirements?			
A.14	Have design risks been identified, analyzed, and responses planned for mitigation?			
A.15	Has a risk assessment report is prepared			
A.16	Have health and safety requirements been considered in the design?			
A.17	Have environmental constraints been considered?			
A.18	Has an environmental impact assessment report been prepared and considered?			
A.19	Have environment-preferred materials and products been considered?			
A 20	Has the Green Building concept been considered?			
A.21	Has energy conservation been considered?			
A.22	Has a daylighting system been considered?			
A.23	Has sensor-controlled lighting been considered?			
A.24	Are the systems energy efficient?			
A.25	Are sustainability requirements considered in the design?			
A.26	Has cost-effectiveness over the entire project life cycle been considered?			
A.27	Have all reasonable design options/systems been considered for the project economy?			
A.28	Does the design have provision for inclusion of facility management requirements?			

(Continued)

TABLE 4.19 CONTINUED
Major Points for Review of Concept Design

Serial Number	Description	Yes	No	Notes
A.29	Does the design have provision for interface with all the low-voltage and control systems?			
A.30	Is the design coordinated with all trades?			
A.31	Have all the regulatory/statutory requirements been taken care of?			
A.32	Does the design support proceed to the next design development stage?			
B: Schedule				
B.1	Is the project schedule practically achievable?			
C: Financial				
C.1	Is the project cost properly estimated?			
D: Resources				
D.1	Has the availability of resources during the construction phase been considered?			
D.2	Has a schedule of equipment and machinery been prepared?			
D.3	Have locally available resources been considered?			
E: Reports				
E.1	Are the reports complete and do they include adequate information about the project?			
E.2	Are the reports prepared for all the trades mentioned in the TOR?			
E.3	Is the report properly formatted and does it have a Table of Contents for each report?			
F: Drawings, Sketches				
F.1	Are the drawings, sketches for all trades prepared as per the TOR?			
F.2	Is the number of drawings as per the TOR requirements?			
G: Models				
G.1	Do the models meet the design objectives? (as applicable)			
H: Submittals				
H.1	Is the numbers of sets prepared as per the TOR?			

4.4.1.16 Finalize the Concept Design

The final design is prepared, incorporating the comments, if any, found during analysis and review of the drawings and documents for submission to the owner/client

4.4.2 PRELIMINARY DESIGN

The preliminary design is mainly a refinement of the elements in the conceptual design phase. The preliminary design is also known as Schematic Design. Preliminary design is a design intent which quantifies functional performance expectations and parameters for each system to be commissioned. It is traditionally labeled as 30% design. The preliminary design adequately describes information about all the proposed project elements in sufficient detail to obtain regulatory approval, necessary permits, and authorization. The central activity of the preliminary design is the architect's design concept of the owner's objective, which can help to make the detailed engineering and design of the required facility. At this phase, the project is planned to the level where sufficient details are available for the initial schedule and cost. This phase also includes the initial preparation of all documents necessary to build the facility/construction project. The primary goal of this phase is to develop a clearly defined design based on the client's requirements. Figure 4.13 illustrates the logic flowchart for the Preliminary Design Phase.

Figure 4.14 illustrates the major activities relating to the Preliminary Design Process developed based on the Project Management Process Groups Methodology.

Preliminary design is a subjective process of transforming ideas and information into plans, drawings, and specifications of the facility to be built. Component/

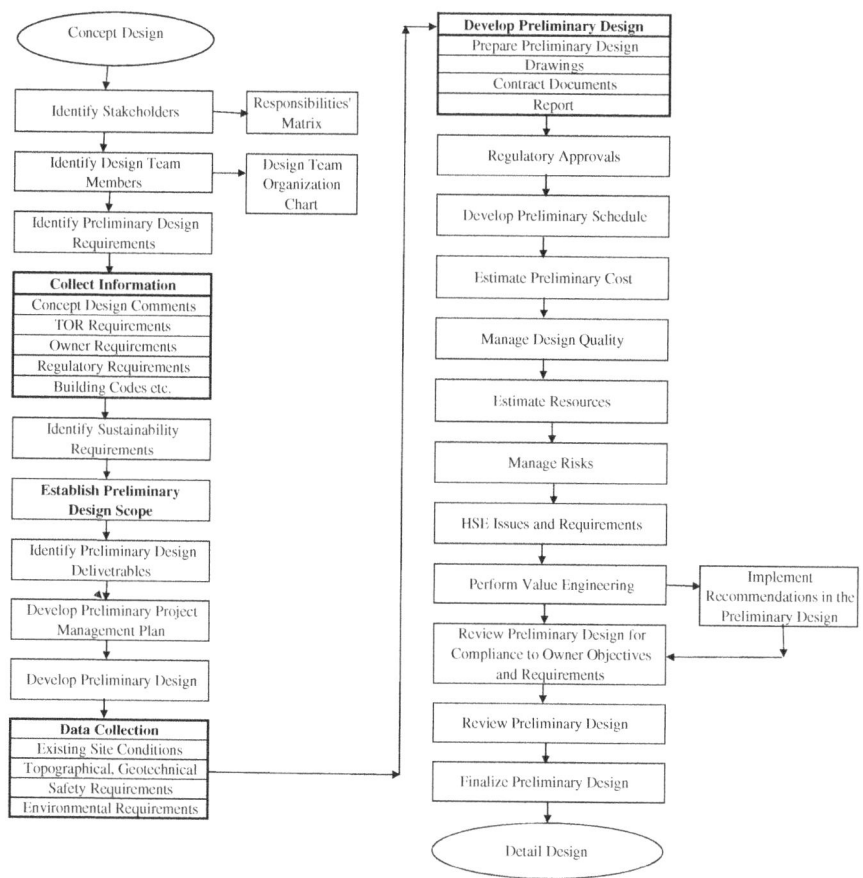

FIGURE 4.13 Logic flowchart for preliminary design phase.

Preliminary/Schematic Design Phase

Management Processes	Project Management Process Groups				
	Initiating Process	Planning Process	Execution Process	Monitoring & Controlling Process	Closing Process
Integration Management	Concept Design Deliverables	Preliminary Project Management Plan	Develop General Layout of Project (Site Plans)	Building Code Requirements	Preliminary Design Deliverables
	Concept Design Comments		Architectural Plans	Concept Design Comments	
	TOR Requirements		Structural Scheme Plans	Existing Conditions	
	Authorities' Requirements		Electromechanical Services	Design Calculations	
	Preliminary Design Requirements		Landscape and Infrastructure	Compliance to Regulatory/ Athorities' Requirements	
			Develop Preliminary Design	Constructability	
Stakeholder Management	Identify Stakeholders	Stakeholders Requirements	Owner's Need Considerations	Stakeholders Requirements	
	Identify Design Team		RegulatoryApprovals	Design Performance	
	Regulatory Approval			Regulatory Approval	
Scope Management	Preliminary Design Scope	Design Deliverables	Design Drawings	Compliance to Owner's Needs	
		Energy Conservation Requirements	Outline Specifications/ Contract Documents	Compliance to TOR	
		Technical and Functional Capability	System Schematics	Stakeholders Approval	
		Sustainability Requirements	Value Engineering	Implements VE Comments	
Schedule Management		Project Schedule (CPM/Bar Chart)	Preliminary Schedule	Project Schedule	
Cost Management		Cost of Activities	Preliminary Cost	Project Cost Estimate	
		Cost of Resources			
		Preliminary Estimate			
Quality Management		Design Criteria	Design Coordination with All Disciplines	Compliance to codes, standards, and Authorities	
		Codes and Standards			
		Authorities Requirements			
Resource Management		Assign Project Design Team	Manage Team Members of Difference Discipline	Performance of Team Members	Assign New Phase/Project
		Estimate Resources			
Communication Management		Design Progress Information	Liaison and Coordination with All Parties	Design Status Information	
			Coordination Meetings		
Risk Management		Management of Design Risk		Control Design Risk	
Contract Management		Contract Terms and Conditions	Preliminary Contract Documents	Check for Contracting System	
HSE Management		Safety Considerations in Design	Environmental Requirements	Life Safety in Design Requirements	
Financial Management		Designer/Consultant Payment			Designer/Consultant Payment
Claim Management		Design Change Payments		Control Changes	Settle Claim by Designer

FIGURE 4.14 Major activities relating to preliminary design process groups. Note: These activities may not be strictly sequential; however, the breakdown allows implementation of project management function to be more effective and easily manageable at different stages of the project phase.

equipment configurations, material specifications, and functional performance are decided during this stage. At this stage, the owner can alter the scope and consider alternatives. The owner seeks to optimize certain facility features within the constraints of other factors, such as cost, schedule, vendor capabilities, and so on.

Design is a complex process. Before design is begun, the scope must adequately define deliverables, that is, what will be furnished. These deliverables are design drawings, contract specifications, type of contracts, construction inspection record drawings, and reimbursable expenses.

Preliminary design is the basic responsibility of the architect (designer/consultant or A/E). In the case of building construction projects, a preliminary design determines:

Management of Quality for Sustainability

1. The general layout of the facility/building/project
2. The required number of buildings/number of floors in each building/area of each floor
3. The different types of functional facilities required, such as offices, stores, workshops, recreation, training centers, parking, etc.
4. The type of construction, such as reinforced concrete or steel structure, precast, or cast *in situ*
5. The type of electromechanical services required
6. The type of infrastructure facilities inside the facilities area
7. The type of landscape and plantation

The designer has to consider the following points while preparing the preliminary design:

1. Concept design deliverables
2. Calculations to support the design
3. System schematics for the electromechanical system
4. Coordination with other members of the project team
5. Authorities' requirements
6. Availability of resources
7. Constructability
8. Health and safety
9. Reliability
10. Energy conservation issues
11. Energy-efficient systems
12. Environmental issues
13. Selection of systems and products that support the functional goals of the entire facility
14. Sustainability requirements
15. Requirements of all stakeholders
16. Optimized life cycle cost (value engineering)

4.4.2.1 Identify Stakeholders

During the preliminary design phase, the following are the stakeholders having direct involvement in the project:

- Owner
- Consultant
- Designer
- Regulatory authorities
- Project/construction manager (if the owner decided to engage one during this phase, depending on the type of project delivery system)

Table 4.20 illustrates the responsibilities of various participants during the preliminary design phase

TABLE 4.20
Responsibilities of Various Participants (Design-Bid-Build Type of Contracts) During the Schematic Design Phase

	Responsibilities		
Phase	Owner	Designer	Regulatory Authorities
Preliminary Design	• Approval of preliminary design	• Develop general layout/scope of facility/project • Regulatory approval • Schedule • Budget • Contract terms and conditions • Value Engineering	• Approval of project submittals

4.4.2.2 Identify the Design Team Members

Design team members are selected based on the organizational structure and the relevant skills required to perform the job. Normally, the design team consists of:

1. Project manager
2. Design managers (one for each trade)
3. Quality manager
4. Team leader (principal engineer) (one for each trade)
5. Team members (engineers, CAD technicians for each discipline)
6. Quantity surveyor (cost engineer)

Figure 4.15 illustrates an example of a structural/civil engineering design team organization chart.

4.4.2.3 Identify the Preliminary Design Requirements

In order to identify the requirements to develop preliminary design, the designer has to gather comments made by the owner/project manager on the submitted concept design, collect the TOR requirements, regulatory requirements, and other related data to ensure that the developed design is error free with minimal omissions.

4.4.2.3.1 Collect Information

The designer has to collect the following information to identify the preliminary design requirements:

1. Concept design comments
2. TOR requirements
3. Owner's requirements
4. Regulatory requirements
5. Building codes and standards

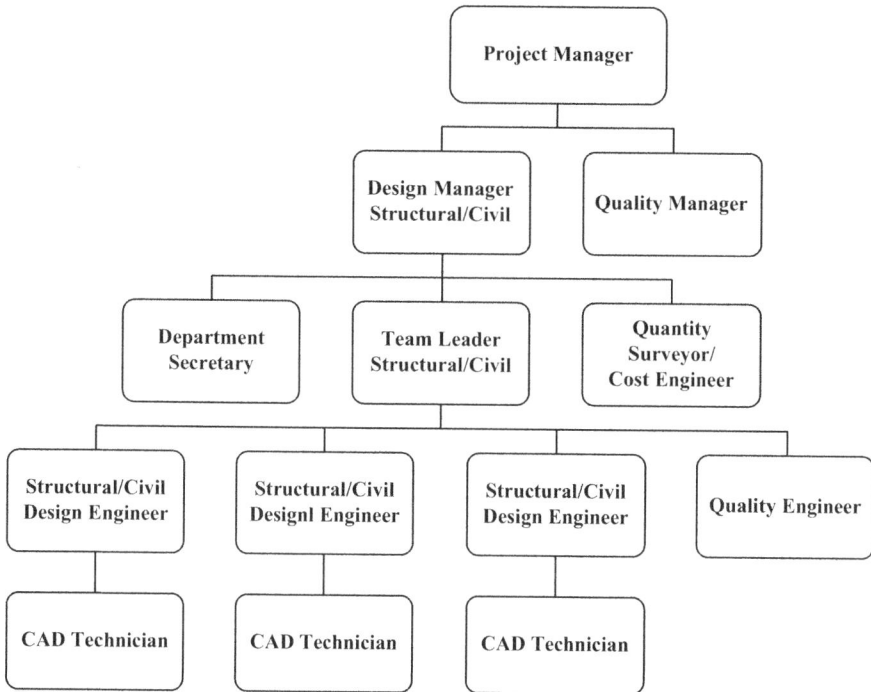

FIGURE 4.15 Structural/civil design team organization chart.

4.4.2.3.1.1 Gather the Concept Design Comments
The objective of the preliminary design is to refine and develop a clearly defined design, based on the client's requirements. While developing the preliminary design, the designer has to review the submitted concept design drawings and reports, and take into consideration the comments, if any, made on the concept design. The designer can discuss in detail with the owner/project manager and incorporate all their requirements to develop the preliminary design.

4.4.2.3.1.2 Identify the TOR Requirements
Normally, the TOR lists the Requirement Guidelines to develop the preliminary design. It consists mainly of the following:

1. Complete preliminary drawings for the Selected/Approved Concept Design
2. Preliminary design report
3. Preliminary models

The TOR details that, during the preliminary design phase, the following items are to be developed, taking into consideration the approved concept design and comments:

A. Drawings
   A.1 Site plan

    A.2 Architectural design
    A.3 Interior design
    A.5 Structural design
    A.6 Conveyance system
    A.7 Mechanical design
        A.7.1 Public health
        A.7.2 Fire suppression
    A.8 HVAC design
    A.9 Electrical design
    A.10 Smart Building Systems (BMS)
    A.11 Special systems
    A.12 Landscape design
    A.13 External works
B. Outline specifications
C. Preliminary contract documents
D. Project schedule (preliminary)
E. Cost estimate (preliminary)
F. Authorities' approvals
G. Value engineering report
H. Preliminary model

4.4.2.3.1.3   Collect the Owner's Requirements
The owner's requirements, discussed under Section 4.4.1.4.2, are to be reviewed and updated, if there are any changes by the owner, and to be further developed in detail in order to be considered in the preliminary design.

4.4.2.3.1.4   Collect the Regulatory Requirements
During this phase, the preliminary design drawings are submitted to the regulatory bodies for their review and approval, for compliance with the regulations, codes, and licensing procedures. Any comments on the drawings are to be incorporated onto the drawings and resubmitted, if required. Normally, for building projects, the following preliminary drawings are submitted for approval by the concerned authorities:

1. Architectural drawings approval by the municipality.
2. Fire department
   a. Escape route (stairs)
   b. Fire alarm system
   c. Fire suppression (sprinkler) system
   d. Smoke ventilation system
   e. Inflammable material and fuel system
   f. Conveyance system (elevator)
   g. Exit and emergency lighting system
3. Public works for utilities connections, such as water supply
4. Public works for storm water, sanitary, and drainage system
5. HVAC System for compliance with energy conservation

Management of Quality for Sustainability 191

6. Electricity agency for approval of Total Connected Load (electrical)
7. Electrical substation location
8. Fuel point for building

For infrastructure and road works, the following approvals are required

1. Environmental impact
2. Sanitary, storm water, drainage system discharge
3. Landscape and plantation
4. Traffic signage and markings
5. Right-of-way
6. Street lighting
7. Substation for streetlights

The requirements differ depending upon the particular country's applicable rules and regulations.

4.4.2.3.1.5 Codes and Standards
The designer has to identify the applicable codes and standards to develop the preliminary design.

4.4.2.3.2 Identify the Sustainability Requirements
The designer has also to identify sustainability requirements for the development of the preliminary design. The following are the basic elements that have to be taken into account by the designer to develop the preliminary design:

- The project budget, as approved by the owner
- Green building concept
- Energy-efficient equipment, systems
- Renewable/alternative energy
- Recycling systems
- Efficient resources
- Ecofriendly material
- Green material and specifications
- Air quality
- Air circulating system
- Daylighting system
- Sensor control lighting in certain areas
- Pollution control
- Stormwater management
- Waste management
- Accessibility and ease of maintenance
- Aesthetic
- Landscape
- Plantation in the project area

- Value engineering (project optimization and economical)
- Regulatory requirements

4.4.2.3.2.1 LEED Requirements

Leadership in Energy and Environmental Design (LEED) is an internationally recognized green building certification system, providing verification by a third party that the building is designed and constructed to the standards and requirements outlined by the LEED rating system. The design team and contractor have to integrate all the required features in their design to ensure that the building is more durable, healthy, and energy efficient.

In order to construct a building that merits LEED Certification, the designer has to consider the following:

- Optimize the site selection
- Orientation of the building to take maximum advantage of sunlight
- Achieve energy-efficient use
- Achieve water-efficient use
- Indoor environmental quality
- Indoor air quality
- Ventilation
- Outdoor airflow monitoring
- Maximum use of alternative (renewable) energy
- Minimize wastewater
- Sustainable material selection
- Regional priority
- Innovation in design

4.4.2.4 Establish the Preliminary Design Scope

The Preliminary Design scope is developed, taking into consideration features identified under the preliminary design requirements as discussed under Section 4.4.2.3. These are as follows:

- Concept design comments
- Approved concept design documents
- TOR requirements
- Owner's preferred requirements
- Regulatory requirements
- Codes and standards
- Approved preliminary project schedule
- Approved preliminary cost estimate

The purpose of the development of the preliminary design is to provide sufficient information to identify the work to be performed and to allow the detailed design to proceed without significant changes that may affect the project schedule and budget. The scope of work during the preliminary design phase mainly comprises:

Management of Quality for Sustainability

- Preparation of preliminary drawings
- Outline specifications
- Preliminary schedule
- Preliminary cost estimate
- Regulatory approvals
- Narrative reports

4.4.2.4.1 Identify the Preliminary Design Deliverables

Based on the scope of the preliminary design, the following are the major deliverables to be developed during the preliminary design:

1. Design drawings
   1.1 Architectural
   1.2 Structural
   1.3 Conveying system
   1.4 Mechanical (public health and fire suppression)
   1.5 HVAC
   1.6 Electrical
   1.7 Landscape
   1.8 External
2. Outline specifications
3. Narrative reports
4. Preliminary schedule
5. Preliminary cost estimate
6. Regulatory approvals
7. Value engineering
8. Models

Table 4.21 lists the Preliminary Design Deliverables to be developed during the Schematic Design phase.

4.4.2.5 Develop the Preliminary Management Plan

The preliminary project management plan developed during the concept design phase is updated based on additional information collected during the schematic design phase. Typical contents of the Project Management Plan are as follows:

1. Project description
2. Project objectives
3. Project organization
   3.1 Organization chart
   3.2 Responsibility matrix
   3.3 Project directory
4. Project (scope) deliverables
5. Project schedule
6. Project budget

TABLE 4.21
Schematic Design Deliverables

Serial Number	Deliverables
1	General
1.1	a. Preliminary/outline specifications
	b. Zoning
	c. Permits and regulatory approvals
	d. Energy code requirements
	e. Construction methodology narration
	f. Descriptive report of environmental, health, and safety requirements
	g. Estimate the construction period (preliminary schedule)
	h. Estimated cost
	i. Value engineering suggestions and resolutions
	j. Life safety requirements
	k. Sketches/perspective
	i. Interior
	ii. Exterior
	m. Graphic presentation
2	Preliminary design drawings
2.1	Architectural
2.1.1	Overall site plans
	a. Existing site plans
	b. Location of building, roads, parking, access, and landscape
	c. Project boundary limits
	d. Site utilities
	e. Water supply, drainage, stormwater lines
	f. Zoning
	g. Reference grids and axis
2.1.2	Floor plans
	a. Floor plans of all floors
	b. Structural grids
	c. Vertical circulation elements
	d. Vertical shafts
	e. Partitions of various types
	f. Doors
	g. Windows
	h. Floor elevations
	i. Designation of rooms
	j. Preliminary finish schedule
	k. Door schedule
	l. Hardware schedule
	m. Ceiling plans
	n. Services closets
	o. Raised floor, if required

(Continued)

TABLE 4.21 CONTINUED
Schematic Design Deliverables

Serial Number	Deliverables
2.1.3	Roof plans
	a. Roof layout
	b. Roof material
	c. Roof drains and slopes
2.1.4	Elevations
	a. Wall cladding
	b. Curtain wall
	c. Stones cladding
	d. Exterior Insulation and Finishing System (EIFS)
	e. Sections at various locations
2.2	Structural
	a. Building structure
	b. Floor grade and system
	c. Foundation system
	d. Tentative size of columns, beams
	e. Stairs
	f. Elevations and sections through various axis
	g. Roof
2.3	Elevator
	a. Traffic studies
	b. Elevator/escalator location
	c. Equipment room
2.4	Plumbing and fire suppression
	a. Sprinkler layout plan
	b. Piping layout plan
	c. Water system layout plan
	d. Water storage tank location
	e. Clean water supply system
	f. Development of preliminary system schematics
	g. Location of mechanical room
2.5	HVAC
	a. Ducting lay out plan
	b. Piping layout plan
	c. Development of preliminary system schematics
	d. Calculations to allow preliminary plant selections
	e. Establishment of primary building services distribution routes
	f. Establishment of preliminary plant location and space requirements
	g. Determine heating and cooling requirements based on heat dissipation of equipment, lighting loads, type of wall, roof, glass, ... etc.
	h. Estimation of HVAC electrical load
	i. Development of BMS schematics showing interface
	j. Location of plant room, chillers, cooling towers
	k. Clean air system/ventilation

(Continued)

TABLE 4.21 CONTINUED
Schematic Design Deliverables

Serial Number	Deliverables

2.6	Electrical
	a. Lighting layout plan
	b. Daylighting
	c. Power layout plan
	d. System design schematic without any sizing of cables and breakers
	e. Substation layout and location
	f. Total connected load
	g. Location of electrical rooms and closets
	h. Location of MLTPs, MSBs, SMBs, EMSB, SEMBs, DBs, EDBs, … etc.
	i. Location of starter panels, MCC panels, … etc.
	j. Location of generator, UPS
	k. Raceway routes
	l. Riser requirements
	m. Information and Communication Technology (ICT)
	   a) Information Technology (Computer Network)
	   b) IP Telephone System (Telephone Network)
	   c) Smart Building System
	n. Loss Prevention Systems
	   a) Fire alarm system
	   b) Access control security system layout
	   c) Intrusion system
	o. Public address, audiovisual system layout
	p. Schematics for FA and other loss prevention systems
	q. Schematics for ICT system, and other low-voltage systems
	r. Location of LV equipment

2.7	Renewable/alternative energy system
2.8	Waste management system
2.9	Pollution control system
3.0	Landscape a. Green area layout b. Plantation b. Selection of plants c. Irrigation system
3.1	External a. Street/road layout b. Street lighting c. Bridges (if any) d. Security system e. Location of electrical panels (feeder pillars) f. Pedestrian walkways

Source: Abdul Razzak Rumane (2013). Quality Tools for Managing Construction Projects. Reprinted with permission of Taylor & Francis Group.

Management of Quality for Sustainability

7. Project quality plan
8. Project resources
9. Risk management
10. Communication matrix
11. Contract management
12. HSE management
13. Project finance management
14. Claim settlement

4.4.2.6 Develop the Preliminary Design

The purpose of the preliminary design is to provide sufficient information to identify the works to be performed and to allow detailed design to proceed without significant changes that may adversely affect the project budget and schedule.

At the preliminary design stage, identified deliverables established in the preliminary design scope must be furnished. It should include a schedule of dates for delivering drawings, specifications, calculations and other information, forecasts, estimates, contracts, materials, and construction. The designer develops preliminary design with the plan, elevation, and other related information that meet the owner's requirements. The designer also develops a concept of how various systems, such as heating and cooling systems, communication systems, etc., will fit into the system.

In order to develop the preliminary design, the designer has to collect the following information:

1. Project-related data/information
    The following data are to be collected to develop the preliminary design for a building project:
  - Needs of the owner
  - Building/project usage
  - Space program
  - Technical and functional capability requirements
  - Zoning requirements
  - Esthetics requirements
  - Fire protection requirements
  - Indoor air quality
  - Lighting/daylighting requirements
  - Conveying system traffic analysis
  - Health and safety features
  - Environmental compatibility requirements
  - Energy conservation requirements
  - Sustainability requirements
  - Facility management requirements
  - Regulatory/authority requirements (permits)
  - Codes and standards to be followed
  - Social responsibility requirements
  - Project constraints

- Ease of constructability
- Number of drawings to be produced
- Milestone for development of each phase of design
- Disabled (special needs) access requirements

2. Site investigation
   - Soil profile and laboratory test of soil
   - Topography of the project site
   - Hydrological information
   - Wind load, seismic load, dead load, and live load
   - Existing services passing through the project site
   - Existing roads, structures surrounding the project site
   - Shoring and underpinning requirements with respect to the adjacent area/structure
3. Topographical details
4. Geotechnical details
5. Health and Safety requirements
6. Environmental requirements

4.4.2.6.1 Prepare the Preliminary Design

During the preliminary design phase, several alternative schemes are reviewed and one scheme which meets the owner's objectives and TOR is selected.

4.4.2.6.2 Prepare the Drawings

Following schematic drawings for building project are generally prepared during this phase;

1. Architectural
   - Site location in relation to the existing environment
   - Overall site plans
   - Floor plans
   - Roof plans
   - Sections
   - Elevations
2. Structural
   - Building structure
   - Foundation system
   - Floor grade and systems
   - Stairs
   - Roof
3. Interior finishes
4. External/interior walls and partitions
5. Conveying system
   - Elevator location
   - Machine room
6. MEP

Management of Quality for Sustainability

- General arrangement of each system
- Single line riser diagram
- Electrical schematic diagram
- Layout of equipment rooms
- Layout of electrical substation
- Electrical rooms
- Vertical shafts

7. Security system
8. External works
9. Landscape plans
10. Specialty items
11. Functional/esthetic aspect of the project
12. Description of critical details
13. Compatibility with the surrounding environment
14. Compliance with sustainability requirements

The designer can use the Building Information Modeling (BIM) tool for designing the project

- **Building Information Modeling**

Building Information Modeling (BIM) is an innovative process of generating a digital database for collaboration and managing building data during its life cycle and preserving the information for reuse and additional industry-specific applications. Building Information Modeling is Autodesk's strategy for the application of information technology to the building industry. It helps better visualization and clash detection. It is an excellent tool to develop project staging plans, study phasing, and coordination issues during a construction project life cycle, preparation of As-Built, and also during maintenance of the project.

4.4.2.6.3 Develop the Contract Documents

During this phase, the following documents are developed in line with the type of the contracting system:

i. Preliminary specifications
ii. Preliminary contract documents

1. **Preliminary specifications (Outline specifications)**

Outline specifications, indicating project-specific features of major equipment, systems, and material, are prepared during this phase. Normally, specifications are prepared per MasterFormat® contract documents produced jointly by the Construction Specification Institute (CSI) and the Construction Specifications Canada (CSC), which are widely accepted as standard practice for the preparation of contract documents.

2. Contract Documents

The consultant/designer team is responsible for developing a set of contract documents that meet the owner's needs, and specifies the required level of quality, schedule, and budget. There are numerous combinations of contract arrangements for handling construction projects; however, Design-Bid-Build is predominantly used in most construction project contracts. This delivery system has been chosen by owners for many centuries and is called the traditional contracting system. In the traditional contracting system, the detailed design for the project is completed before tenders for construction are invited.

Based on the type of contracting arrangements, and how the owner would like to handle the project, necessary documents are prepared by establishing a framework for the execution of the project. Generally, FIDIC (Federation International des Ingénieurs Counseils) models of the conditions for international civil engineering contracts are used as a guide to prepare these contract documents.

4.4.2.6.4 The Preliminary Design Report

The preliminary design phase reports mainly consist of the following:

- Systems
- Materials
- Finishes
- Design features describing the selected option
- Construction methodology
- Project risk
- HSE requirements

4.4.2.7 Regulatory Approval

Once the preliminary design is approved, the requisite drawings are to be submitted to regulatory bodies for their review and approval for compliance with the regulations, codes, and licensing procedure.

4.4.2.8 Develop the Preliminary Schedule

The project schedule is developed using top-down planning and key elements (refer Table 4.15)

4.4.2.9 Estimate the Preliminary Cost

Based on the preliminary design, the preliminary cost is prepared by estimating the cost of activities and resources. Estimation of the preliminary cost is an important activity that results in a timed phased plan summarizing the expected expenses toward the contract and also the income or the generation of funds necessary to achieve the milestone. The estimated cost for a construction project is the maximum amount the owner is willing to spend for design and construction of the facility that meets the owner's need. The preliminary cost is determined by estimating the cost of activities and resources and is related to the schedule of the project. If the cash flow or resulting budget is not acceptable, the project schedule should be modified. It is required that, while preparing the budget, the risk assessment of the project is also performed.

Management of Quality for Sustainability

The cost estimate during this phase is based on an elemental parametric tool/methodology and is also known as the budgetary cost (refer to Table 4.16)

4.4.2.10 Manage the Design Quality

In order to minimize design errors, minimize design omissions, and reduce rework during the preliminary design, the designer has to plan quality (planning of design work), perform quality assurance, and control quality for preparing schematic design. This will mainly consist of the following:

1. Plan Quality:
   - Establish the owner's requirements
   - Identify the requirements listed under the TOR
   - Identify the quality standards and codes to be complied with
   - Identify the regulatory requirements
   - Establish the design criteria
   - Establish the quality organization with a responsibility matrix
   - Determine the number of drawings to be produced
   - Identify the sustainability requirements
   - Establish the scope of work
   - Develop the design (drawings and documents) review procedure
   - Establish the submittal plan
   - Establish the design review procedure
2. Quality Assurance:
   - Collect the data
   - Investigate the site conditions
   - Prepare the preliminary drawings
   - Prepare the preliminary/outline specifications
   - Ensure the functional and technical compatibility
   - Coordinate with all the disciplines
   - Select material to meet the owner's objectives
3. Control Quality:
   - Check the design drawings
   - Check the specifications/contract documents
   - Check for regulatory compliance
   - Check the preliminary schedule
   - Check the cost of the project (preliminary cost)

The BIM tool can be used to control the quality of the project.

4.4.2.11 Estimate the Resources

The designer has to estimate the resources required to complete the project. At this stage, more details about the activities and works to be performed during the construction, testing, and commissioning phase are available, and the designer has to update the earlier estimated resources and prepare a manpower histogram. Also, the designer can estimate the total number of design team members to develop the design and construction documents.

4.4.2.12 Manage the Risks

The following are typical risks which normally occur during the preliminary design phase:

- Concept design deliverables and review comments are not taken into consideration while preparing the preliminary design
- All the TOR requirements are not taken into consideration
- Sustainability requirements are not taken into consideration
- Regulatory authorities' requirements are not taken into consideration
- Preliminary design scope of the work is incomplete
- The related project data and information collected is incomplete
- The related project data and information collected is likely to be incorrect and wrongly estimated
- Site investigations for existing conditions are not carried out
- Fire and safety considerations
- Environmental consideration
- Incomplete design
- Prediction of possible changes in design during the construction phase
- Inadequate and ambiguous specifications
- Wrong selection of materials and systems
- Undersized HVAC equipment selection
- Incorrect water supply requirements
- Estimated total electrical load is much lower than the expected actual consumption
- Errors in calculating the traffic study for the conveying system
- Errors in estimating the project schedule
- Errors in cost estimation
- Number of drawings not as per the TOR requirements

The designer has to take into account the above-mentioned risk factors while developing the preliminary design.

Furthermore, the designer has to consider the following risks while planning the duration for completion of the schematic phase:

- Impractical schematic design preparation schedule
- Delay obtaining regulatory approval
- Delay in the site investigations
- Delay in the data collection

4.4.2.13 HSE Issues and Requirements

The designer has to identify HSE issues and requirements to manage the activities during the preliminary design stage. The designer has to carry out an Environmental Impact Assessment study and take it into consideration, while developing the preliminary design.

4.4.2.14 Perform Value Engineering

Value engineering (VE) studies can be conducted at various phases of a construction project; however, the studies conducted in the early stage of a project tend to provide the greatest benefit. In most projects, VE studies are performed during the preliminary phase of the project. At this stage, the design professionals have considerable flexibility to implement the recommendations made by the VE team, without significant impacts on the project's schedule or design budget. In certain countries, for a project over US$5 million, a VE study must be conducted as part of the preliminary design process. The team members who perform the VE study depend on the client's/owner's requirements. It is advisable that a SAVE internationally registered certified value specialist be assigned to lead this study. Figure 4.16 illustrates VE activities.

4.4.2.15 Review the Preliminary Design

Table 4.22 illustrates major points for review of the preliminary design.

4.4.2.16 Finalize the Preliminary Design

The final preliminary design is prepared, incorporating the comments, if any, found during analysis and review of the drawings and documents for submission to the owner/client.

Normally, the following items are submitted to the owner for their review and approval in order to proceed with the development of the detail design:

1. Preliminary design drawings
2. Outline specifications
3. Preliminary contract documents
4. Preliminary schedule
5. Budgetary cost estimate
6. Preliminary design reports
7. Model (if applicable)

4.4.3 Detail Design

Detail design follows the preliminary design phase and takes into consideration the configuration and the allocated baseline derived during the preliminary design phase. Detail design is also known as Design Development/Detailed engineering. During this phase, all suggested changes are reevaluated to ensure that the changes will not detract from meeting the project design goals/objectives. Detail design involves the process of successively breaking down, analyzing, and designing the structure and its components so that it complies with the recognized codes and standards of safety and performance, while rendering the design in the form of drawings and specifications that will tell the contractors exactly how to build the facility to meet the owner's needs. During this phase, detail design of the work, detail plan, budget, estimated cash flow, contract documents, regulatory approval, and tender/bidding documents are prepared. Depending on the type of contract the owner would like

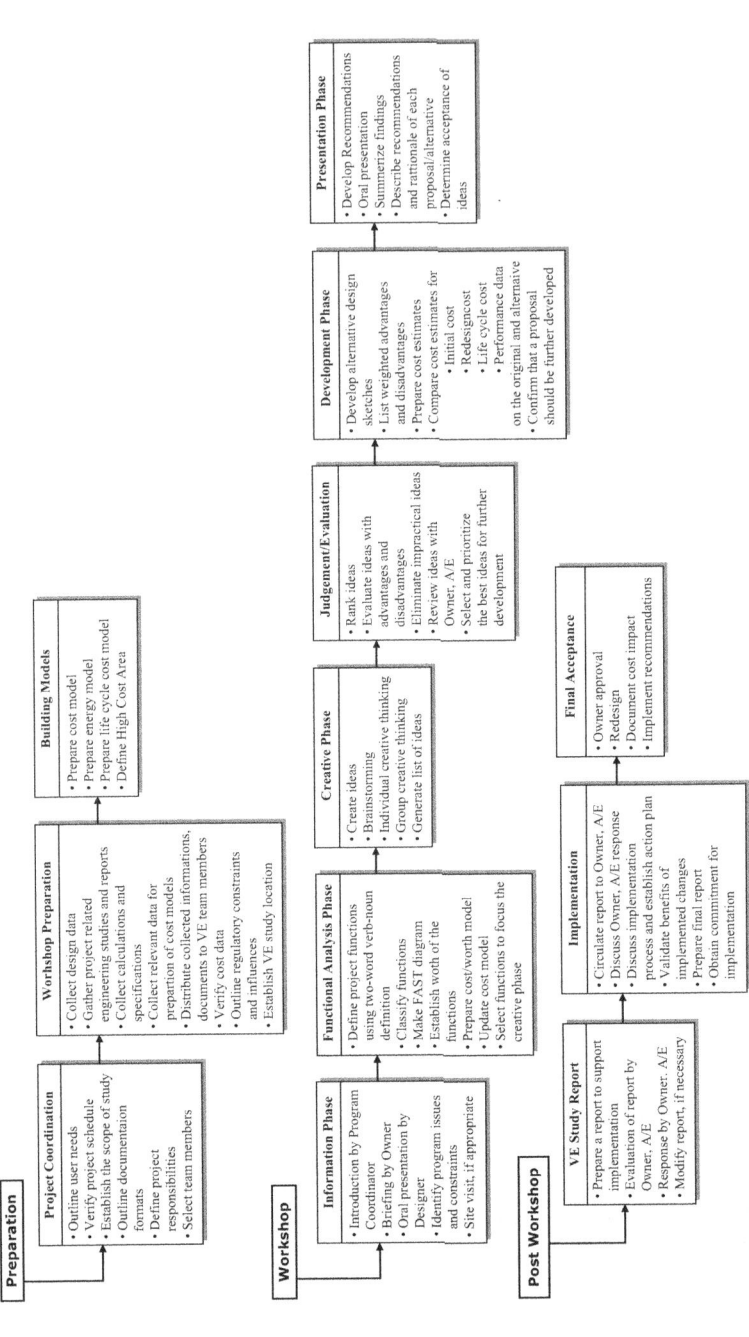

FIGURE 4.16 Value engineering study process activities. Source: Abdul Razzak Rumane. (2017). *Quality Management in Construction Projects*, Second Edition. Reprinted with permission from Taylor & Francis Group Company.

TABLE 4.22
Major Points to Review for the Preliminary Design

Serial Number	Description	Yes	No	Notes
A: Preliminary Design (General)				
A.1	Does the design support the owner's project goals and objectives?			
A.2	Does the design meet all the elements specified in the TOR?			
A.3	Are comments on Concept Design taken care of while preparing the Schematic Design?			
A.4	Have regulatory approvals been obtained?			
A.5	Does the design meet all the performance requirements?			
A.6	Has constructability been taken care of?			
A.7	Has technical and functional capability been considered?			
A.8	Have health and safety requirements been considered in the design?			
A.9	Does the design meet fire and egress requirements?			
A.10	Have design risks been identified, analyzed, and responses planned for mitigation?			
A.11	Have environmental constraints been considered?			
A.12	Is energy conservation considered in the design?			
A.13	Are the sustainability requirements considered in the design?			
A.14	Is cost-effectiveness over the entire project life cycle considered?			
A.15	Does the design meet LEED requirements?			
A.16	Is accessibility considered?			
A.17	Does the design result in ease of maintenance?			
A.18	Does the design have provision for the inclusion of facility management requirements?			
A.19	Does the design support proceed to the next development stage of the design?			
A.20	Are all the drawings numbered?			
B: Architectural				
B.1	Are the drawings coordinated with other disciplines?			
B.2	Is the grid system established?			
B.3	Is zoning considered?			
B.4	Do the overall site plans show all the major areas?			
B.5	Do the floor plans show overall dimensions?			
B.6	Are roof plans shown?			
B.7	Are preliminary elevations shown?			
B.8	Are all the rooms numbered?			
B.9	Are all entrances, stairways, lobbies, and corridors identified?			
B.10	Are all service rooms, equipment rooms, and plant rooms identified?			
B.11	Is there a finishes schedule/requirement prepared?			

(*Continued*)

TABLE 4.22 CONTINUED
Major Points to Review for the Preliminary Design

Serial Number	Description	Yes	No	Notes
C: Structural				
C.1	Is the preliminary foundation plan shown?			
C.2	Are the structural systems identified?			
C.3	Is the preliminary building structure prepared?			
C.4	Are there preliminary framing plans for all floors and the roof?			
C.5	Are the relevant codes regarding seismic zone and wind speed considered for structural load calculations?			
C.6	Is slab loading for equipment considered?			
D: Elevator				
D.1	Is traffic analysis performed?			
D.2	Are elevator locations shown?			
D.3	Is the equipment room location shown?			
E: Mechanical				
E.1	Are the preliminary plans for toilets/rest rooms and pantry shown?			
E.2	Are the main water supply, sanitary, and stormwater system shown?			
E.3	Is the riser diagram/single line diagram for the plumbing system shown?			
E.4	Are plumbing fixtures identified?			
E.5	Does the fire protection system comply with regulatory requirements?			
E.6	Is a single line diagram prepared for the fire protection system (sprinkler)?			
E.7	Is the equipment room (plant room) location shown?			
E.8	Is a special fire suppression system considered for the electrical substation or generator room?			
F: HVAC				
F.1	Are single line diagrams prepared for all related systems?			
F.2	Are shaft locations and approximate pipe sizes and duct size shown?			
F.3	Is the location of the plant room considered?			
F.4	Is the approximate load for the chiller and pump sizes considered from available data?			
F.5	Are gross HVAC zoning and typical individual space zoning considered while preparing the design?			
G: Electrical				
G.1	Are preliminary lighting plans prepared?			
G.2	Are preliminary power layout plans prepared?			
G.3	Is the substation location shown?			
G.4	Is the electrical room located?			

(Continued)

TABLE 4.22 CONTINUED
Major Points to Review for the Preliminary Design

Serial Number	Description	Yes	No	Notes
G.5	Are the cable tray and other raceway routes shown?			
G.6	Is the riser shaft for cables and bus ducts considered?			
G.7	Is a system schematic diagram prepared?			
G.8	Does the fire alarm system comply with regulatory requirements?			
G.9	Are preliminary plans for all low-voltage systems (communication, public address, audiovisual, access control, and security system) prepared?			
G.10	Is the emergency diesel generator considered?			
H: Landscape				
H.1	Are landscape plans prepared?			
H.2	Are plants selected?			
H.3	Is an irrigation system layout prepared?			
I: External				
I.1	Are site layout plans, showing roads, walkways, and parking areas, prepared?			
I.2	Are street lighting plans prepared?			
I.3	Are project site boundaries properly marked and demarcated?			
J: Financial				
J.1	Is the project cost properly estimated?			
K: Schedule				
K.1	Is the project schedule practically achievable?			
L: Value Engineering				
L.1	Is a value engineering study performed and recommendations considered?			
M: Reports				
M.1	Is the narrative description complete and includes adequate information about the project?			
M.2	Do the outline specifications include all the works			
M.3	Are preliminary contract documents considered with respect to all the TOR requirements			
N: Drawings, Sketches				
N.1	Are drawings, sketches for all trades prepared as per TOR?			
O: Models				
O.1	Do the models meet the design objectives?			
P: Submittals				
P.1	Is the number of sets prepared as per TOR?			

to have for completing the facility, the designer (consultant) can start preparing the detailed design. The success of a project is closely correlated with the quality and depth of the engineering plans prepared during this phase.

Figure 4.17 illustrate the logic flowchart for the Detail Design Phase.

Figure 4.18 illustrates major activities relating to the Detail Design Process developed based on Project Management Process Groups Methodology.

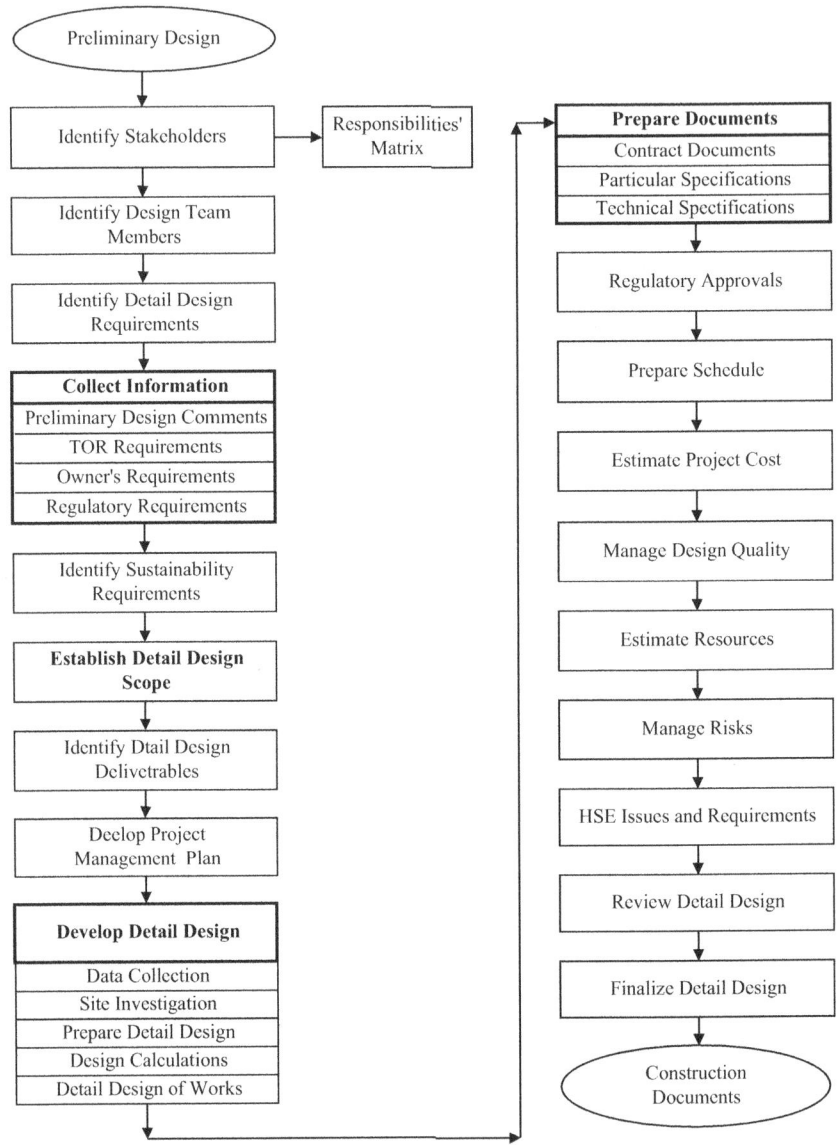

FIGURE 4.17 Logic flowchart for detail design phase.

Management of Quality for Sustainability

Design Detail Design Phase

Management Processes	Process Management Groups				
	Initiating Process	Planning Process	Execution Process	Monitoring & Cintrolling Process	Closing Process
Integration Management	Schematic Design Deliverables	Project Management Plan	Data Collection/ Site Investigations	Design Calculations	Detail Design Drawings
	Comments on Preliminary Design		Detail Design Drawings	Interdisciplinary Coordination	
	TOR Requirements		Bill of Quantities	Compliance to TOR	
	Authorities' Requirements		Model		
Stakeholder Management	Identify Design Team	Identify Stakeholders		Design Progress	
		Stakeholders Requirements			
		Stakeholders Matrix			
Scope Management		Owner's Needs, Project Goals and Objectives	Schematic/Riser Diagram	Authorities Approval	
		Design Development	Detail Calculations	Compliance to Owner's Need	
		Design Documents	Project Specifications	Stakeholders Approval	
		Sustainability Requirements	Contract Documents		
		Design Deliverables			
		Bill of Quantity			
Schedule Management		Activity Duration	Project Schedule	Design pProgress	
		Precedence Diagram			
		Construction Schedule			
Cost Management		Price Analysis	Project Budget	Control Budget	
		Bill of Quantities			
		Resources			
		Detail Estimate			
Quality Management		Codes and Standards	Design Coordination with All Disciplines	Design Compliance to Owner's Goals and Objectives	
		Regulatory Requirements	Assure Design Quality	Coordination with all Disciplines	
		Design Crieteria	Develop Project Quality	Control Design Quality	
		Well Defined Specifications			
		Plan Design Quality			
Resource Management		Estimate Project Resources	Manage Team Members from All Discipline	Performance of Team Project Members	
Communication Management		Communication Matrix	Liaison with All Disciplines	Design Status Information	
			Coordination Meetings		
Risk Management		Identification of Risk during Bidding, Construction, Testing and Commissioning		Design Risk Control	
Contract Management		Bidding and Tendering Documents		Check for Contracting System	Contract Documents
HSE Management		Safety in Design	Safety Requirements		
		Environmental Compatibility	Environmental Requirements	HSE Compliance in Design	
Financial Management		Designer/Consultant Payment		Payment to Designer/Consultant	
Claim Management		Design Change Payment		Control Changes	Settle Designer's Claim

FIGURE 4.18 Major activities relating to detail design process. Note: These activities may not be strictly sequential; however, the breakdown allows implementation of project management function to be more effective and easily manageable at different stages of the project phase.

4.4.3.1 Identify the Stakeholders

The following stakeholders have direct involvement in the project during the detail design phase:

- Owner
- Consultant
- Designer
- Regulatory authorities
- Project/construction manager (if the owner decided to engage one during this phase, depending on the type of project delivery system)

Table 4.23 illustrates the responsibilities of the various participants during the detail design phase.

4.4.3.2 Identify Design Team Members

Figure 4.19 illustrates the Design Management Team and their major responsibilities.

Each of the managers has many other team members. These members are selected based on the organizational structure and suitable skills required to perform the job. These include

1. Team leader (principal engineer) (one for each trade)
2. Team members (engineers and CAD technicians for each discipline)
3. Quantity surveyor (cost engineer)
4. Owner's representative
5. End-user

TABLE 4.23
Responsibilities of Various Participants (Design--Bid-Build Type of Contracts) during the Detail Design Phase

	Responsibilities		
Phase	Owner	Designer	Regulatory Authorities
Detail Design	• Approval of design • Approval of time schedule • Approval of budget	• Development of detail design • Submission for Authorities approval • Detail plans • Schedule • Budget • BOQ • Specifications • Contract documents • Verification of design	• Review and approval of project submittals

Management of Quality for Sustainability

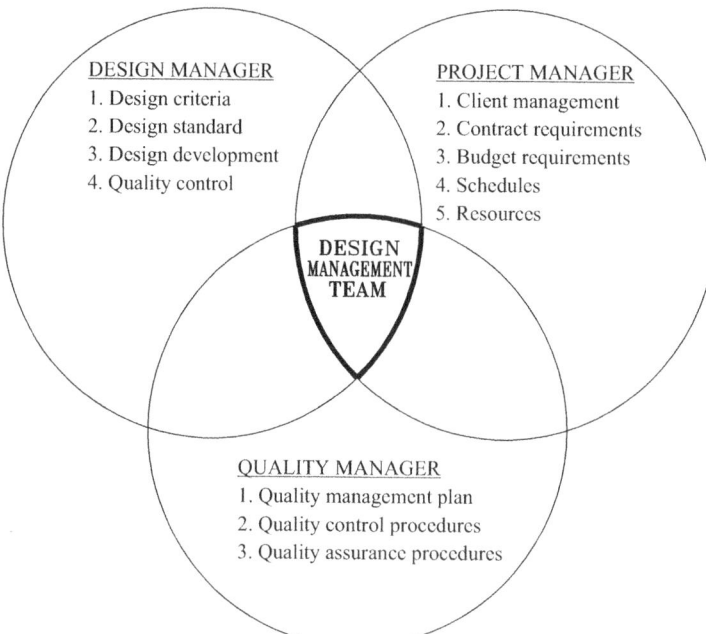

FIGURE 4.19 Design management team. Source: Abdul Razzak Rumane. (2013). *Quality Tools for Managing Construction Projects*. CRC Press, Boca Raton, FL. Reprinted with permission of Taylor & Francis Group.

Accuracy in the project design is a key consideration in the life cycle of the project; therefore, it is required that the designer/consultant be not only an expert in the technical field but should also have a broad understanding of engineering principles, construction methods, and value engineering. The designer must know the availability of the latest products on the market and to use proven technology, methods, and materials to meet the owner's objectives. He or she must refrain from using a monopolistic product, unless its use is important or critical for proper functioning of the system. He or she must ensure that at least two or three sources are available on the market, producing the same type of product that complies with all its required features and intent of use. This will help the owner get competitive bidding during the tender stage.

The design professional should be:

- Fully qualified to provide services contractually undertaken and required
- Having the skill to develop the scope of design works and other services
- Having the appropriate skill to develop the design works
- Having the ability to avoid conflict of interest
- Having the knowledge of applicable codes, standards, and regulatory requirements

- Having the knowledge of schedule development
- Having full knowledge of cost estimation
- Having knowledge to achieve project quality
- Having information on the availability of resources
- Aware of applicable contracting documents

4.4.3.3 Identify the Detail Design Requirements

Detail design involves enhancement of the work carried out during the schematic design phase. During this phase, a comprehensive design of works with a detailed work breakdown structure of design, drawings, specifications, and contract documents is prepared. The detail design phase is the realm of design professionals, including architects, interior designers, landscape architects, and experts in several other disciplines, such as civil, mechanical, electrical, and other engineering professionals, as needed.

During this phase, detailed plans, sections, and elevations are drawn to scale, principal dimensions are noted, and design calculations are checked to conform the accuracy of design and its compliance with the codes and standards.

4.4.3.3.1 Collect Information

In order to identify detail design requirements, the designer has to collect the following information:

1. Preliminary design comments
2. TOR requirements
3. Owner's requirements
4. Regulatory requirements

4.4.3.3.1.1 Gather the Preliminary Design Comments

The objective of detail design is the enhancement of the preliminary design and to develop a clearly defined design based on the client's requirements. While developing the detail design, the designer needs to review the approved preliminary design drawings and reports, and take them into consideration in the comments, if any, made on the preliminary design.

4.4.3.3.1.2 Identify the TOR Requirements

Normally, the TOR lists the Requirements Guidelines to develop the detail design. It mainly consists of the following:

1. Detail Design drawings based on the approved Preliminary Design
2. Project schedule
3. Definitive estimate
4. Bill of Quantities
5. Project specifications
6. Contract documents
7. Format, scale, and size of reports and drawings

Management of Quality for Sustainability

The TOR lists the following items to be developed during the Detail Design, taking into consideration the approved Preliminary Design and comments:

A. Drawings
    A.1 Project title
    A.2 Drawing index, legends, and symbols
    A.3 Existing site condition
    A.4 Site plans
    A.5 Architectural design
    A.6 Interior design
    A.7 Structural design
    A.8 Mechanical design
        A.8.1 Public health
        A.8.2 Fire suppression
    A.9 HVAC design
    A.10 Electrical design
    A.11 Smart Building Systems
    A.12 Special systems
    A.13 Landscape design
    A.14 External works
B. Project schedule
C. Definitive cost estimate
D. Project specifications
E. Contract documents
F. Authorities' approvals

4.4.3.3.1.3   Collect the Owner's Requirements
Any additional requirements/changes to those considered during the development of the schematic design are considered while developing the detail design. This may include additional systems or changes in the preferred requirements.

4.4.3.3.1.4   Collect the Regulatory Requirements
Government agency regulatory requirements have a considerable impact on precontract planning. Some agencies require that the design drawings are submitted for their preliminary review and approval to ensure that the designs are compatible with local codes and regulations. These include:

- Submission of drawings to electrical authorities showing the anticipated electrical load required for the facility.
- Fire alarm and firefighting system drawings.
- Water supply and drainage system.
- HVAC drawings to ensure compliance with energy conservation codes.
- Technical details of conveying system.

4.4.3.3.2 Identify the Sustainability Requirements

The designer also has to identify the sustainability requirements for developments of the detail design. The following are the basic elements that have to be taken care of by the designer to develop the preliminary design:

- Project cost within the budget
- Efficient resources
- Efficient and durable materials
- Ecofriendly materials
- Green materials and specifications
- Detailed calculations to select the structure
- Detailed calculation to select the MEP Systems
- Schematics/riser diagrams (applicable systems)
- Energy-efficient MEP systems, equipment
- Water conservation
- Renewable/alternative energy
- Recycling systems
- Air quality
- Air circulating system
- Comfortable lighting
- Pollution control
- Waste management
- Stormwater management
- Accessibility and ease of maintenance
- Green Building concept
- Aesthetic
- Landscape compatible with the surrounding area
- Plantation in the project area
- Regulatory requirements
- Optimized design

4.4.3.4 Establish the Detail Design Scope

Design development scope is developed, taking into considerations requirements established under Section 4.4.3.3.1, which are as follows:

- Approved preliminary design documents
- TOR requirements
- Owner's preferred requirements
- Regulatory requirements
- Approved preliminary project schedule
- Approved preliminary cost estimate
- Codes and standards

The purpose of the development of the detail design is to provide sufficient information and detail to ensure that the work to be performed by the contractor is properly

Management of Quality for Sustainability

identified and correctly addressed, taking necessary steps to mitigate errors and omissions in the design. The scope of work during the detail design phase mainly comprises:

- Design development drawings
- Project schedule
- Detail cost estimate
- Bill of Quantities
- Technical specifications
- Authorities' approvals
- Design development report

4.4.3.4.1 Identify the Detail Design Deliverables

Based on the scope for design development, the following are the major deliverables to be developed during the detail design phase:

1. Design Drawings
    1.1 Site plans
    1.2 Architectural
    1.3 Interior
    1.4 Structural
    1.5 Conveying system
    1.6 Mechanical (public health and fire suppression)
        1.6.1 Clean water supply system
    1.7 HVAC
        1.7.1 Clean air supply system
    1.8 Electrical
        1.8.1 Alternative energy system
        1.8.2 Comfortable lighting
    1.9 Low-voltage system
    1.10 Waste management system
    1.11 Pollution control system
    1.12 Landscape
    1.13 External

2. Specifications
3. Project schedule
4. Definitive cost estimate
5. Regulatory approvals

Table 4.24 lists the detail design deliverables to be developed during development of the detail design phase.

4.4.3.5 Develop the Project Management Plan

The preliminary project management plan, developed during the preliminary design phase, is updated based on additional information collected during the detail design

TABLE 4.24
Detail Design Deliverables

Serial Number	Deliverables
1	General
1.1	a. Project specifications
	b. Project schedule
	c. Project estimate
	d. Bill of Quantities
	e. Detailed calculations (all trades)
	f. Schematic/riser diagram (all applicable systems)
	g. Site investigations
	h. Design report
	i. Model
2	Detail Design drawings
2.1	Architectural
2.1.1	Overall site plans
2.1.2	Floor plans
2.1.3	Roof plans
2.1.4	Elevations
2.2	Structural
2.3	Elevator
2.4	Plumbing and fire suppression
2.5	HVAC
2.6	Electrical
2.7	Waste management system
2.8	Pollution control system
2.9	Landscape
3.0	External

phase. The contents of the project management plan are the same as that of the preliminary design, but additional information is added. A typical Table of Contents (TOC) of the project management plan is given below:

1. Project description
2. Project objectives
3. Project organization
   a. Organization chart
   b. Responsibility matrix
   c. Project directory
4. Project (scope) deliverables
5. Project schedule
6. Project budget
7. Project quality plan

Management of Quality for Sustainability

8. Project resources
9. Risk management
10. Communication matrix
11. Contract management
12. HSE management
13. Project finance management
14. Claim settlement

4.4.3.6 Develop the Detail Design

Detail design activities are similar, although more in-depth than the design activities in the preliminary design stage. The size, shape, levels, performance characteristics, technical details, and requirements of all the individual components are established and integrated into the design. Design engineers of different trades have to take into consideration all these at least while preparing the scope of works. The range of design work is determined by the nature of the construction project.

4.4.3.6.1 Data Collection

The designer has to collect all the missing data to ensure that the detail design is developed without any errors or omissions.

4.4.3.6.2 Site Investigation

The designer has to revisit the site to ensure that there are not many changes to the earlier performed investigations.

4.4.3.6.3 Prepare the Detail Design

The following detail drawings for the building project are generally prepared during this phase:

- Architectural
    - Site plans
    - Floor plans
    - Roof plans
    - Partitions
    - Sections
    - Elevations
    - Reflected ceiling plan
    - Finishes schedule
    - Door schedule
    - Typical windows details
    - Furnishings
- Structural
    - Building structure
    - Foundation system
    - Footings
    - Stairs

- Roof
- Sections
- Structural floor plans
- Conveying system
  - Elevator location
  - Machine room
- MEP
  - General arrangement of each system
  - Single line riser diagrams
  - Electrical schematic diagram
  - Layout of equipment rooms
  - Layout of electrical substation
  - Electrical rooms
  - Vertical shafts
- Low-voltage systems
- Waste management system
- Pollution control system
- External works
- Landscape and plantation plans

4.4.3.6.3.1 Calculations
The designer has to submit the following calculations along with the detail design:

- Calculation of all structural elements (structural works)
- Analysis and selection of the HVAC System
- Analysis and selection of mechanical systems
- Calculation of lighting (Isolux calculations)
- Short circuit calculations for cables
- Lightning protection system
- Earthing (grounding) System
- Sound system (audiovisual system)

4.4.3.6.4 Prepare Drawings
Detail design activities are similar, although more in-depth than the design activities in the preliminary design stage. The size, shape, levels, performance characteristics, technical details, and requirements of all the individual components are established and integrated into the design. Design engineers of different trades have to take into consideration all these at a minimum while preparing the scope of works. The range of design work is determined by the nature of the construction project.

4.4.3.6.4.1 Detail Design of Work
The following are the aspects of work to be considered by design professionals while preparing the detail design. These can be considered as a base for the development of design to meet customer requirements and will help achieve the qualitative project.

Management of Quality for Sustainability

Architectural Design

- Intent/use of building/facility
- Property limits
- Aesthetic look of the building
- Environmental conditions
- Elevations
- Plans
- Axis, grids, and levels
- Room size to suit the occupancy and purpose
- Zoning as per usage/authorities requirements
- Identification of zones, areas, and rooms
- Modules to match with structural layout/plan
- Number of floors
- Ventilation
- Thermal insulation details
- Stairs and elevators (horizontal and vertical transportation)
- Fire exits
- Ceiling height and details
- Reflected ceiling plan
- Internal finishes
- Internal cladding
- Partition details
- Masonry details
- Joinery details
- Schedule of doors and windows
- Utility services
- Toilet details
- Required electromechanical services
- External finishes
- External cladding
- Glazing details
- Finishes schedule
- Door schedule
- Windows schedule
- Hardware schedule
- Special equipment
- Fabrication of items, such as space frame, steel construction, retaining wall, having a special importance for appearance/finishes
- Special materials/products to be considered, if any
- Any new material/product to be introduced
- Conveying system core details
- Maintenance access for equipment/services requirements
- Ramp details

- Hard and soft landscape
- Parking areas
- Provision for future expansion (if required)

Concrete Structure

- Property limits/surrounding areas
- Type of foundation
- Energy-efficient foundation
- Design of foundation based on field and laboratory tests of the soil investigation which gives the following information:
   a) Subsurface profiles, subsurface conditions, and subsurface drainage
   b) Allowable bearing pressure and immediate and long-term settlement of footing
   c) Coefficient of sliding on foundation soil
   d) Degree of difficulty of excavation
   e) Required depth of stripping and wasting
   f) Methods of protecting below-grade concrete members against the impact of soil and ground water (water and moisture problems, termite control, and radon, where appropriate)
   g) Geotechnical design parameters such as angle of shear resistance, cohesion, soil density, modulus of deformation, modulus of sub-grade reaction, and predominant soil type
   h) Design loads such as dead load, live load, wind load, and seismic load
- Footings
- Grade and type of the concrete
- Size of the bars for reinforcement and the characteristic strength of the bars
- Clear cover for reinforcement for
   a) Raft foundation
   b) Underground structure
   c) Exposed to weather: structures such as columns, beams, slabs, walls, or joists
   d) Not exposed to weather: columns, beams, slabs, walls, or joists
- Reinforcement bar schedule, stirrup spacing
- Expansion joints
- Concrete tanks (water)
- Insulation
- Services requirements (shafts, pits)
- Shafts and pits for the conveying system
- Location of columns in coordination with architectural requirements
- Number of floors
- Height of each floor
- Beam size and height of beam
- Openings for services
- Substructure

Management of Quality for Sustainability

    (a) Columns
    (b) Retaining walls
    (c) Walls
    (d) Stairs
    (e) Beams
    (f) Slabs
- Superstructure:
    (a) Columns
    (b) Stairs
    (c) Walls
    (d) Beams
    (e) Slabs
- Consideration of waterproofing requirements for roof slab against water leakage
- Deflection, which may cause fatigue of structural elements, crack, or failure of fixtures, fittings, or partitions, or discomfort of occupants
- Movement and forces due to temperature
- Equipment vibration criteria
- Load sensors to measure deflection
- Reinforcement bar schedule, stirrup spacing
- Building services to fit in the building
- Environmental compatibility
- Parapet wall
- Excavation
- Dewatering
- Shoring
- Backfilling

Elevator Works

- Type of elevator
- Loading capacity
- Speed
- Number of stops
- Travel height
- Cabin, cabin accessories, cabin finishes, and car operating system
- Door, door finishes, and door system
- Safety features
- Drive, size, and type of motor
- Floor indicators, call button
- Control system
- Cab overhead dimensions
- Pit depth
- Hoist way
- Machine room
- Operating system

Fire Suppression System

The fire suppression system provides protection against fire to life and property. The system is de-signed taking into consideration the local fire code and NFPA standards. The system includes the following:

- Sprinkler system for fire suppression in all the areas of the building
- Hydrants (landing valve) for professional fighting
- Hose reel for public use throughout the building
- Gaseous fire protection system for communication rooms
- Fire protection system for the diesel generator room
- Size of fire pumps and controls
- Water storage facility
- Interface with other related systems

Plumbing Works

- Maximum working pressure to have adequate pressure and flow of water supply
- Maximum design velocity
- Maximum probable demand
- Demand weight of fixture in fixture units for public uses
- Friction loss calculation
- Maximum hot water temperature at fixture outlet
- Water heater high-temperature outlets
- Providing isolating valves to ensure that the system is easily maintained
- Hot water system
- Central water storage capacities
- Size of pumps and controls
- Location of storage tank
- Schematic diagram for water distribution system

Drainage System

While designing the drainage system, the schedule of foul drainage demand units and frequency factors for the following items should be considered for sizing the piping system, number of manholes, capacity of sump pump, and capacity of sump pit:

- Washbasins
- Showers
- Urinals
- Restrooms
- Kitchen sinks
- Other equipment such as dishwashers and washing machines
- Waste management system

HVAC Works

- Environmental conditions
    - Outdoor design conditions

Management of Quality for Sustainability

- Indoor design conditions
- Indoor air quality
- Air conditioning calculations
  - Cooling load calculations
  - Heating load calculations
  - Space temperature and humidity at required set point
  - Occupancy load
  - Lighting load
- Room pressurizing and leakage calculations
- Energy consumption calculations
- Air conditioning calculations for IT equipment room(s) based on heat emission of equipment
- Air distribution system calculations
- Smoke extractor ventilation calculations
- Exhaust ventilation calculations
- Ductwork sizing calculations
- Selection of the ductwork components, such as balancing dampers, constant volume boxes, variable air volume boxes, attenuators, grilles and diffusers, fire dampers, pressure relief dampers, etc.
- Pipe work sizing calculations
- Selection of the inline pipe work components, e.g., valves, strainers, air vents, commissioning sets, flexible connections, sensors, etc.
- Selection of boilers, pressurization units, air conditioning calculations
- Pipe work and duct work insulation selection
- Details of grilles and diffusers, control valves, etc.
- Selection of the duct work systems plant and equipment, e.g., air handling units, fan coil units, filters, coils, fans, humidifiers, duct heaters, etc.
- Selection of chillers, cooling towers
- Selection of pumps
- Selection of fans
- Equipment system calculations
- Space requirements for chillers, cooling towers, pumps, and other equipment (plant room)
- Mechanical room location and access
- Preparation of the plan and section layouts, and plant room drawings
- Electrical load calculations
- Comparison of electrical consumption with electrical conservation code
- Preparation of equipment schedules
- HVAC-related electrical works
- Control details
- Starter panels, MCC Panels, schematic diagram of MCC
- Selection of program equipment
- Preparation of point schedule for Building Management System (BMS)
- Schematic diagram for BMS

Electrical System

- Lighting calculations for different areas based on the illumination level recommended by CIE/CEN/CIBSE and Isolux diagrams
- Selection of light fittings, types of lamps
- Daylighting system
- Selection of control gear for light fixture
- Environmental consideration for selection of light fixture and control gear for comfortable lighting
- EXIT/EMERGENCY lighting system
- Circuiting references, normal as well as emergency
- Sizing of conduits
- Power for wiring devices
- Power supply for equipment (HVAC, PH&FF, conveying system, others)
- Sizing of cable tray
- Sizing of cable trunking
- Selection (type and size) of wires and cables
- Voltage drop calculations for wires and cables
- Selection of upstream and downstream breakers
- De-rating factor
- Sensitivity of breakers (degree of protection)
- Selection of isolators
- IP ratings (degree of ingress protection) of panels, boards, isolators
- Schedule of Distribution boards, switchboards, and main low-tension boards
- Cable entry details
- Location of Distribution Boards, switchboards, and low-tension panels
- Short circuit calculations
- Sizing of diesel generator set for emergency power supply
- Sizing of ATS (automatic transfer switch)
- Generator room layout
- Sizing of capacitor bank
- Provision for solar system integration
- Schematic diagrams
- Sizing of transformers
- Substation layout
- Calculations for grounding (earthing) system
- Grounding system layout
- Calculations for lightning protection system
- Lightning protection system layout
- Renewable energy/alternative energy system

Fire Alarm System

A fire alarm system is designed, taking into consideration the local fire code and NFPA standards. The system includes the following:

Management of Quality for Sustainability

- Conduiting and raceways
- Type of system: analog/digital/addressable
- Type of detectors, based on the area and spacing between the detectors and the walls
- Break glass/pull station
- Type of horns/bells
- Voice evacuation system, if required
- Type of wires and cables
- Mimic panel, if required
- Repeater panel, if required
- Main control panel
- Interface with other systems such as HVAC, elevator
- Riser diagram

Solar System

The designer should consider the following points while developing a solar power system:

- Avoid shading from trees, buildings, etc. (especially during peak sunlight hours).
- Check the proposed plan for the proposed site to ensure that future, neighboring construction will not cast shade on the array.
- Determine where the solar array can be placed (roof, carport, facade, curtain wall, boundaries, double skin, or elsewhere).
- Keep the south-facing section obstruction-free if possible; if the roof is sloped, the south-facing section will optimize the system performance.
- Minimize rooftop equipment to maximize the available open area for the solar collector placement.
- Ensure that the design of the type of roof will have adequate space to install the solar system at a later stage to optimize the cost of installing the solar system at a later stage.
- Ensure the roof is capable of carrying the load of the solar equipment.
- Analyze wind loads on rooftop solar equipment to ensure that the roof structure is sufficient.
- Add additional safety equipment for the solar equipment access and installation.

Information and Communication Technology

- Structured cabling, considering the type and size of the cable: copper, fiber optic
- Type and size of the cables
- Racks
- Wiring accessories/devices
- Access/distribution switches

- Internet switches
- Core switch
- Access gateway
- Router
- Network management system
- Servers
- Telephone handsets

Public Address System

- Conduiting and raceways
- Type of system: analog/digital/IP-based
- Types of wires and cables
- Types of speakers
- Distribution of speakers
- Required noise level in different areas
- Calculations for sound pressure level
- Zoning of system, if required
- Size and type of premixer
- Size and type of amplifier
- Microphones
- Paging system
- Message recorder/player
- Interface with other systems

Audiovisual System

- Conduiting and raceways
- Type of system: analog/digital/IP-based
- Types of wires and cables
- Racks
- Type, size, and brightness of projectors
- Type and size of speakers and sound pressure level
- Type and size of screens
- Microphones
- Cameras (visualizers)
- CD/DVD players-recorders
- Control processors
- Video switch matrix
- Mounting details of equipment

Security System/CCTV

- Type of system: digital/IP-based
- Conduiting and raceways

Management of Quality for Sustainability

- Wires and cabling network
- Level of security required
- Type and size of cameras
- Types of monitors/screens
- Video/event recording
- Video servers
- Database server
- System software
- Schematic diagram
- System console

Security System Access Control

- Conduiting and raceways
- Wires and cabling network
- Proximity RFID reader
- Fingerprint and proximity combined reader
- Magnetic lock
- Release button
- Door contact
- RFID card
- Reader control panel
- Server
- Multiplexer
- Monitors
- Workstation
- Metal detector

Landscape Works

As a landscape architect, the following points are to be considered while designing the landscape system:

- Property boundaries
- Size and shape of the plot
- Shape and type of dwelling
- Integration with surrounding areas
- Orientation to the sun and wind
- Climatic/environmental conditions
- Ecological constraints (soil, vegetation, etc.)
- Location of pedestrian paths and walkways
- Pavement
- Garage and driveway
- Vehicular circulation
- Location of sidewalk
- Play areas and other social/community requirements

- Outdoor seating
- Location of services, positions of both under- and aboveground utilities and their levels
- Location of existing plants, rocks, or other features
- Site clearance requirements
- Foundation for paving, including front drive
- Top soiling, or top soil replacement
- Soil for planting
- Planting of trees, shrubs, and ground cover
- Grassed area
- Sowing grass or turfing
- Lighting poles/bollards
- Special features, if required
- Signage, if required
- Surveillance, if required
- Installation of irrigation system
- Marking out the borders
- Storage for landscape maintenance material

External Works (Infrastructure and Road)
External works are part of the contract requirements of a project that involves construction of a service road and other infrastructure facilities to be connected to the building and also includes care of existing services passing through the project boundary line. The designer has to consider the following while designing external works:

- Grading material
- Asphalt paving for road or street
- Pavement
- Pavement marking
- Precast concrete curbs
- Curbstones
- External lighting
- Cable routes
- Piping routes for water, drainage, storm water system
- Sump pump(s) for drainage, storm water
- Trenches or tunnels
- Bollards
- Manholes and hand holes
- Traffic marking
- Traffic signals
- Boundary wall/retaining wall, if required

Bridges
Designer should use relevant authorities' design manual and standards, and consider the following points while designing bridges:

Management of Quality for Sustainability

- Soil stability
- Alignment with road width, property lines
- Speed
- Intersections/interchanges
- No. of lanes, width
- Right-of-way lines
- Exits, approaches, and access
- Elevation datum
- Superelevation
- Clearance with respect to railroad, roadway, navigation (if applicable)
- High and low levels of water (if applicable)
- Utilities passing through the bridge length
- Slopes
- Number and length of span
- Live loads, bearing capacity
- Water load, wind load, earthquake effect (seismic effect)
- Bridge rails, protective screening, guard rails, barriers
- Shoulder width
- Footings, columns, and piles
- Abutment
- Beams
- Substructure
- Superstructure, deck slab
- Girders
- Slab thickness
- Reinforcement
- Supporting components, deck hanger, tied arch
- Expansion and fixed joints
- Retaining walls, crash wall
- Drainage
- Lighting
- Esthetic
- Sidewalk, pedestrian, and bike facilities
- Signage, signals
- Durability
- Sustainability

Highways

Designer should use relevant authorities' design manual and standards, and consider the following points while designing highways.

- Type of highway
- Soil stability
- Speed
- Number of lanes, width

- Shoulder width
- Gradation
- Type of pavement and thickness
- Right-of-way lines
- Exits, approaches, access, and ramp
- Superelevation
- Slopes, curvature, turning
- Median, barriers, curb
- Sidewalks, driveways
- Pedestrian accommodation
- Bridge roadway width
- Drainage
- Gutter
- Special conditions, such as snow and rain
- Pump(s) for drainage, rain water
- Lighting
- Signage, signals
- Durability
- Sustainability

Furnishings/Furniture (Loose)
In building construction projects, loose furnishings/furniture is tendered as a special package and is normally not part of the main contract. In order to express all the features of the furnishing/furniture products in the specification, the descriptive feature of the product is not enough. In order to give enough information and understanding of the product, the product specifications are accompanied with the pictorial view/cut-out sheet/photo of the product and the furniture layout.

4.4.3.7 Develop the Contract Documents

The designer/consultant is responsible for preparing detail specifications and contract documents that meet the owner's needs and specify the required level of quality, schedule, and budget.

Generally, the contract documents include all the details, as well as references to generally accepted quality standards published by international standards organizations. Proper specifications and contract documentations are extremely important as these are used by the contractor as a measure of quality compliance during the construction process.

4.4.3.7.1 Prepare the Contract Documents

Construction contract documents are the written documents that constitute a set of contract documents developed to meet the owner's needs, the required level of quality, schedule, and budget. Based on the type of project delivery system and the type of contracting arrangements, necessary documents are prepared by establishing a framework for the execution of the project to satisfy the owner's requirement.

Management of Quality for Sustainability

It is essential that contract documents are clearly written in simple language that is unambiguous and convenient to understand by all the concerned parties. These documents mainly consist of the following:

1. Bidding
   - Tendering procedures
   - Bid bond
   - Performance bond
   - Agreement forms
2. Contracting
   - General conditions
   - Particular conditions
3. Construction
   - General specifications
   - Particular specifications
   - Drawings
   - Bill of Quantities
   - Price analysis schedule
   - Forms
   - Reports
   - Soil test reports
   - Survey reports
   - Addenda (if any)
   - Special technical requirements

There are organizations/agencies producing different types of construction contract formats which are globally used as a guide to prepare construction contract documents. The following are the best known organizations whose documents are most commonly used to prepare construction contract documents for a specific project:

1. EJCDC – The Engineers Joint Contract Documents Committee (USA)
2. FIDIC – Fédération Internationale des Ingénieurs-Conseils (International Federation of Consulting Engineers)
3. MasterFormat® – Construction Specifications Institute (CSI) and Construction Specifications Canada (CSC)
4. NEC – New Engineering Contract (NEC) or NEC Engineering and Construction Contract (UK; Institution of Civil Engineers)

Based on the type of contracting arrangements which the owner would like to handle the project, necessary documents are prepared by establishing a framework for execution of the project. Generally, FIDIC (Fédération International des Ingénieurs-Conseils) model conditions for international civil engineering contracts are used as a guide to prepare these contract documents.

4.4.3.7.2 Prepare the Specifications

Specifications of the work quality are an important feature of construction project design. Specifications of the required quality and components represent part of the contract documents and are detailed under various sections of the particular specifications.

Particular specifications consist of many sections related to specific topics. Detailed requirements are written in these sections to enable the contractor to understand the product or system to be installed in the construction project. The designer has to interact with the project team members and owner while preparing the contract documents.

Generalized writing of these sections is as follows:

Section Number
Title
Part 1 – General
   1.01 – General reference/related sections
   1.02 – Description of work
   1.03 – Related work specified elsewhere in other sections
   1.04 – Submittals
   1.05 – Delivery, handling, and storage
   1.06 – Spare parts
   1.07 – Warranties
     In addition to the above, a reference is made for items such as preparation of a mockup, quality control plan, and any other specific requirement related to the product or system specified herein.
Part 2 – Product
   2.01 – Materials
   2.02 – Manufacturer's Qualification/List of Recommended Manufacturers
Part 3 – Execution
   3.01 – Installation
   3.02 – Site Quality Control

Preparation of detailed documents and specifications as per the master format is one of the activities performed during this phase of the construction project. The contract documents must specify the scope of works, location, quality, and duration for the completion of the facility. As regards the technical specifications of the construction project, master format specifications are included in the contract documents. Normally, construction documents are prepared as per the MasterFormat® contract documents produced jointly by the Construction Specification Institute (CSI) and Construction Specifications Canada (CSC), which are widely accepted as standard practice for the preparation of contract document.

Master format is a master list of section titles and numbers for organizing information about construction requirements, products, and activities into a standard sequence. Master format is a uniform system for organizing information in project manuals, for organizing cost data, for filling product information and other technical

Management of Quality for Sustainability

data, for identifying drawing objects, and for presenting construction market data. MasterFormat® (2016 Edition) consists of 48 divisions (49 is reserved).

4.4.3.8 Regulatory Approvals

Government agency regulatory requirements have considerable impact on precontract planning. Some agencies require that the design drawings be submitted for their preliminary review and approval to ensure that the designs are compatible with local codes and regulations. These include submission of drawings to electrical authorities showing the anticipated electrical load required for the facility, approval of fire alarm and fire suppression system drawings, and approval of drawings for water supply and the drainage system. Technical details of the conveying system are also required to be submitted for approval from the concerned authorities.

4.4.3.9 Develop the Project Schedule

The project schedule is developed using bottom-up planning details, using key events. It is also known as Class 1 Schedule or Schedule Level 3 (please refer to Figure 4.11 and Table 4.15).

4.4.3.10 Estimate the Project Cost

The cost estimate during this phase is based on elemental parametric methodology. It is also known as Detailed Costing (Detailed Estimate) (please refer Table 4.16).

4.4.3.11 Manage the Design Quality

In order to reduce errors and omissions, it is necessary to review and check the design for quality as assurance by the quality control personnel from the project team through itemized review checklists to ensure that the design drawings fully meet the owner's objectives/goals. It is also required to review the design with the owner to ensure a mutual understanding of the build process. The designer has to ensure that the installation/execution specification details are comprehensively and correctly described and also that the installation quality requirements for systems are specified in detail.

The designer has to plan quality (planning of design work), perform quality assurance, and control quality for preparing the detail design. This will mainly consist of the following:

4.4.3.11.1 Plan Quality
- Review the comments on the preliminary design
- Determine the number of drawings to be produced
- Establish the scope of work for preparation of the detail design
- Identify the requirements listed under the TOR
- Identify quality standards and codes to be complied with
- Establish the design criteria
- Identify the regulatory requirements
- Identify the environmental requirements
- Establish the quality organization with the responsibility matrix

- Develop the Design (Drawings and Documents) review procedure
- Establish the submittal plan
- Establish the design review procedure

4.4.11.3.2   Quality Assurance
- Collect the data
- Investigate the site conditions
- Prepare the design drawings
- Prepare the detailed specifications
- Prepare the contract documents
- Prepare the Bill of Quantities
- Ensure functional and technical compatibility
- Ensure the design is constructible
- Ensure the operational objectives are met
- Ensure the drawings are fully coordinate with all disciplines
- Ensure the design is cost-effective
- Ensure the selected/recommended material meets the owner's objectives
- Ensure that the design fully meets the owner's objectives/goals

4.4.3.11.3   Control Quality
- Check the quality of the design drawings
- Check the accuracy and correctness of the design
- Verify the Bill of Quantities
- Check the specifications
- Check the contract documents
- Check for regulatory compliance
- Check the project schedule
- Check the project cost
- Check the interdisciplinary requirements
- Check the required number of drawings prepared drawing

Table 4.25 illustrates the mistake-proofing chart to eliminate design errors.

4.4.3.12   Estimate the Resources

The designer has to estimate the resources required to complete the project. During the design development phase, detailed information is available to estimate the manpower resources during the construction phase.

4.4.3.13   Manage the Risks

The following are typical risks which normally occur during the design development phase:

- Preliminary design deliverables and review comments are not taken into consideration while preparing the detail design
- Regulatory authorities' requirements are not taken into consideration

TABLE 4.25
Mistake-Proofing for Eliminating Design Errors

Serial Number	Items	Points to be Considered to Avoid Mistakes
1	Information	1. Terms of Reference (TOR)
		2. Client's preferred requirement matrix
		3. Data collection
		4. Regulatory requirements
		5. Codes and standards
		6. Historical data
		7. Organizational requirements
2	Mismanagement	1. Compare production with actual requirements
		2. Inter disciplinary coordination
		3. Application of different codes and standards
		4. Drawing size of different trades/ specialist consultants
3	Omission	1. Review and check design with TOR
		2. Review and check design with client's requirements
		3. Review and check design with regulatory requirements
		4. Review and check design with codes and standards
		5. Check for all required documents
4	Selection	1. Qualified team members
		2. Available material
		3. Installation methods

Source: Abdul Razzak Rumane (2013). Quality Tools for Managing Construction Projects. Reprinted with permission of Taylor & Francis Group.

- The detail design scope of work is not properly established and is incomplete
- The related project data and information collected are incomplete
- The related project data and information collected are likely to be incorrect and wrongly estimated
- Site investigations for existing conditioned are not verified
- Fire and safety considerations recommended by the authorities are not incorporated in the design
- Environmental consideration
- Incomplete design drawings and related information
- Inappropriate construction method
- Conflict with different trades
- Interdisciplinary coordination not done
- Wrong selection of materials and systems
- Undersized HVAC equipment selection
- Incorrect water supply requirements
- Estimated total electrical load is much lower than the expected actual consumption

- Traffic study for the conveying system is not verified, taking into consideration the final load
- Prediction of possible changes in design during the construction phase
- Inadequate and ambiguous specifications
- Project schedule not updated as per detailed data and project assumptions
- Errors in detail cost estimation
- Number of drawings not as per the TOR requirements

The designer has to take into account the above-mentioned risk factors while developing the detail design.

Furthermore, the designer has to consider the following risks while planning the duration for completion of the design development phase.

- Impractical design development/design preparation schedule
- Duration to obtain authorities' approval

4.4.3.14 HSE Issues and Requirements

The designer has to develop the HSE plan and identify issues to manage the activities during the detail design stage. The designer has to carry out an Environmental Impact Assessment study and take it into consideration while developing the detail design.

While developing the design, the designer has to consider the following:

- Hazardous properties of the materials, products used in the project
- Hazardous emissions
- Pollution and its impact
- Impact on health
- Safety in design (safe design)
- Safety in operation of MEP systems
- Waste management system
- Environmental compatibility
- Regulatory and other environmental protection agencies' requirements

The designer has to prepare an Environmental Impact Assessment Report for submission to the owner along with other documents while submitting the concept design package. The typical contents of an environmental impact assessment report are as follows:

1. Purpose of the report
2. Description of the project (goals and objectives)
3. References (relevant codes, standards, owner references)
4. Scope of work (during all the life cycle phases of the project)

The designer is also required to submit an HSE Management Plan along with other documents during this phase.

Management of Quality for Sustainability

4.4.3.15 Review the Detail Design

The success of a project is highly correlated with the quality and depth of the engineering design prepared during this phase. Coordination and conflict resolution is an important factor during the development of design to avoid omissions and errors. The designer has to review the detail design for accuracy of the drawings, interdisciplinary coordination, and documents before these are submitted to the owner/project manager for subsequent preparation of construction documents.

4.4.3.15.1 Review the Drawings

Figure 4.20 illustrates the design review steps for the detail design.

4.4.3.15.2 Perform Interdisciplinary Coordination

Table 4.26 illustrates the major points to perform interdisciplinary coordination.

4.4.3.15.3 Review the Documents

The designer has to review the contract documents and ensure that all the requirements listed under the TOR are taken account of. The contract documents are prepared for the approved type of project delivery system and contract pricing method. Table 4.27 lists the items to be checked for detail design by the designer before submission to the owner/project manager.

4.4.3.16 Finalize the Detail Design

The final detail design is prepared by incorporating the comments, if any, found during analysis, review, and interdisciplinary coordination of the drawings and documents for submission to the owner/client

Normally, the following items are submitted to the owner for their review and approval in order to proceed with development of construction documents for the bidding/tendering purpose;

1. Detail design drawings
2. Contract documents
3. Project specifications
4. Bill of Quantities
5. Project schedule
6. Detailed cost estimate
7. Detail design report
8. Calculations
9. Site investigations, survey reports
10. Model (if applicable)

4.4.4 CONSTRUCTION DOCUMENTS

During the construction documents phase, the drawings and specifications prepared during the design development phase are further developed into the working drawings. All the drawings, specifications, documents, and other related elements

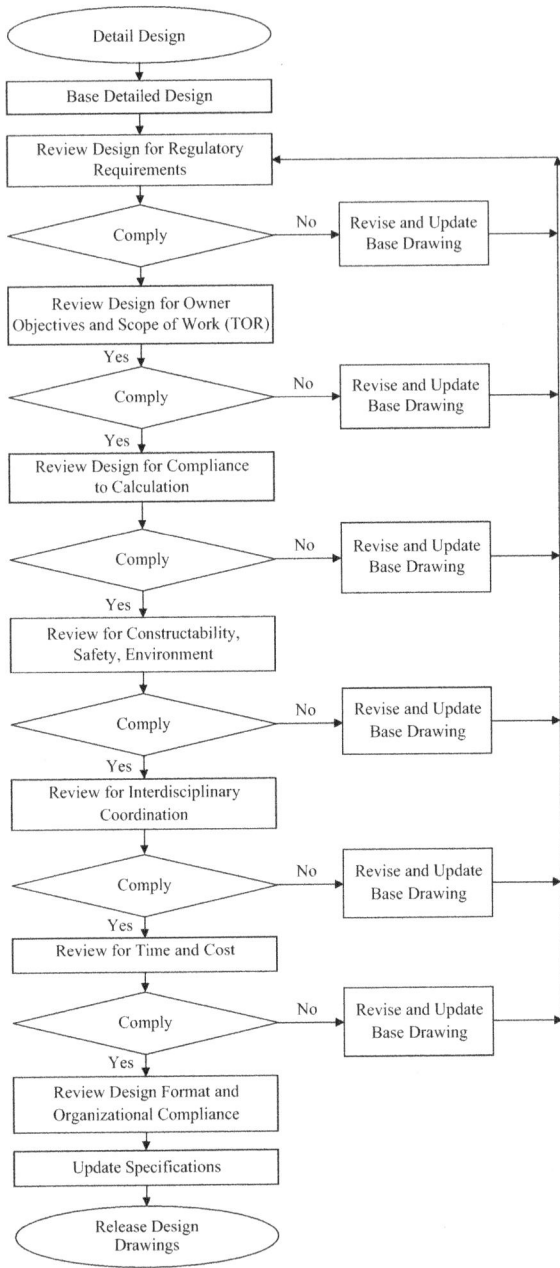

FIGURE 4.20 Design review steps. Source: Abdul Razzak Rumane. (2013). *Quality Tools for Managing Construction Projects*. Reprinted with permission of Taylor & Francis Group.

Management of Quality for Sustainability

TABLE 4.26
Interdisciplinary Coordination

Serial Number	Discipline	Architectural	Structural	Mechanical	HVAC	Electrical	External Works
1	ARCHITECTURAL		1. Structural framing plans,	1. Pump Room location and size of room	1. Plant room location and size	1. Location and size of substation and door sizing	1. Property limits
			2. Axis, grids, levels	2. Void above false ceiling for piping	2. Void above false ceiling for HVAC equipment, duct and piping	2. Trenches for cables in substation, electrical room and generator room	2. Location of outdoor equipment
			3. Location of columns, beams	3. Sprinkler in false ceiling	3. Access for maintenance of equipment	3. Location and size of electrical room and Closets	3. Location of plants
			4. Modules to match with structural plan	4. Location of sanitary fixtures and accessories	4. K-value of thermal insulation, type of external glazing and U-value,	4. Location of electrical devices	4. Location of seating/relax area
			5. Location of stirs, fire exits	5. Location of fire hydrant, cabinet and landing valves	5. Location of louvers, grills, diffusers	5. Location of light fittings in the false ceiling	5. Location of maintenance room/area
			6. Expansion joints	6. Location of water tank	6. Location of thermostat and other devices	6. Void above false ceiling for cable tray and trunking	6. Location of manholes
			7. Building dimensions	7. Shaft for water supply, sanitary and drainage pipes	7. Staircase pressurization system with respect to HVAC	7. Cable tray, cable trunking route	7. Location of generator exhaust pipe

(Continued)

TABLE 4.26 CONTINUED
Interdisciplinary Coordination

Serial Number	Discipline	Architectural	Structural	Mechanical	HVAC	Electrical	External Works
				8. Location of fuel filling point for fuel carrying tanker 9. Location of manholes	8. Location of HVAC equipment on roof 9. HVAC shaft requirement	8. Location and size of low-voltage rooms 9. Location and size of generator room 10. Ventilation of substation and generator room	
2	STRUCTURAL	1. Structural framing plans 2. Axis, grids, levels 3. Location of columns, beams 4. Modules to match with structural plan 5. Location of stairs, fire exits 6. Expansion joints		1. Opening for pipe crossing in the walls and slab 2. Shaft for pipe risers (water supply, sanitary, drainage) 3. Opening for roof drain 4. Openings/sleeves for piping 5. Opening for main circulation drain 6. Water tank inlet location	1. Shaft for piping and duct 2. Openings/sleeves for duct and pipings 3. Operating weight of all HVAC equipment 4. Floor height to accommodate equipment 5. Expansion joints requirements 6. Pump room equipment loads with HVAC equipment	1. Base for Transformers 2. Base for generator 3. Trenches for electrical cables 4. Openings/sleeves for cable tray, electrical bus duct 5. Shaft for cable trays 6. Foundation for light poles	1. Manholes 2. Foundation for light poles 3. Manhole/foundation for electrical panels 4. Manhole/foundation for feeder pillars 5. Underground services tunnel

(Continued)

TABLE 4.26 CONTINUED
Interdisciplinary Coordination

Serial Number	Discipline	Discipline					
		Architectural	Structural	Mechanical	HVAC	Electrical	External Works
3	MECHANICAL	1. Pump room location and size of room	1. Opening for pipe crossing in the walls and slab	7. Sanitary manholes	1. Make up water requirements for HVAC	7. Manhole/foundation for electrical panels	1. Irrigation system with external works
		2. Void above false ceiling for piping	2. Shaft for pipe risers (water supply, sanitary, drainage)		2. Connection of chilled water for plumbing works	1. Power supply for pumps and other equipment	2. Area drain, road gully with external/asphalt work
		3. Sprinkler in false ceiling	3. Opening for roof drain		3. Interface with building management System	2. Location of isolators for power supply	3. Stormwater manholes with external works
		4. Location of sanitary fixtures and accessories	4. Openings/sleeves for piping		4. HVAC/AHU drain with drainage system	3. Interface with fire alarm system	4. External services to be hooked up with the municipality route
		5. Location of fire hydrant, cabinet and landing valves	5. Opening for main circulation drain				
		6. Location of water tank	6. Water tank inlet location				
		7. Shaft for water supply, sanitary and drainage pipes	7. Sanitary Manholes				

(Continued)

TABLE 4.26 CONTINUED
Interdisciplinary Coordination

Serial Number	Discipline	Architectural	Structural	Mechanical	HVAC	Electrical	External Works
		8. Location of fuel filling point for fuel carrying tanker 9. Location of manholes					
4	HVAC	1. Plant room location and size	1. Shaft for ducts and piping	1. Make up water requirements for HVAC		1. Power supply for chillers, pumps, AHUs, and other equipment	1. Access for underground services
		2. Void above false ceiling for HVAC equipment, duct, and piping	2. Opening for ducts and piping in the wall and roof	2. Connection of chilled water for plumbing works		2. Location of isolators for power supply	2. Location of exhaust for underground ventilation system
		3. Access for maintenance of equipment		3. Interface with building management system		3. Heat dissipation from lighting and other electrical panels	
		4. K-value of thermal insulation, type of external glazing and U-value,		4. HVAC/AHU drain with drainage system		4. 3 Phase/single phase power requirements	
		5. Location of louvers, grills, diffusers				5. Power supply load during summer/winter	

(Continued)

Management of Quality for Sustainability

TABLE 4.26 CONTINUED
Interdisciplinary Coordination

Serial Number	Discipline	Architectural	Structural	Mechanical	HVAC	Electrical	External Works
		6. Location of thermostat and other devices				6. Electrical power supply for equipment connected to generator	
		Staircase pressurization system with respect to HVAC				Interface with fire alarm system	
		Location of HVAC equipment on roof				8. Interface with Building Management System	
		9. HVAC shaft requirement					
5	ELECTRICAL	1. Location and size of substation and door sizing	1. Base for transformers	1. Power supply for pumps and other equipment	1. Power supply for chillers, pumps, AHUs, and other equipment		Location of lighting poles
		2. Trenches for cables in substation, electrical room and generator room	2. Base for Generator	2. Location of isolators for power supply	2. Location of isolators for power supply		2. Location of earth pits
		3. Location and size of electrical room and closets	3. Trenches for electrical cables	3. Interface with fire alarm system	3. Heat dissipation from lighting and other electrical panels		Location of electrical manholes, handholes

(Continued)

TABLE 4.26 CONTINUED
Interdisciplinary Coordination

Serial Number	Discipline	Architectural	Structural	Mechanical	HVAC	Electrical	External Works
		4. Location of electrical devices	4. Openings/sleeves for cable tray, electrical bus duct		4. Three-phase/single phase power requirements		Underground cable routes
		5. Location of light fittings and other devices in the false ceiling	5. Shaft for cable trays		Power supply load during summer/winter		5. Location of bollards
		6. Void above false ceiling for cable tray and trunking	6. Foundation for light poles		6. Electrical power supply for equipment connected to generator		6. Location of electrical panels, feeder pillars
		7. Cable tray, cable trunking route	7. Manhole/foundation for electrical panels		Interface with fire alarm system		
		8. Location and size of low-voltage rooms			Interface with Building Management System		
		9. Location and size of generator room					
		10. Ventilation of sustation and generator rom					
6	LANDSCAPE/ EXTERNAL	1. Property limits	1. Manholes	1. Irrigation system with external works	1. Access for underground services	1. Location of lighting poles	
		2. Location of outdoor equipment	2. Foundation for light poles	2. Area drain, road gully with external/asphalt work	2. Location of exhaust for underground ventilation system	2. Location of earth pits	

(Continued)

TABLE 4.26 CONTINUED
Interdisciplinary Coordination

Serial Number	Discipline	Discipline						
		Architectural	Structural	Mechanical	HVAC	Electrical	External Works
		3. Location of plants	3. Manhole/foundation for electrical panels	3. Stormwater manholes with external works		Location of electrical manholes, handholes	
		4. Location of seating /relax area	4. Manhole/foundation for feeder pillars	4. External services to be hooked up with municipality route		Underground cable routes	
		5. Location of maintenance room/ area	5. Underground services tunnel			Location of bollards	
		6. Location of manholes				Location pf Electrical Panels, Feeder Pillars	
		7. Location of generator exhaust pipe				Location of generator exhaust pipe	

Source: Abdul Razzak Rumane (2013). Quality Tools for Managing Construction Projects Reprinted with permission of Taylor & Francis Group.

TABLE 4.27
Checklist for Detail Design Review

Serial Number	Items to be Checked
1	Does the design meet the owner's requirements and the complete scope of work (TOR)?
2	Are designs prepared using authenticated and approved software?
3	Are design calculation sheets included in the set of documents?
4	Is the design fully coordinated for conflict between different trades?
5	Has design been taken into consideration for relevant collected data requirements?
6	Have reviewer's comments on preliminary design been responded to?
7	Has regulatory approval been obtained and comments, if any, incorporated and all review comments responded to?
8	Does the design have environmental compatibility?
9	Are energy efficiency measures considered?
10	Are sustainability requirements considered when selecting the equipment/systems?
11	Is design constructability considered?
12	Does design match with property limits?
13	Do legends match with layout?
14	Are the design drawings properly numbered?
15	Do the design drawings have the owner's logo, designer logo as per standard format?
16	Is the design format of different trades uniform?
17	Is the project name and contract reference shown on the drawing?

Source: Abdul Razzak Rumane (2013). Quality Tools for Managing Construction Projects. Reprinted with permission of Taylor & Francis Group.

necessary for construction of the project are assembled as construction documents and subsequently released for bidding and tendering. Figure 4.21 illustrates a logic flowchart for the construction documents phase.

Figure 4.22 illustrates major activities relating to the Construction Documents Process developed based on the Project Management Process Groups methodology.

4.4.4.1 Identify the Stakeholders

The following stakeholders have direct involvement in the construction document phase:

- Owner
- Consultant
- Designer
- Project/construction manager (if the owner decided to engage one during this phase, depending on the type of project delivery system)

Table 4.28 illustrates the responsibilities of various participants during the construction document phase.

Management of Quality for Sustainability

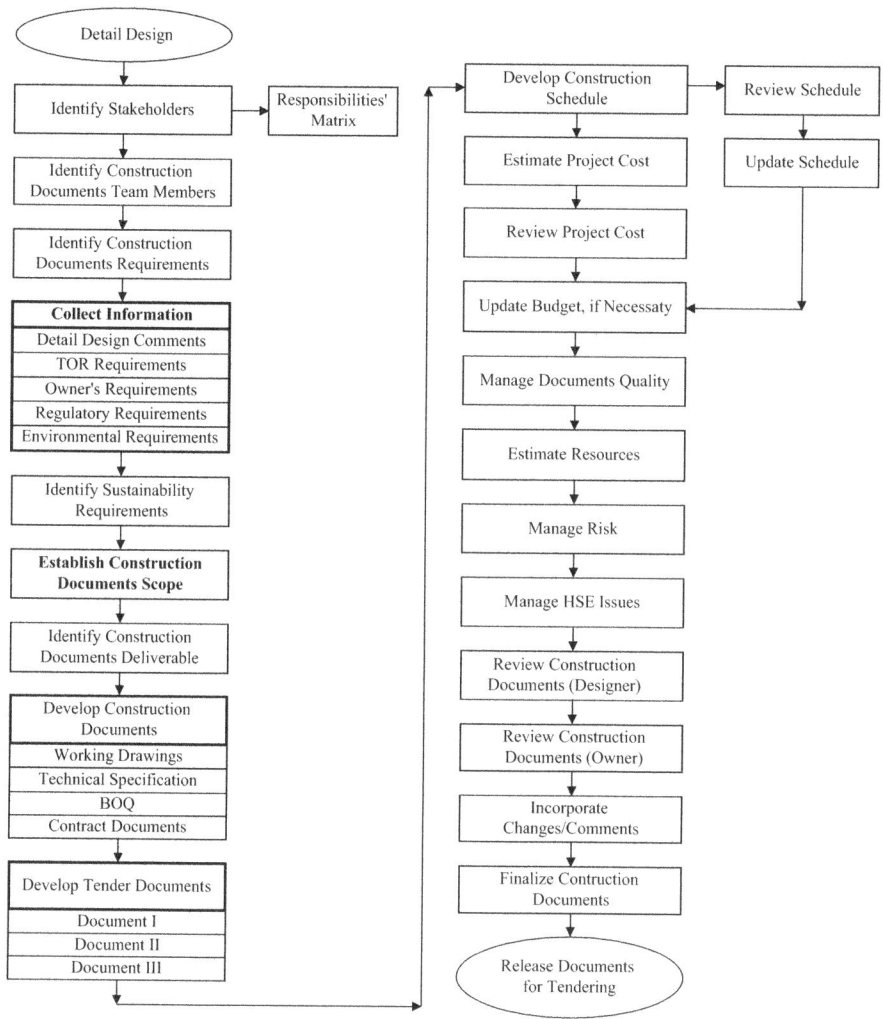

FIGURE 4.21 Logic flowchart for construction documents phase.

4.4.4.2 Identify the Construction Documents Team Members

The following project team members have direct involvement in the construction document phase:

- Owner
- Consultant
- Designer
  - Quantity surveyor (contract administrator)
- Project/construction manager (if the owner decided to engage one during this phase, depending on the type of project delivery system)

Construction Document Phase

Management Processes	Process Management Groups				
	Initiating Process	Planning Process	Execution Process	Monitoring & Cintrolling Process	Closing Process
Integration Management	Detail Design Deliverables	Project Management Plan	Working Drawings	Design Calculations	Working Drawings
	Comments on Detail Design		Project Specifications	Interdisciplinary Coordination	Spectications
	Authorities' Requirements		Bill of Quantities	Compliance to TOR	Contract Documents
	Environmental Requirements		Contract Documents	Sustainability Requirements	Tender Documents
	TOR Requirements		Tender Documents		
Stakeholder Management		Project Stakeholders			
		Stakeholders Requirements		Working Drawings Progress	
		Construction Documents		Authorities Approval	
		Sustainability Requirements		Compliance to Owner's	
Scope Management		Construction Documents Deliverables		Stakeholders Approval	
		Tender Documents			
Schedule Management		Project Schedule		Project Schedule	
		Construction Document Schedule Deliverables		Construction Document Schedule	
Cost Management		Bill of Quantities Price Analysis		Project Budget	
		Definitive Estimate			
Quality Management		Codes and Standards	Design Coordination with All Disciplines	Design Compliance to Owner's Goals and	
		Regulatory Requirements	Assure Design Quality	Coordination with all Disciplines	
		Design Quality		Control Design Quality	
		Well Defined Specifications			
		Documents Quality			
Resource Management		Supervision Team Requirements	Manage Team Members from All Discipline	Performance of Team Project Members	Assign New Project
		Contractor's Core Team			
		Contractor's Manpower			
Communication Management		Communication Matrix	Liaison with All Disciplines	Design Status Information	
			Coordination Meetings		
Risk Management		Identification of Risk during Bidding, Construction, Testing and Commissioning		Design Risk Control	
Contract Management		Bidding and Tendering Documents		Check for Contracting System	Contract Documents
					Tender Documents
HSE Management		Safety in Design	Safety Requirements		
		Environmental Compatibility	Environmental Requirements	HSE Compliance in Design	
Financial Management		Designer/Consultant Payment		Payment to Designer/Consultant	
Claim Management		Design Change Payment		Control Changes	Settle Designer's Claim

FIGURE 4.22 Major activities relating to construction documents process groups. Note: These activities may not be strictly sequential; however, the breakdown allows implementation of project management function to be more effective and easily manageable at different stages of the project phase.

During this phase, the quantity surveyor/contract administrator has great responsibilities. His/her team, under the leadership of the project manager (design), is responsible for coordinating and assembling all the required documents.

4.4.4.3 Identify the Construction Documents Requirements

The construction document phase provides a complete set of working drawings of all the disciplines, site plans, technical specifications, Bill of Quantities (BOQ), schedule (except for the standards specifications, documents for insertions normally added

Management of Quality for Sustainability

TABLE 4.28
Responsibilities of Various Participants (Design-Bid-Build Type of Contracts) During the Construction Documents Phase

Phase	Responsibilities		
	Owner	Designer	Regulatory Authorities
Construction Document	• Approval of working drawings • Approval of tender documents • Approval of time schedule • Approval of budget	• Development of working drawings • Development of specifications • Development of contract documents • Project schedule • Project budget • BOQ • Development of tender documents • Review of construction documents	• Review and approval of project submittals

during the bidding and tendering phase), and related graphic and written information to bid the project. It is necessary that the utmost care is taken to develop and assemble all the documents and ensure the accuracy and correctness to meet the owner's objectives.

4.4.4.3.1 Data/Information

In order to identify the requirements to assemble the contract documents, the designer has to gather the comments on the submitted detail design by the owner/project manager, collect the TOR requirements, the owner's requirements, regulatory requirements, environmental requirements, and all other related information to ensure that nothing is missed.

4.4.4.3.1.1 Gather the Detail Design Comments

The designer has to review the comments on the detail design and coordinate with all the disciplines/trades and incorporate the same while preparing the construction documents.

4.4.4.3.1.2 Identify the TOR Requirements

The designer must identify all the requirements listed under the TOR. This includes:

- Final drawings to be prepared to the required scales, format with the necessary logo, client's name, location map, north orientation, project name, designer's name, drawing title, drawing number, contract reference number,

date of drawing, revision number, drawing scale, and duly signed by the designer.
- Technical specifications
- Bill of Quantities and schedule of rates
- Contract documents
- Project schedule
- Cost estimates
- Summary report

4.4.4.3.1.3 Collect the Owner's Requirements

The designer has to discuss with the owner to ascertain that there are no changes to the earlier requirements established by the owner and, if there are any additional requirements or changes, then the scope is to be updated and incorporated in the final design and documents before sending them for tendering.

4.4.4.3.1.4 Identify the Regulatory Requirements

The designer has to ensure that there are no changes to the existing regulatory requirements. If there are updates to regulatory requirements, then the designer has to incorporate the same as any changes during construction will have an adverse effect on the project.

4.4.4.3.1.5 Identify the Environmental Requirements

Environmental agencies always update their requirements to protect the environment. The designer has to verify that there are no changes to the requirements/assumptions considered during the detail design development phase.

4.4.4.3.2 Identify the Sustainability Requirements

The designer also has to identify the sustainability requirements for development of the construction documents. The following are the basic elements that have to be taken account of by the designer to develop construction documents:

- Meeting/satisfying the owner's requirements
- Project value within the approved budget
- Efficient resources
- Efficient and durable material
- Working drawings with all the details
- Schematics of MEP systems
- Energy-efficient MEP systems, equipment
- Green Building concept
- Aesthetic
- Landscape compatible with the surrounding area
- Regulatory requirements
- Applicable codes and standards
- Contract documents clearly written in simple language that is unambiguous and easy to understand

Management of Quality for Sustainability

4.4.4.4 Establish the Construction Documents Scope

The scope for development of the construction documents is prepared, taking into consideration requirements established under Section 4.4.4.3.1 as follows:

- Approved detail design phase documents
- TOR requirements
- Owner's preferred requirements
- Regulatory requirements
- Environmental requirements
- Approved project schedule
- Approved detail cost estimate
- Codes and standards

The purpose of construction documents is to provide sufficient information and detail to ensure that the bidders will be able to submit the definitive cost for the project. There must be no ambiguity in the drawings and specifications, and the work to be performed by the contractor is properly identified and correctly addressed, taking necessary measures to mitigate errors and omissions in the design. The scope of work during the construction document phase mainly comprises of:

- Preparation of the working (final) drawings
- Technical specifications
- Bill of Quantities
- Project schedule
- Definitive cost estimate
- Authorities' approvals
- Existing site conditions/site plans
- Site surveys
- Design calculations

4.4.4.4.1 Identify the Construction Documents Deliverables

Table 4.29 lists the Construction Document deliverables to be developed during this phase.

4.4.4.5 Develop the Construction Documents

The following items are mainly developed during the Construction Documents phase:

1. Working drawings
2. Technical specifications
3. Contract documents
4. Tender documents

4.4.4.5.1 Working Drawings

All the drawings prepared during the Detail Design phase are reviewed to ensure that all the related information and adjustments are carried out. The following is the

TABLE 4.29
Construction Document Deliverables

Serial Number	Deliverables	
1	Document I	
1.1	Tendering Procedure	
	i. Invitation to tender	
	ii. Instructions to bidders	
	iii. Forms for tender and appendix	
	iv. List of equipment and machinery	
	v. List of contractor's staff	
	vi. Contractor's certificate of work statement	
	vii. List of subcontractor(s) or specialist(s)	
	viii. Initial bond	
	ix. Final bond	
	x. Forms of agreement	
2	Document II	
2.1	Conditions of contract	
	II-1	General conditions
	II-2	Particular conditions
	II-3	Public tender laws
3	Document III	
	III-1	General specifications
	III-2	Particular specifications
	III-3	Drawings with schematics
	III-4	Schedule of Rates and Bill of Quantities
	III-5	Analysis of prices
	III-6	Addenda
	III-7	Tender requirements (if any) and any other instructions issued by the owner

list of major disciplines in building construction projects for which working drawings are developed:

1. Architectural drawings
2. Structure works drawings
3. Elevator
4. Fire suppression
5. Plumbing
6. Drainage
7. HVAC works
8. Electrical system (light and power)

Management of Quality for Sustainability

9. Fire alarm system
10. Information and communication system
11. Public address system
12. Audiovisual system
13. Security system/CCTV
14. Security system/access control
15. Satellite/main antenna system
16. Integrated automation system
17. Landscape and plantation
18. External works (infrastructure and road)
19. Furnishings/furniture (loose)

Each of these trade drawings shall have:

- Detail drawings produced at different scales and formats
- Plans
- Sections
- Elevations
- Schedule
- Drawing index

The following information will be included on all the drawings

- Client name
- Client logo
- Location map
- North orientation
- Project name
- Drawing title
- Drawing number
- Date of drawing
- Revision number
- Drawing scale
- Contract reference number
- Signature block
- Signed by the designer for check and approval

4.4.4.5.2 Specifications

The designer has to prepare comprehensive technical specifications as per the Division and Section, taking into consideration the related drawings. It is essential to have close coordination between working drawings and specifications. MasterFormat® specification documents are used to prepare specifications for building projects. These Divisions and sections are divided into numbers of volumes for ease of reference. The Bill of Quantities (BOQ) for the project activities is prepared corresponding to these Divisions and Sections.

4.4.4.5.3 BOQ
BOQ developed during the detail design phase is reviewed to ensure that it matches with the working drawings.

4.4.4.5.4 Contract Documents
Contract documents are prepared, taking into consideration the contract format suitable for a specific type of project, from any of the following organizations that are producing different types of contracting systems:

1. EJCDC – The Engineers Joint Contract Documents Committee (USA)
2. FIDIC – Féderation Internationale des Ingénieurs-Conseils (International Federation of Consulting Engineers)
3. MasterFormat® – Construction Specifications Institute (CSI) and Construction Specifications Canada (CSC)
4. NEC – New Engineering Contract (NEC) or NEC Engineering and Construction Contract (UK) (Institution of Civil Engineers)

4.4.4.5 Develop the Tender Documents
The following documents are prepared during this phase:

1. Complete set of working (construction) drawings duly coordinated with other disciplines and technical specifications
2. Detailed Bill of Quantities
3. Technical specifications for all the activities shown on the drawings
4. Schedule
5. Cost estimate
6. Legal and contractual information
7. Contractor bidding requirements
8. Contract conditions
9. General specifications
10. Schedules
11. Reports

The above listed documents are used as guidelines by the designer to prepare the tender documents. These are as follows:

Document-I
1. Tendering procedure, consisting of:
    I. Invitation to tender
    II. Instruction to bidders
    III. Forms for tender and appendix
    IV. List of equipment and machinery
    V. List of contractor's staff
    VI. Contractor certificate of work statement
    VII. List of subcontractor(s) or specialist(s)

Management of Quality for Sustainability

VIII. Initial bond
IX. Final bond
X. Form of Agreement
XI. List of tender documents

Document-II

1. II-1 General conditions
2. II-2 Particular conditions
3. II-3 Public tender laws

Document-III

1. III-1 General specifications
2. III-2 Particular specifications (Division 1–49)
3. III-3 Drawings
4. III-4 Schedule of rates and Bill of Quantities
5. III-5 Analysis of prices

4.4.4.6 Develop the Project Schedule
The project schedule is developed using bottom-up planning details and key activities/events. It is also known as Class 1 Schedule or Schedule Level 4

4.4.4.7 Estimate the Project Cost
The cost estimate during this phase is based on the detail costing methodology. During this phase, all the project activities are known, and a detailed Bill of Quantities is available for costing purposes. It is also known as Detailed Costing (Definitive Estimate)

4.4.4.8 Manage the Construction Documents Quality
In order to reduce errors and omissions, it is necessary to review and check the design for quality assurance by the quality control personnel from the project team through itemized review checklists to ensure that the working drawings are suitable for construction. The designer has to ensure that the installation/execution specification details are comprehensively and correctly described and coordinated with the working drawings, and also the installation quality requirements for systems are specified in detail.

The designer has to plan quality, perform quality assurance, and control quality for preparing contract documents. This will mainly consist of the following:

4.4.4.1 Plan Quality
- Review comments on the detail design package
- Determine the number of drawings to be produced
- Establish the scope of work for preparation of construction documents
- Identify the requirements listed under the TOR

- Identify the quality standards and codes to be complied with
- Identify regulatory requirements
- Identify environmental requirements
- Establish the quality organization with the responsibility matrix
- Develop a review procedure for the working drawings produced
- Develop a review procedure for the specifications and contract documents
- Establish a submittal plan for construction documents

4.4.4.8.2 Quality Assurance
- Prepare working drawings
- Prepare detailed specifications
- Prepare contract documents
- Prepare Bill of Quantities and Schedule of Rates
- Ensure functional and technical compatibility
- Ensure the design is constructible
- Ensure the operational objectives are met
- Ensure the drawings are fully coordinated with all disciplines
- Ensure the design is cost-effective
- Ensure selected/recommended materials meet the owner's objectives
- Ensure that the design fully meets the owner's objectives/goals
- Ensure that the construction documents match with approved project delivery system
- Ensure the type of contracting/pricing as per adopted methodology

4.4.4.8.3 Control Quality:
- Check the quality of the design drawings
- Check the accuracy and correctness of the design
- Verify the Bill of Quantities for correctness as per the working drawings
- Check that the complete specifications are prepared and coordinated to match the working drawings and BOQ
- Check the contract documents as per the project delivery system
- Check for regulatory compliance
- Check the project schedule
- Check the project cost
- Check the calculations
- Review the studies and reports
- Check the accuracy of the design
- Check interdisciplinary requirements
- Check the required number of prepared drawing

Before the drawings are released for bidding and tendering, it is necessary to check the drawings for formatting, annotation, and interpretation. Table 4.30 lists the items to be checked for the quality (correctness) of the working drawings.

TABLE 4.30
Quality Check for Working Drawings

Serial Number	Points to be Checked	Yes/No
1	Check for use of approved version of AutoCAD	
2	Check drawing for: • Title frame • Attribute • North orientation • Key plan • Issues and revision number	
3	Client name and logo	
4	Designer (consultant name)	
5	Drawing title	
6	Drawing number	
7	Contract reference number	
8	Date of drawing	
9	Drawing scale	
10	Annotation: • Text size • Dimension style • Fonts • Section and elevation marks	
11	Layer standards including line weights	
12	Line weights, line type (continuous, dash, dot … etc.)	
13	Drawing continuation reference and match line	
14	Plot styles (CTB-color dependent plot style tables)	
15	Electronic CAD file name and project location	
16	XREF (X Reference) attachments (if any)	
17	Image reference (if any)	
18	Section references	
19	Symbols	
20	Legends	
21	Abbreviations	
22	General notes	
23	Drawing size as per contract requirements	
24	List of drawings	

Source: Abdul Razzak Rumane (2013). Quality Tools for Managing Construction Projects. Reprinted with permission of Taylor & Francis Group.

4.4.4.9 Estimate the Resources
At this stage, the designer can estimate the resources with accuracy as more details are available to estimate the exact resources.

4.4.4.10 Manage the Risks
The following are the typical risks which normally occur during the construction document phase:

- The design development deliverables and review comments are not taken into consideration while preparing the construction documents
- The scope of work to produce the construction documents is not properly established and is incomplete
- Documents do not match as per the project delivery system
- Documents not as per type of contract/pricing methodology
- Regulatory authorities' requirements are not taken into consideration
- The latest environmental considerations are not taken into account
- Conflict among different trades
- Conflict between the working drawings and the specifications
- Prediction of possible changes in the design during the construction phase
- Inadequate and ambiguous specifications
- The project schedule is not updated as per the detailed data and project assumptions
- Errors in definitive cost estimation
- The number of drawings is not as per the TOR requirements
- It is likely that owner-supplied items, if any, are not included in the documents

Designer has to take into account the above-mentioned risk factors while construction documents are being prepared.

Furthermore, the designer has to consider the following risk factors while planning the duration for completion of the construction document phase.

- Impractical construction document preparation

4.4.4.11 HSE Issues and Requirements
The designer has to consider HSE issues and requirements while preparing the construction documents.

4.4.4.12 Review Construction Documents (Designer)
Construction documents are to be reviewed by the designer for accuracy and meeting/satisfying the owner's requirements. Table 4.31 illustrates the items to be reviewed for construction documents.

4.4.4.13 Construction Documents Review (Owner)
The construction documents need to be reviewed by the owner/project manager (owner's representative) for compliance with their requirements

TABLE 4.31
Items to be Reviewed for Construction Documents

Serial Number	Items to be Reviewed	Yes/No
1	Are there construction elements that are impossible or impractical to build?	
2	Does the design follow industry standards and practices?	
3	Is the structural design according to site conditions, soil conditions and bearing capacity?	
4	Are the site conditions verified and suitable with respect to access, availability of utility services?	
5	Will all the specified material be available during the construction phase?	
6	Is the specified material available from a single source or multiple sources and brands?	
6	Is the design suitable for construction using the specified material, equipment?	
8	Is the design suitable for construction using a recommended method statement?	
9	Are the available labor resources capable of building the facility as per contract drawings and contracted methods and practices?	
10	Is the design fully coordinated with technical specifications and CSI format divisions?	
11	Do the specifications cover the material considered in the design?	
12	Does the design fully meet regulatory requirements?	
13	Are the drawings coordinated with all the trades and are cross-references indicated wherever applicable?	
14	Is the design coordinated with adjacent land and its accessibility?	
15	Are requirements of the general public and persons of special needs considered?	
16	Are the construction schedule and the milestone practical to achieve?	
17	Can application of QA/QC requirements be complied with?	
18	Has environmental impact and its mitigation been considered?	
19	Is there space for temporary office facilities and parking space for workforce vehicles?	
20	Has availability of storage space for construction material been considered?	
21	Is the design sustainable?	

Source: Abdul Razzak Rumane (2013). Quality Tools for Managing Construction Projects. Reprinted with permission of Taylor & Francis Group.

4.4.4.14 Finalize the Construction Documents

The designer has to implement/incorporate the comments made by the owner/owner's representative before submitting the documents for bidding and tendering.

The final construction documents package is prepared, taking into consideration the review comments and risks identified by the designer and comments from the owner/project manager.

4.4.4.15 Release for Bidding and Tendering

Normally, the following items are submitted to the owner for their review and approval in order to proceed to the Bidding and Tendering phase of the project:

1. Working drawings
2. Project specifications
3. Bill of Quantities
   a. Priced
   b. Unpriced
4. Project schedule
5. Definitive cost estimate
6. Project summary report
7. Tender Documents comprising:
   I. Document-I Tendering procedure
   II. Document-II Condition of contract
   III. Document-III consisting of:
   III-1 General specifications
   III-2 Particular specifications
   III-3 Drawings
   III-4 Bill of Quantities and schedules of rates
   III-5 Analysis of prices
8. Soft copy of construction documents

4.5 SUSTAINABILITY AT THE BIDDING AND TENDERING STAGE

The Bidding and Tendering Stage is a competitive bidding method. During this phase, tender documents are released for bidding and a contract is awarded to the successful bidder. In many countries, it is a legal requirement that government-funded projects employ the competitive bidding method. This requirement gives an opportunity to all qualified contractors to participate in the tender, and normally the contract is awarded to the lowest bidder. Private-funded projects have more flexibility in evaluating the tender proposal. Private owners may adopt the competitive bidding system, or the owner may select a specific contractor and negotiate the contract terms. Negotiated contract systems have the flexibility of pricing arrangement as well as the selection of the contractor based on their expertise or the owner's past experience with the contractor successfully completing one of his or her projects.

Figure 4.23 illustrates the logic flow process for the Bidding and Tendering Phase.

Figure 4.24 illustrates the major activities relating to the Bidding and Tendering Phase developed based on the Project Management Process Groups methodology.

4.5.1 IDENTIFY THE STAKEHOLDERS

The following stakeholders have direct involvement in the Bidding and Tendering Phase;

- Owner
- Tender Committee

Management of Quality for Sustainability

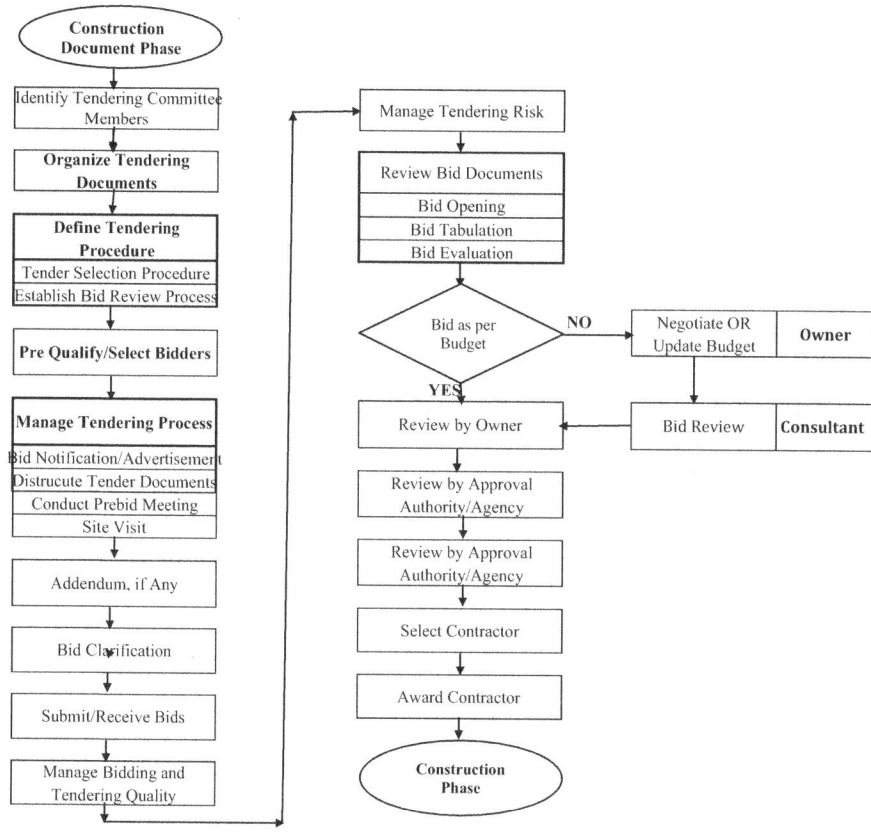

FIGURE 4.23 Logic flowchart for bidding and tendering phase.

- Designer (Consultant)
- Project/construction manager (if the owner decided to engage one during this phase, depending on the type of project delivery system)
- Bidders/contractors

Table 4.32 illustrates the responsibilities of the various participants during the construction document phase.

4.5.2 Organize the Tendering Documents

The owner hands over the approved construction documents/tender documents to the Tender Committee for further action. The bid documents are prepared as per the procurement method and contract strategy adopted during the early stage of the project. The tendering procedure documents submitted by the designer are updated and necessary owner-related information is inserted into the tender documents. The bid advertisement material is prepared and, upon approval from the owner, the bid notification is announced through different media as per the Organization's/Agency's policy.

Management Processes	Process Management Groups				
	Initiating Process	Planning Process	Execution Process	Monitoring & Cintrolling Process	Closing Process
Integration Management	Construction Documents	Organize Tender Documents	Tendering Documents	↑	Award Contract
	TOR Requirements				
	Owner's Requirements				
	Authorities Requirements				
Stakeholder	Identify Tendering Team		Contractor Selection		
Scope Management		Identify Bidders	Bid Review	Addendum	
		Bidder Selection Procedure			
		Bid Review Procedure			
Schedule Management		Bid Period		Monitor Bid Duration	
		Bid Review Duration		Monitor Review Duration	
Cost Management		Estimate Bid Price		Control Bid Value	
Quality Management					
Resource Management					
Communication Management		Advertize Tender	Conduct Meetings		
Risk Management			Manage Risk	Control Risk	
Contract Management		Select Bidders	Prepare Construction Contract		Signed Contract
HSE Management					
Financial Management				Update Project Finances	
Claim Management					

FIGURE 4.24 Major activities relating to bidding and tendering process groups. Note: These activities may not be strictly sequential; however, the breakdown allows implementation of project management function to be more effective and easily manageable at different stages of the project phase.

TABLE 4.32
Responsibilities of Various Participants (Design-Bid-Build Type of Contracts) during the Bidding and Tendering Phase

	Responsibilities		
Phase	Owner/Tender Committee	Consultant (Designer)	Bidder/Contractor
Bidding and Tendering	• Advertise bids • Distribute bids • Collect bids (proposal) • Negotiation • Approve contractor • Award contract	• Review/Evaluate bid • Bid conference/Meeting • Bid clarification • Recommend successful bidder • Prepare contract documents	• Collection of bid documents • Preparation of proposal • Submission of proposal

Management of Quality for Sustainability

4.5.3 Identify the Tendering Procedure

The owner has to identify the tendering procedure. There are different types of tendering methods based on the organization's procurement strategy. The following are the common procurement methods for the selection of the project teams/contractors;

1. Low Bid
   - Selection is based solely on the price.
2. Best Value
   a) Total Cost
      − Selection is based on total construction cost and other factors
   b) Fees
      − Selection is based on a weighted combination of fees and qualification.
3. Qualification-Based Selection (QBS)
   - Selection is based solely on qualification
4. Bidders/contractors commitment to sustainability policies

4.5.3.1 Define the Bidder Selection Procedure

Figure 4.25 illustrates the shortlisting/selection of bidders/contractors.

4.5.3.2 Establish the Bid Review Process

The designer (consultant) establishes the bid evaluation procedure in consultation with the owner. Table 4.33 lists the items to be reviewed prior to evaluation of the bid documents.

4.5.4 Identify the Bidders

Identification/shortlisting of bidders/contractors is done by gathering information through the following methods;

- Request for information
- Request for qualification
- Prequalification questionnaires

4.5.4.1 Prequalify the Bidders/Contractors

Table 4.34 lists the prequalification questionnaires to select the bidder/contractor for the Design-Bid-Build type of contracting system.

4.5.5 Manage the Tendering Process

The tendering process involves following activities;

- Bid notification
- Distribution of the tender documents

Quality Management

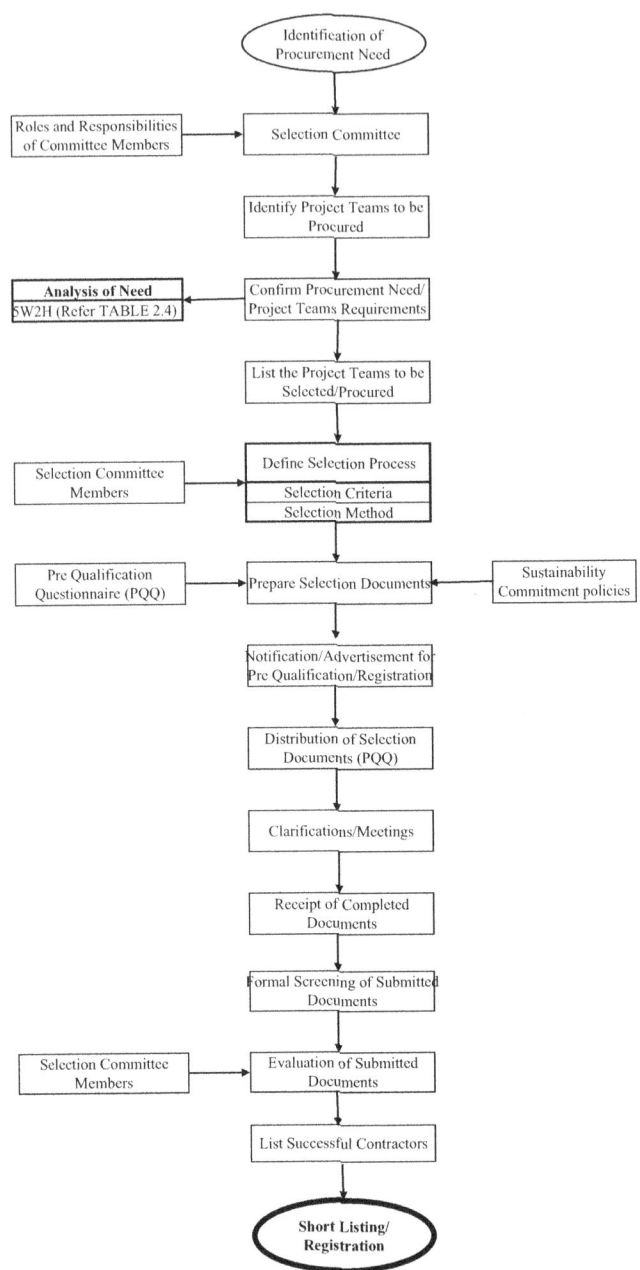

FIGURE 4.25 Project team procurement strategy for short listing/registration.

Management of Quality for Sustainability

TABLE 4.33
Checklist for Bid Evaluation

Serial Number	Description	Yes	No	Notes
A: Documents				
A.1	Bid submitted before closing time on the date specified in the bid documents			
A.2	Bidders identification is verified			
A.3	Bid is properly signed by the authorized person			
A.4	Bid bond is included			
A.5	Required certificates are included			
A.6	Bidders confirmation of the validity period of bid			
A.7	Confirmation to abide by the specified project schedule			
A.8	Bid documents have no reservations or conditions (limitation or liability)			
A.9	Preliminary method statement			
A.10	List of equipment and machinery			
A.11	List of proposed core staff as listed in the tender documents			
A.12	Complete responsiveness to the commercial terms and conditions			
A.13	All the required information is provided (completeness of information)			
A.14	All the supporting documents required to determine technical responsiveness are submitted			
B: Financial				
B.1	All the items are priced			
B.2	Bid amount is clearly described			
B.3	Prices of provision items			

- Pretender meeting(s)
- Site visit
- Bid clarification
- Issuing an addendum, if any
- Bid submission

4.5.5.1 Bid Notification/Advertisement

The tender is announced in different types of media, such as newspapers, magazines, and electronic media as per the organization's/agency's policy.

4.5.5.2 Distribute the Tender Documents

Normally, tender documents are distributed to eligible bidders in response to payment of fee announced in the bid notification, which is nonrefundable.

TABLE 4.34
Pre-Qualification Questionnaires (PQQ) for Selecting Contractor

Serial Number	Question	Answer
1	Name and address of the organization	
2	Organization's registration and licence number	
3	ISO certification	
4	Registration/classification status of the organization	
5	Joint venture with any international contractor	
6	Total turnover of past five years	
7	Audited financial report for past three years	
8	Insurance and bonding capacity	
9	Total experience (years) in construction of the following types of projects	
	9.1 Residential	
	9.2 Commercial (mixed use)	
	9.3 Institutional (governmental)	
	9.4 Industrial	
	9.5 Infrastructure	
10	Size of Project (Maximum Amount Single Project)	
	10.1 Residential	
	10.2 Commercial (mixed use)	
	10.3 Institutional (governmental)	
	10.4 Industrial	
	10.5 Infrastructure	
11	List of Successfully Completed Projects	
	11.1 Residential	
	11.2 Commercial (mixed use)	
	11.3 Institutional (governmental)	
	11.4 Industrial	
	11.5 Infrastructure	
12	List of similar type (type to be mentioned) of projects completed	
	12.1 Project name and contracted amount	
	12.2 Project name and contracted amount	
	12.3 Project name and contracted amount	
	12.4 Project name and contracted amount	
	12.5 Project name and contracted amount	
13	List of subcontractors	
14	Resources	
	14.1 Management	
	14.2 Engineering	
	14.3 Technical	

(Continued)

TABLE 4.34 CONTINUED
Pre-Qualification Questionnaires (PQQ) for Selecting Contractor

Serial Number	Question		Answer
	14.4	Foreman/supervisor	
	14.5	Skilled manpower	
	14.6	Unskilled manpower	
	14.7	Plant and equipment	
15	Current projects		
16	Quality management policy		
17	Health, Safety, and Environment Policy		
	17.1	Number of accidents during the past three years	
	17.2	Number of fires at site	
18	Staff development policy		
19	List of delayed projects		
21	List of failed contract		
22	List of professional awards		
23	Litigation (dispute, claims) on earlier projects		

4.5.5.3 Conduct Prebid Meeting

The owner conducts prebid/pretender meeting to provide an opportunity for the contractors bidding for the project to review and discuss the construction documents and to discuss:

- General scope of the project
- Any particular requirements of bidders that may have been difficult to specify
- Explain details of complex matters
- Engagement of subcontractors and specialist subcontractors
- Particular risks
- Any other matters that will contribute to the efficient delivery of the project

The meeting is attended by the designer (consultant), bidders (contractors), the project management consultant, and Tender Committee members. Queries from the contractors pertaining to contract documents are noted and the designer (consultant) provides written responses to these queries by clarifying all the points. The bidders have to consider the clarification points and incorporate the requirements while calculating the bid price. The responses recorded in the meeting become part of the contract documents (part of the addendum) which is signed by the owner and the successful bidder.

4.5.5.4 Site Visit

A site visit is arranged to get acquainted with the site conditions and the surrounding areas.

4.5.5.5 Bid Clarification

Figure 4.26 illustrates the Bid Clarification Form which becomes part of the contract documents.

4.5.5.6 Addendum, if Any

Any addendum to the advertised contract documents is considered for addition/changes in the contract.

4.5.6 Submit/Receive Bids

Bids are received in accordance with the Instructions to Bidders section of the tender documents. Each bid should be accompanied by an initial bond in favor of the owner/tender committee to be valid for a period mentioned in the tendering procedures. All the bids received are documented and notified. The tender, which is submitted as a sealed document, is opened as mentioned in the tendering procedures.

4.5.7 Manage the Bidding and Tendering Quality

In order to minimize errors and omissions during the Bidding and Tendering phase, the designer has to plan quality (planning of bidding work), perform quality assurance, and control quality for preparing the bidding and tendering activities. This will mainly consist of the following:

1. **Plan Quality**
   - Organize the tender documents
   - Identify regulatory requirements
   - Identify bidders
   - Bid notification
   - Develop a tender documents distribution system

Serial Number	Name of Contractor	Item No. and Clause Reference	Queries	Owner/Consultant's Clarification	Remarks

Project Name
Project Number
Bid Clarification Form

SAMPLE FORM

FIGURE 4.26 Bid clarification.

- Establish the responsibility matrix
- Bid clarification method
- Bid collection method
- Bid evaluation procedure
- Contract award procedure

2. **Quality Assurance**
   - Collect regulatory information
   - Organize owner's approval for bidding
   - Distribution of bid documents
   - Pretender meetings
   - Site visits
   - Bid clarifications
   - Bid collection

3. **Control Quality**
   - Bid documents sent to prequalified/shortlisted/registered bidders
   - Contents of bid documents as required
   - Bid document distribution as per announced date
   - Check for specified number of documents
   - Check bid bond amount
   - Bid evaluation

4.5.8 Manage the Tendering Risk

The following are typical risks likely to occur during this phase:

- Not all the qualified bidders take part in bidding for the project
- Bidders noticing errors and omissions in the construction documents, resulting in delay in submission of bids
- BOQ not matching with working (design) drawings
- Amendment to construction documents
- Addendum
- Delay in submission of bids than the notified one
- Bid value exceeding the estimated definitive cost (approved budget)
- Successful bidder fails to submit performance bond

The owner/designer has to consider these risks and plan the phase duration accordingly.

Table 4.35 lists the risks the contractor has to manage during the bidding and tendering phase.

4.5.9 Review the Bid Documents

The Tender Committee/designer (consultant) reviews the bids for compliance with tender requirements.

TABLE 4.35
Major Risk Factors Affecting Contractor

Serial Number		Risk Factor
1	**Bidding/Tendering**	
	1.1	Low bid
	1.2	Poor definition of scope of work
	1.3	Overall understanding of project
	1.4	Review of contract specs with Bill Of Quantities
	1.5	Errors in resource estimation
	1.6	Errors in resource productivity
	1.7	Errors in resource availability
	1.8	Errors in material price
	1.9	Improper schedule
	1.10	Quality standards
	1.11	Exchange rate
	1.12	Review of contract document requirements with regulatory requirements
	1.13	Unenforceable conditions or contract clauses

4.5.9.1 Evaluate the Bids

The bids are evaluated as per the bidder selection procedure. Refer to Table 4.33, as discussed earlier under Section 4.5.3.2

4.5.9.2 Select the Contractor

The contractor is selected based on the procurement strategy adopted by the owner.

4.5.10 AWARD THE CONTRACT

Figure 4.27 illustrates the contract award process.

4.6 SUSTAINABILITY IN THE CONSTRUCTION STAGE

Construction involves translating the owner's goals and objectives, by the contractor, to build the facility as stipulated in the contract documents, plans, specifications within budget and on schedule. Construction is the sixth phase of the construction project life cycle and is an important phase in construction projects. The majority of total project budget and schedule is expended during construction. Similar to costs, the time required to construct the project is much higher than the time required for the preceding phases. Construction usually requires large numbers of workers and a variety of activities. Construction activities involve erection, installation, or construction of any part of the project. Construction activities are actually carried out by the contractor's own work force or by subcontractors. Construction therefore

Management of Quality for Sustainability

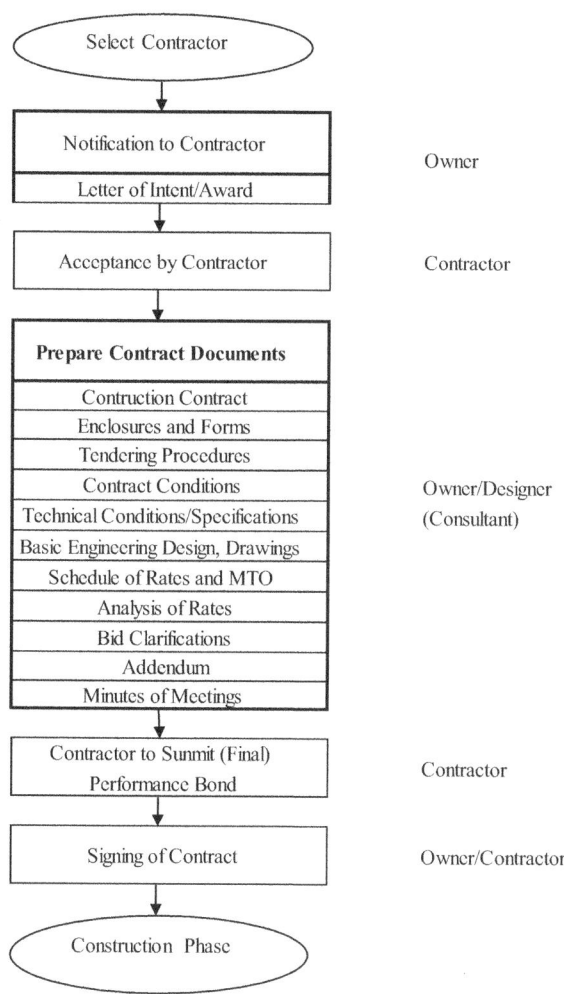

FIGURE 4.27 Contract award procedure. Source: Abdul Razzak Rumane. (2016). *Handbook of Construction Management*. Reprinted with permission of Taylor & Francis Group.

requires more detailed attention being paid to its planning, organization, monitoring, and control of the project schedule, budget, quality, safety, and environmental concerns. Figure 4.28 illustrates a logic flowchart for the Construction Phase.

Figure 4.29 illustrates major activities relating to the Construction Phase based on Project Management Process Groups methodology.

Once the contract is awarded to the successful bidder (contractor), then it is the responsibility of the contractor to respond to the needs of the client (owner) by constructing the project as specified in the contract documents, drawings, and specifications within the specified time and budget. The contractor is given a few weeks to start the construction works after signing of the contract. A letter from the client/

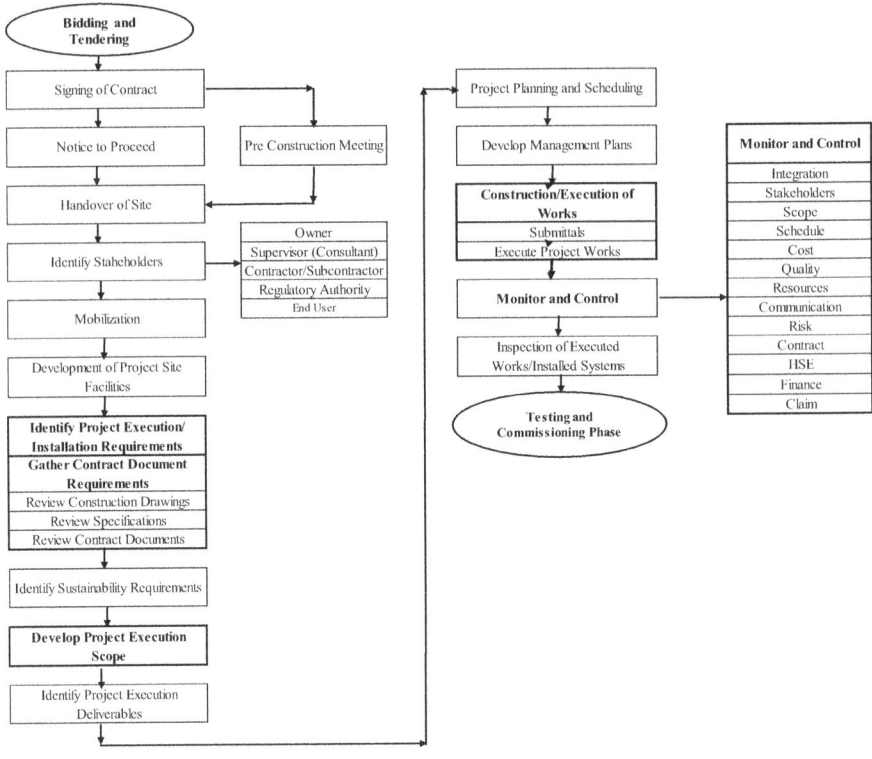

FIGURE 4.28 Logic flowchart for construction phase.

owner is issued to the contractor to begin the project work subject to the conditions of the contract. This letter is known as the "Notice to Proceed" letter.

1. **Notice to Proceed**

The Notice to Proceed authorizes the contractor to proceed with work to construct the project/facility as per the agreement. Prior to issuance of the Notice to Proceed, the owner has to ensure that:

- Necessary permits have been obtained from the relevant authorities/agencies to hand over the construction site to the contractor
- Supervision staff/project manager/construction manager has been selected to supervise the work
- Relevant departments have been informed about the signing of the contract to release availability of funds
- The supervision undertaking guarantee is signed by the Supervision Consultant
- The owner's representative is selected

Management of Quality for Sustainability

Construction Phase

Management Processes	Project Management Process Groups				
	Initiating Process	Planning Process	Execution Process	Monitoring & Controlling Process	Closing Process
Integration Management	Contract Documents	Construction Management Plan	Mobilization	Compliance to Contract Documents	Executed Project
	Tender Documents		Submittals	Change Management	
	Notice to Proceed		Management Plans		
			Execution Process		
			Construction Work		
Stakeholder Management	Owner's Representative	Responsibility Matrix		Project Status/Performance	
	Supervision Team	Stakeholder Requirements		Payments	
	Contrator's Core Staff	Reports		Variation Orders	
	Subcontractor	Meetings		Conflict Resolution	
	Authorities				
Scope Management		Scope Change Management	Design Changes	Authorities Approval	
		Preventive and Corrective Actions		Stakeholders Approval	
		Sustainabilty Requirements		Scope Change Control	
		Execution/Installation Scope		Alternate Material	
				Site Work Instruction	
				Variation Orders	
				Preventive and Corrective Plan Updates	
Schedule Management		Contractor's Construction Schedule		Schedule Monitoring	
				Schedule Control	
				Work Progress Monitoring	
				Submittals Monitoring	
Cost Management		Contracted Value of Project		Cost Control	
		Construction Budget		Cash Flow	
				Progress Payment	
				Variation Orders	
Quality Management		Contractor's Quality Control Plan	Quality Assurance	Quality Control	
			Shop Drawings	Quality Auditing	
			Builders Drawings	Material Inspection	
			Composite Drawings	Work Inspection/Testing	
			Material Approvals	Rework	
			Method Statement	Regulatory Compliance	
Resource Management		Resource Management Plan	Training of Project Team Members	Performance of Team Members	Demobilization of Workforce
		Project Manpower	Manage Project Team	Dispute Resolution	
		Sub contractor			
		Material and Equipment		Performance of Workforce	
Communication Management	Kick off Meeting	Communication Plan	Site Administration Matrix	Meetings	
		Submittals		Submittals Monitoring	
		Documentation			
		Correspondence			
Risk Management		Risk Management Plan		Control Risk	
		Construction Risks Register		Risk Audit	
Contract Management		Contract Management	Contract Documents	Inspection	Finalise Work Performed
		Plan Purchase of Material/ Equipment	Selection of Sub contractor(s)	Check List	Finalise Material/ Equipment Supplier's Contract
			Material, Systems and Equipment		
HSE Management		Safety Management Plan	Site Safety	Accident Prevention Measures	
		Waste Management Plan	Temporary Fire Fighting	Loss Prevention during Construction	
Financial Management		Finance Management Plan	Contractor's Payments	Financial Control	Payment to Consultant
		Contractor Payment	Staff Payment		
		Material and Equipment Payments	Material and Equipment Payments		Payment to Contractor/Sub contractor
			Progress (Interim Payment) Payment		Material and Equipment Payment
Claim Management		Claim Identification	Claim/Dispute Administration	Claim Prevention	Claim Payments
		Claim Quantification		Conflict Resolution	Settle Claims

FIGURE 4.29 Major activities relating to construction process groups. Note: These activities may not be strictly sequential; however, the breakdown allows implementation of project management function to be more effective and easily manageable at different stages of the project phase.

- The Notice to Proceed date is mutually discussed and agreed as per the conditions of the contract
- Copies of the construction documents are distributed to the stakeholders concerned
- The authorization letter to the owner's representative and the engineer's representative has already been issued

Figure 4.30 is a sample Notice to Proceed.

2. Kickoff Meeting

The "Kickoff Meeting" is the first meeting with the owner/client and project team members. It is also called the preconstruction meeting. The Kickoff Meeting provides an opportunity to all project team members to interact with and get to know each other.

Figure 4.31 is a sample Kickoff Meeting agenda.

LETTER HEAD

Ref: --------------------
Date: ----------------------

NOTICE TO PROCEED

To,

Contractor Name: *SAMPLE LETTER*

Address:

Subject: Contract Number ------------

Attention: -------------------

Sir/Madam

You are hereby authorized to proceed with Project No. ------------ in accordance with construction contract dated ----------- This contract calls all the contracted works to be completed within --------- calendar days. The date of enterprise shall be -------------.

Sincerely,

Enclosures:

CC:

FIGURE 4.30 Notice to proceed.

PROJECT NAME

Contract Number:			
Type of Meeting:		Date of Meeting:	
Place of Meeting:		Time of Meeting:	
Owner:			
PMC:			
Contractor:			
Others (As Applicable)			

AGENDA — SAMPLE AGENDA

1.0 Points to be Discussed

1.1 Intoduction

1.2 Project goals and objectives

1.3 Scope of work

1.4 Permit, Bonds, Insurance

1.5 Site handover procedure

1.6 Mobilization

1.7 Contractor's organization chart (Design, Construction)

1.8 Construction Schedule

1.9 Communication and Correspondance

1.10 Transmittals and Submittal Procedure

1.11 Meetings (Progrees, Coordination)

1.12 Construction Management Plan

1.13 Quality Management Plan

1.14 Risk Management Plan

1.15 HSE Management Plan

1.16 Payment

1.7 Nominated sub contractors

2.0 Any other business

Signed by: …………………………. Position: ………………………….
Deate: …………………………..

FIGURE 4.31 Kickoff meeting agenda.

4.6.1 Identify the Stakeholders

The following stakeholders have involvement in the project during the construction phase:

1. Project owner
   - Owner's representative/project manager
2. Construction supervisor
   - Consultant (designer)
   - Specialist consultant
3. Contractor
   - Main contractor
   - Subcontractor(s)
   - Supplier/vendor(s)
4. Regulatory authorities
5. End-user

Table 4.36 illustrates the responsibilities of the various participants during the construction phase.

4.6.1.1 Identify the Owner's Representative/Project Manager

From his/her office, the owner deputes or hires from outside the Owner's Representative (OR) to administer the overall project. The OR should have relevant experience and knowledge about the construction processes of projects of a similar nature. He/she should be able to manage the project with the help of the supervision

TABLE 4.36

Responsibilities of Various Participants (Design-Bid-Build Type of Contracting System) During the Construction Phase

Phase	Responsibilities		
	Owner	Supervisor	Contractor
Construction	• Approve subcontractor(s) • Approve contractor's core staff • Legal/regulatory clearance • SWI • Variation Orders • Payments	• Supervision • Approve plan • Monitor work progress • Approve material • Approve shop drawings • Monitor schedule • Control budget • Recommend payment	• Execution of work • Contract management • Selection of subcontractor(s) • Planning • Resources • Procurement • Quality • Safety

team members. In FIDIC terminology, this person is known as the "Engineer" who is appointed by the "Employer" (owner).

4.6.1.2 Identify the Supervision Team

In a traditional type of contract, the client selects the same firm which has designed the project. The firm, known as the "Consultant," is responsible for supervising the construction process and achieving the quality goals of the project. The firm appoints a representative, who is acceptable to and approved by the owner/client, to be on site and is often called the Resident Engineer (RE). The Resident Engineer, along with the supervision team members, is responsible for supervising, monitoring, and controlling, implementing the procedure specified in the contract documents and ensuring the completion of the project within a specified time, budget, and per defined scope of work.

In order to ensure the smooth flow of supervision activities, the RE has to follow the organization's supervision manual and contractual requirements. Depending on the type and size of the project, the supervision team usually consists of the following personnel:

1. Resident engineer
2. Contract administrator/quantity surveyor
3. Planning/scheduling engineer
4. Engineers from different trades such as architectural, structural, mechanical, HVAC, electrical, low-voltage system, landscape, and infrastructure
5. Inspectors from different trades
6. Interior designer
7. Document controller
8. Office secretary

The construction phase consists of various activities such as mobilization, planning and scheduling, execution of works/installation of systems, control and monitoring, management of resources/procurement, quality, and inspection. Table 4.37 illustrates the major activities to be performed by the supervisor (consultant) during the construction phase.

4.6.1.3 Identify the Contractor's Core Staff

Contract documents normally specify a list of a minimum number of core staff to be available on-site during the construction period. The contractor's core staff requirements and their qualifications are listed under the tendering procedures. The absence of these staff from the project site without prior permission attracts a penalty to the contractor. Normally, the penalty amount is specified in the contract documents. Upon signing of the contract, the contractor has to submit the names of the staff for the positions described in the contact documents for approval from the owner/consultant to work on the project. The contractor has to select an appropriate candidate to propose for the specified position. Figure 4.32 describes the Site Staff Selection Procedure and Figure 4.33 described the Project Staffing Process.

TABLE 4.37
Responsibilities of Supervision Consultant

Sr. No.	Description
1	Achieving the quality goal as specified
2	Review contract drawings and resolve technical discrepancies/errors in the contract documents
3	Review construction methodology
4	Approval of contractor's construction schedule
5	Monitoring and controlling construction time
6	Monitoring and controlling construction expenditure
7	Approval of contractor's quality control plan
8	Regular inspection and checking of executed works
9	Review and approval of construction materials
10	Review and approval of shop drawings
11	Inspection of construction material
12	Conduct progress and technical coordination meetings
13	Coordination of owner's requirements and comments related to site activities
14	Project-related communication with contractor
15	Coordination with regulatory authorities
16	Evaluating risk and making decisions related to unforeseen conditions
17	Maintaining project record
18	Processing of site work instruction for owner's action
19	Evaluation and processing of variation order/change order
20	Recommendation of contractor's payment to owner
21	Approval of contractor's safety management plan
22	Monitor safety at site and HSE requirements
23	Supervise testing, commissioning, and handover of the project
24	Issue substantial completion certificate

The following is a typical list of contractor's minimum core staff needed during the construction period for the execution of work of a major building construction project.

1. Project manager
2. Senior engineer for civil works
3. Senior engineer for architectural works
4. Senior engineer for interior works
5. Senior engineer for elevator/conveying system works
6. Senior engineer for plumbing/public health works
7. Senior engineer for fire suppression works
8. Senior engineer for HVAC works
9. Senior engineer for electrical works

Management of Quality for Sustainability

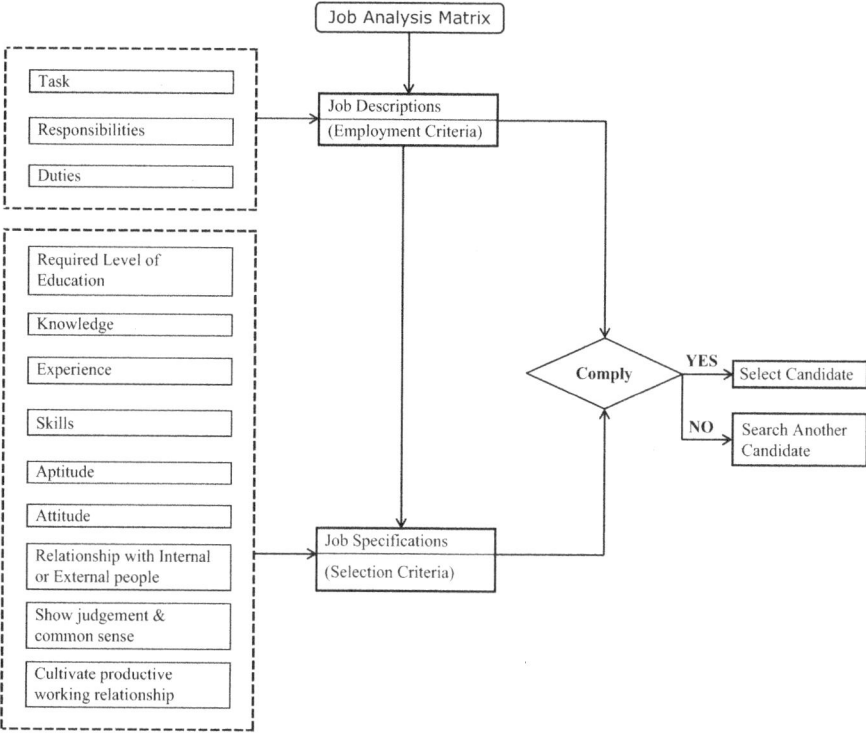

FIGURE 4.32 Site staff selection procedure. Source: Abdul Razzak Rumane. (2016). *Handbook of Construction Management: Scope, Schedule, and Cost Control.* Reprinted with permission of Taylor & Francis Group.

10. Senior engineer for information and communication works
11. Senior engineer for low-voltage systems
12. Senior engineer for infrastructure works
13. Planning and control engineer
14. QA/QC manager/engineer
15. Senior quantity surveyor/cost engineer
16. Contract administrator
17. Senior engineer for landscape works
18. Senior safety engineer/officer
19. Laboratory technician
20. Foremen (different trades)

4.6.1.4 Identify the Subcontractors

In most construction projects, the contractor engages special subcontractors to execute certain parts of the contracted project works. Areas of subcontracting are generally listed in the particular conditions of the contract document. Generally,

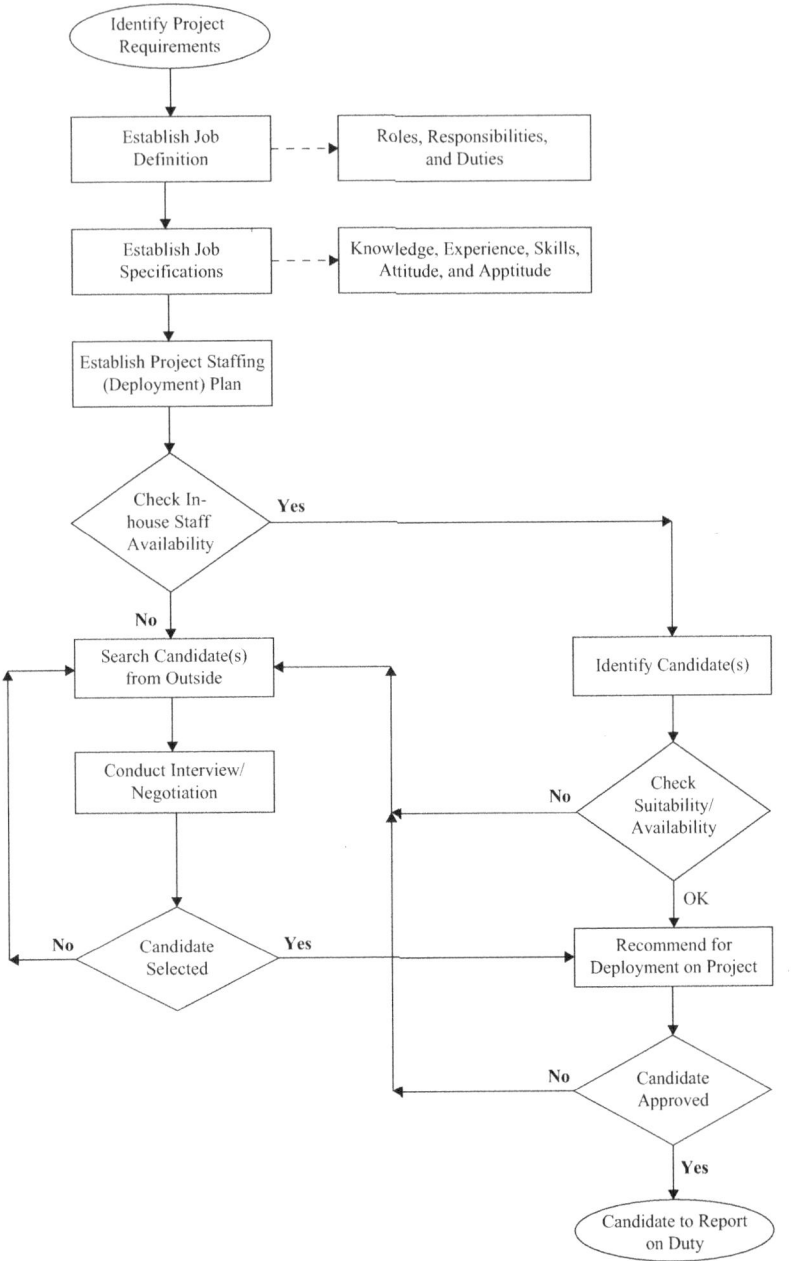

FIGURE 4.33 Project staffing process. Source: Abdul Razzak Rumane. (2016). *Handbook of Construction Management: Scope, Schedule, and Cost Control*. Reprinted with permission of Taylor & Francis Group.

the contractor has to submit subcontractors/specialist contractors to execute the following types of works:

1. Precast concrete works
2. Metal works
3. Space frame, roofing works
4. Wood works
5. Aluminum works
6. Internal finishes such as painting, false ceilings, tiling, cladding
7. Furnishings
8. Waterproofing and insulation works
9. Mechanical works
10. HVAC works
11. Electrical works
12. Low-voltage systems/Smart Building Systems
13. Landscape
14. External works
15. Any other specialized works

The contractor has to submit their names for approval to the owner prior to their engagement to perform any work at the site. Table 4.38 is an example of a subcontractor selection questionnaire.

4.6.1.4.1 Nominated Subcontractors

In certain projects, the owner/client select a contractor, known as a nominated subcontractor, to carry out certain parts of the work on behalf of the owner/client. Normally, the nominated subcontractor is imposed on the main contractor after the main contractor is appointed. The nominated subcontractor may be a specialist supplier or a specialist contractor nominated in accordance with the contract to be employed by the contractor for the supply of materials or services, or to execute the project work.

4.6.1.5   Identify the Regulatory Authorities

In certain countries, there is a regulation to submit electrical, mechanical, and HVAC drawings for review and approval by the authorities. The contractor has to identify which drawings/documents are to be submitted to the authorities during the construction phase. The necessary letter to the regulatory authorities/agencies is issued by the owner upon request for such a letter from the contractor.

4.6.2   Mobilization

The activities to be performed during the mobilization period are defined in the contract documents. During this period, the contractor is required to perform many of the activities before the beginning of actual construction work at the site. Necessary permits are obtained from the relevant authorities to start the construction work at

TABLE 4.38
Subcontractor Prequalification Questionnaire

Instructions

Please type or write all of your replies legibly. Attach additional sheets, if required.

PART I

I.1 Company Information

I.1.1 Name of organization:

I.1.2 Commercial registration no.:

I.1.3 Year of establishment:

I.1.4 Type of company:

I.1.5 Company address:

I.1.6 Affiliate company name(s) and address:

I.2 Subcontract Works (Please tick mark all interested)

Sr. no.	Work description	Detail design	Preparation of shop drawing	Construction	Inspection/auditing
1	Process engineering	☐			
2	Piping				
3	Mechanical				
4	Civil				
5	Instrumentation				
6	Electrical				
7	Buildings				
8	Low-voltage systems				
9	Landscape work				
10	External works				
11					
12					

(Continued)

TABLE 4.38 CONTINUED
Subcontrator Prequalification Questionnaire

PART II
II.1 Financial Information
II.1.1 Provide copy of audit balance sheet:
II.1.2 Provide bonding capacity:
II.1.3 Provide insurance capacity:
II.1.4 Provide bank reference:

PART III
III.1 Organization Details
III.1.1. Core business area:
III.1.2. Organization chart:
III.1.3. ISO certification:
III.1.4. Years of experience:

III.2 Project Details
III.2.1. Project History for past ten years

Sr.No.	Name of Project	Type of Work	Value	Peak Workforce	Start Date	Finish Date
1						
2						
3						
4						

III.2.2 Current Projects

Sr.No.	Name of Project	Type of Work	Value	Peak Workforce	Start Date	Expected Finish Date
1						
2						
3						

(Continued)

TABLE 4.38 CONTINUED
Subcontractor Prequalification Questionnaire

PART IV

IV.1 Management Staff

IV.1.1 Provide List of Project Managers, Project Engineers, Engineers

IV.2 Workforce

Sr.No.	Work Description	Technicians	Foreman	Skilled	Unskilled
1	Process engineering				
2	Piping				
3	Civil				
4	Mechanical				
5	Instrumentation				
6	Electrical				
7	Buildings				
8	Low-voltage systems				
9	Landscape work				
10	External works				
11					
12					

PART V

V.1 Quality Management System

V.1.1 Provide copy of ISO Certificate:

V.I.2 Person in charge of QA/QC activities:

V.I.3 Number of quality auditors:

PART VI

VI.1 HSE System

VI.1 Does the company have an HSE Policy?

VI.1.2 Provide site accident records for past two years

(Continued)

TABLE 4.38 CONTINUED
Subcontractor Prequalification Questionnaire

Declaration

We hereby declare that the information provided herein is true to our knowledge.

Note:
All relevant documents attached.

Signature of Authorised Person

the site. After being granted access to the construction site by the owner, the contractor starts mobilization work, which consists of the preparation of site offices/field offices for the owner, supervision team (consultant), and for the contractor. This includes all the necessary on-site facilities and services necessary to carry out specific tasks. Mobilization activities usually occur at the beginning of a project but can occur anytime during a project when specific on-site facilities are required. During this time, the project site is handed over to the contractor. The contractor performs site survey and testing of soil, etc., to facilitate the start of construction work.

In anticipation of the award of the contract, the contractor begins the following activities far in advance, but these are part of the contract documents, and the contractor's action is required immediately after signing of the contract in order to start construction:

- Mobilization of construction equipment and tools
- Workforce required to execute the project

4.6.2.1 Bonds, Permits, and Insurance

In order to proceed with the project execution activities, and as per contract documents, the contractor has to:

1. Submit an advance payment guarantee
2. Obtain a permit from the local authority (municipality)
3. Obtain insurance policies covering the following areas:
    a. Contractor's all risks and third-party insurance policy
    b. Contractor's plant and equipment insurance policy
    c. Workmen's compensation insurance policy
    d. Site storage insurance policy

Normally, the submitted originals are retained by the owner and the copies are kept by the resident engineer or project/construction manager (consultant).

4.6.3 Development of the Project Site Facilities

The requirements to set up temporary facilities are specified in the contract. The contractor has to submit layout plans, dimensions, and other pertinent details for temporary facilities to be constructed. Upon approval of the plans, the contractor proceeds with construction of the temporary facilities and necessary utilities. These include:

1. Site offices for owner, supervisor (consultant)
2. Storage facilities
3. Toilets and washrooms
4. Sanitary and drainage system
5. Drinking water facility
6. Safety and healthcare facilities
7. Site electrification
8. Temporary firefighting system

9. Site fence
10. Site access road
11. Necessary utilities for construction
12. Communication system
13. Signage
14. Fuel storage area
15. Guard room
16. Testing laboratory
17. Waste dumping area
18. Area for storage of hazardous material

The contractor has to designate an authority-approved dumping area for waste material.

4.6.4 Identify the Project Execution/Installation Requirements

Once the contract is awarded to the successful bidder (contractor), then it is the responsibility of the contractor to respond to the needs of the client (owner) by building the facility as specified in the contract documents, drawings, and specifications within the allotted budget and time.

4.6.4.1 Gather the Contract Document Requirements

In order to develop the contract requirements, the contractor has to review all the construction documents which are part of the contract that the contractor has signed with the owner of the project. These include:

1. Construction drawings
2. Specification
    - General specifications
    - Particular specification
    - Bill of Quantities (BOQ)
3. Contract documents
    - General conditions
    - Particular conditions
4. Other related documents

4.6.4.1.1 Review the Construction Drawings

The contractor has to review the construction/working drawings to understand the project requirements. Each trade engineer has to review the drawings and understand the construction procedure to be followed to avoid omission and rework. The trade engineers have to;

- Prepare a list of issues to be resolved and information to be obtained from the designer (consultant)
- Prepare a list of conflicting items in the drawings
- Compare the estimated bid activities and actual material for proper execution of the project

- Identify high-risk items
- Check constructability as per the specified method statement
- Identify where the execution process can be simplified
- Identify discrepancies between the drawings and the specifications
- Identify any missing drawings required to execute the project
- Identify the value engineering change proposal
- Identify the complete scope of work
- Prepare the construction material takeoff (BOQ)

4.6.4.1.2 Review the Specifications

The contractor has to review specifications and check for the following major points:

- Matching of construction drawings and specification requirements
- Any missing specifications
- Issues that need to be resolved
- If there are discrepancies between specifications and drawings
- Codes and standards to be followed
- Submittal requirements
- Quality requirements
- Safety requirements
- Environmental requirements
- Installation procedure for owner-supplied items

4.6.4.1.3 Review the Contract Documents

The contractor has to carefully study all the clauses of the contract and identify:

- High-risk clauses
- Items having price differences between the bid price and the actual price for the items to be installed for specified performance of the work/system
- Items which need to be resolved by raising a Request for Information (RFI)
- Clauses having conflict with regulatory requirements
- Schedule constraints
- Ambiguous clauses
- Clauses that favor the owner
- Coordination among all the parties involved in the project
- Any hidden clause that will entitle the owner for a compensation claim from the contractor
- Priorities among the various construction documents
- Discrepancies and conflicting clauses
- Change order process
- Claim for extra work
- Payment procedure
- Damage and penalty for delay clauses
- *Force Majeure* clause

Management of Quality for Sustainability 289

4.6.5 IDENTIFY THE SUSTAINABILITY REQUIREMENTS

The contractor has to also identify sustainability requirements for the construction/execution of the project. The following are the basic elements that have to be taken care of by the contractor to develop the construction project:

- Meeting/satisfying the owner's requirements
- Completion of the execution/installation works within the agreed schedule
- Completion of the execution/installation of the project works within the budget
- Quality management
- Use of energy-efficient equipment, systems
- Water conservation systems
- Use of local resources (manpower)
- Efficient and durable material
- Ecofriendly material
- Material handling and utilization
- Optimized material storage
- Supply chain management
- Just-in-time delivery
- Follow systematic work process
- Use efficient method of execution/installation
- Follow the Green Building concept
- AeEsthetic
- Install energy-efficient MEP systems
- Waste treatment system
- Stormwater management
- Hazardous material management
- Clean air/ventilation system
- Daylighting
- Sensor-controlled lighting in certain areas
- Install a waste treatment plant
- Risk management
- Landscape compatible with the surrounding area
- Plantation in the project area
- Regulatory requirements
- Applicable codes and standards

4.6.6 DEVELOP THE PROJECT EXECUTION SCOPE

Contract requirements are established, taking into consideration items identified under Sections 4.6.4 and 4.6.5. The following are the major requirements:

- Construction of the facility/project as per the specifications
- Complete the facility within the specified schedule

- Complete the facility within the contracted budget
- Complete the facility as per the quality requirements

The following is the scope of works to be executed during the construction phase:

A. **Execution of Works**
   - Site work such as cleaning and excavation of the project site
   - Construction of foundations including footings and grade beams
   - Construction of columns and beams
   - Forming, reinforcing, and placing of the floor slab
   - Laying up masonry walls and partitions
   - Installation of roofing system
   - Finishes
   - Furnishings
   - Conveying system
   - Installation of fire suppression system
   - Installation of water supply, plumbing, and public health system
   - Installation of heating, ventilating, and air conditioning system
   - Integrated automation system
   - Installation of electrical lighting and power system
   - Emergency power supply system
   - Fire alarm system
   - Communication system
   - Electronic security and access control system
   - Landscape works
   - External works

B. **Monitoring and Control during Construction**
   - Stakeholder's requirements
   - Project execution scope
   - Project schedule
   - Project cost
   - Project quality
   - Safety during construction

C. **Inspection of Executed Works**

D. **Regulatory Approvals**

4.6.6.1 Identify the Construction Phase Deliverables

The deliverables of construction phase are executed and the construction project is complete and duly inspected with the following works:

A. **Construction-Installation/Execution and Inspection**
   1. Site work such as cleaning and excavation of project site
   2. Grading works
   3. Earth work
   4. Structural work

5. Architectural
   6. Internal finishes
   7. External finishes
   8. Elevator/conveying system
   9. Fire suppression works
   10. Mechanical/public health
   11. HVAC works
   12. Electrical works
   13. Lightning protection system
   14. Emergency power/generator system
   15. Alternative energy system
   16. Information and communication
   17. Low-voltage systems
   18. Landscape work
   19. External work
   20. Parking
   21. Furnishings
B. **Planning and Scheduling**
   1. Construction schedule
   2. S-Curve
   3. Contractor's quality control plans
   4. Management plans
C. **Monitor and Control of Construction:**
   1. Stakeholder's requirements
   2. Project scope
   3. Project schedule
   4. Project cost
   5. Project quality
   6. Resources
   7. Contract
   8. Safety during construction
D. **Inspection of Executed Works**
   1. Inspection of executed works
   2. Validate executed works
E. **Drawings**
   1. Record drawings/update working drawings (shop drawings)
F. **Specifications**
   1. Update specifications

4.6.7 Project Planning and Scheduling

Project planning is a logical process to ensure that the work of the project is carried out:

- In an organized and structured manner
- Reducing uncertainties to a minimum

- Reducing risk to a minimum
- Establishing quality standards
- Achieving results within budget and scheduled time

Prior to the start of the execution of a project or immediately after the actual project starts, the contractor prepares the project construction plans based on the contracted time schedule of the project. Detailed planning is needed at the start of construction to decide how to use resources such as laborers, plant, materials, finance, and subcontractors economically and safely to achieve the specified objectives. The plan shows the periods for all sections of the works and activities, indicating that everything can be completed by the date specified in the contract and ready for use or for installation of equipment by other contractors.

Effective project management requires planning, measuring, evaluating, forecasting, and controlling all aspects of project quality and quality of work, cost, and schedules. The purpose of the project plan is to successfully control the project to ensure completion within the budget and schedule constraints. Project planning is the evolution of the time and efforts to complete the project.

Upon signing of the contract, the contractor has to submit the following to the construction supervisor (consultant) for their review and approval:

1. Contractor's construction schedule
2. S-Curve
3. Contractor's quality control plan

Among these plans, the construction schedule is the most important. This is the first program that the contractor has to submit for approval. The contractor cannot proceed with construction unless the preliminary construction schedule is approved. In certain cases, the progress payment is involved with approval of the contractor's construction schedule. The contractor is not paid unless the contractor's construction schedule is approved.

The contractor has to prepare the following plans and submit them along with the construction schedule:

- Resource management plan
- Project S-Curve

Apart from the above plans, the contractor also has to prepare and submit the following plans:

- Stakeholder management plan
- Resource management plan
- Communication management plan
- Risk management plan
- Contract management plan
- HSE management plan

Management of Quality for Sustainability

4.6.7.1 Develop the Contractor's Construction Schedule

Depending on the size of the project, the project is divided into multiple zones, and relevant activities are considered for each zone to prepare the construction program. While preparing the program, the relationships between project activities and their dependency and precedence are considered by the planner. These activities are connected to their predecessor and successor activity based on the way the task is planned to be executed. There are basically four possible relationships that exist between various activities:

 i. Finish-to-start relationship
 ii. Start-to-start relationship
 iii. Finish-to-finish relationship
 iv. Start-to-finish relationship

Once all the activities are established by the planner and the estimated duration of each activity has been assigned, the planner prepares the detailed program, fully coordinating all the construction activities.

The first step in preparation of the construction program is to establish project activities, while the next step is to sequence the project activities and to establish the estimated time duration of each activity. The deadline for each activity is fixed, but it is often possible to reschedule by changing the sequence in which the tasks are performed, while retaining the original estimated total duration.

Figure 4.34 illustrates the schedule development process.

The activities to be performed during execution of the project are grouped into a number of categories. Each of these categories has a number of activities. The following are the major categories of the construction projects schedule:

A. **General Activities**
 1. Mobilization
 2. Staff approval
 3. Subcontractor approval
B. **Engineering**
 1. Detailed engineering (if applicable)
 2. Procurement of materials submittal and approval
 3. Shop drawing submittal and approval
C. **Procurement**
 1. Approval of vendor/supplier
 2. Submittal and approval of material, equipment, and systems
 3. Procurement of material, equipment, and systems
 4. Inspection of material, equipment, and systems
D. **Construction (a few site activities are listed below)**
 1. Site earthworks
 2. Dewatering and shoring
 3. Excavation and backfilling
 4. Raft works
 5. Retaining wall works
 6. Concrete foundation and grade beams

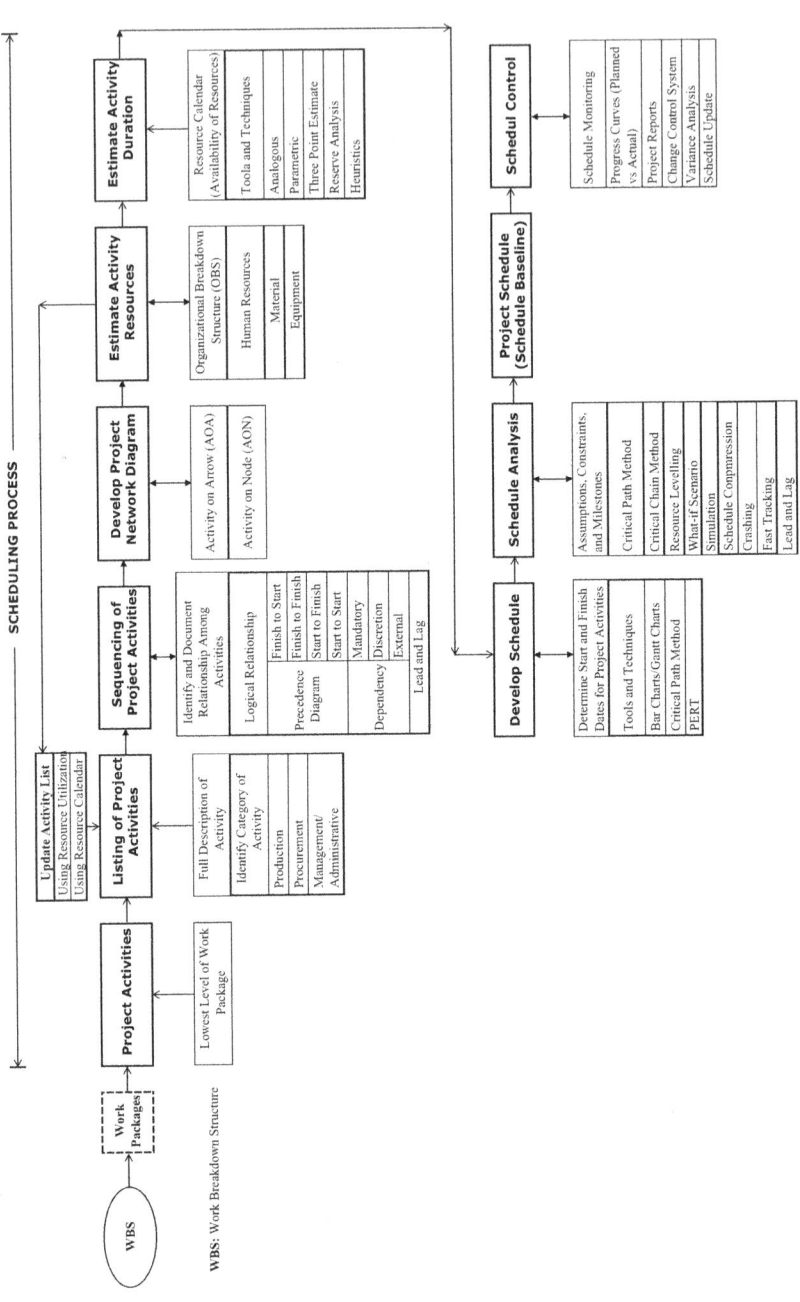

FIGURE 4.34 Schedule development process. Source: Abdul Razzak Rumane. (2016). *Handbook of Construction Management: Scope, Schedule, and Cost Control*. Reprinted with permission of Taylor & Francis Group.

7. Waterproofing
8. Concrete columns and beams
9. Casting of slabs
10. Wall partitioning
11. Interior finishes
12. Furnishings
13. External finishes
14. Equipment
15. Conveying systems works
16. Plumbing and public health works
17. Firefighting works
18. HVAC works
19. Electrical works
20. Fire alarm system works
21. Communication system works
22. Low-voltage system works
23. Landscape works
24. External works

E. **Close Out**
1. Testing and commissioning
2. Completion and handover

The contractor also submits the following along with the construction schedule:
1. Resources (equipment and manpower) schedule
2. Cost loading (schedule of item pricing based on the Bill of Quantities)

4.6.7.2 Develop the Project S-Curve

The S-Curve is a graphical display of cumulative cost, resources, or other quantities plotted against time. The S-Curve is used for forecasting cash flow which is based on the work (activities) the contractor is expected to complete and how much amount (payment) will be received and predicts how much will be spent over the established project schedule (time). The S-Curve is used to measure project performance and predict the expenses over the project duration. The S-Curve helps the owner/client to know the project funding requirements. It is also an indication of the progress of work to be completed in a project. Funding requirements, both total and periodic, are derived from the S-Curve. This also represents the planned progress of a project.

4.6.7.2.1 Cost-Loaded Curve

Conceptually, cash flow is a simple comparison of when revenue will be received and when the financial obligations must be paid. This is obtained by loading each activity in the approved schedule with the budgeted cost in the Bill of Quantities (BOQ). The process of inputting schedule of values is known as Cost Loading. The graphical representation of the above is obtained as a curve and is known as a cost-loaded curve.

4.6.7.3 Develop the Contractor's Quality Control Plan

The contractor's quality control plan (CQCP) is the contractor's everyday tool to ensure meeting the performance standards specified in the contract documents. The adequacy and efficient management of CQCP by the contractor's personnel have a great impact on both the performance of the contract and the owner's quality assurance surveillance of the contractor's performance.

CQCP is the documentation of the contractor's process for delivering the level of construction quality required by the contract. It is a framework for the contractor's process for achieving quality construction. CQCP does not endeavor to repeat or summarize contract requirements. It describes the process which the contractor will use to ensure compliance with the contract requirements. The quality plan is virtually manual, tailor-made for the project and is based on contract requirements

The CQCP is prepared based on the project's specific requirements, as specified in the contract documents. The plan outlines the procedures to be followed during the construction period to attain the specified quality objectives of the project, fully complying with the contractual and regulatory requirements.

In the quality plan, the generic documented procedures are integrated with any necessary additional procedures specific to the project in order to attain specified quality objectives. Application of various quality tools, methods, and principles at different stages of construction projects is necessary to make the project qualitative and economical and to meet the owner's needs/specification requirements.

Based on the contract requirements, the contractor prepares his quality control plan and submits the same to the consultant for their approval. Figure 4.35 illustrates the logic flow for the Contractor's Quality Control Plan. This plan is followed by the contractor to maintain the project quality.

The quality control plan outlines the procedures to be followed during the construction period to attain the specified quality objectives of the project, fully complying with the contractual and regulatory requirements. The plan provides the mechanism by which to achieve the specified quality by identifying the procedures, controls, instructions, and tests required during the construction process to meet the owner's objectives.

Table 4.39 illustrates contents of Contractor's Quality Control Plan. However, the contractor has to take into consideration the requirements listed under the contract documents, depending on the nature and complexity of the project.

4.6.8 Develop Management Plans

Apart from the construction schedule, S-Curve, and Contractor's Quality Control Plan, the contractor has to submit the following plans to the construction supervisor (consultant) for their review and approval:

1. Stakeholder management plan
2. Resource management plan
3. Communication management plan
4. Risk management plan
5. Contract management plan
6. HSE management plan

Management of Quality for Sustainability

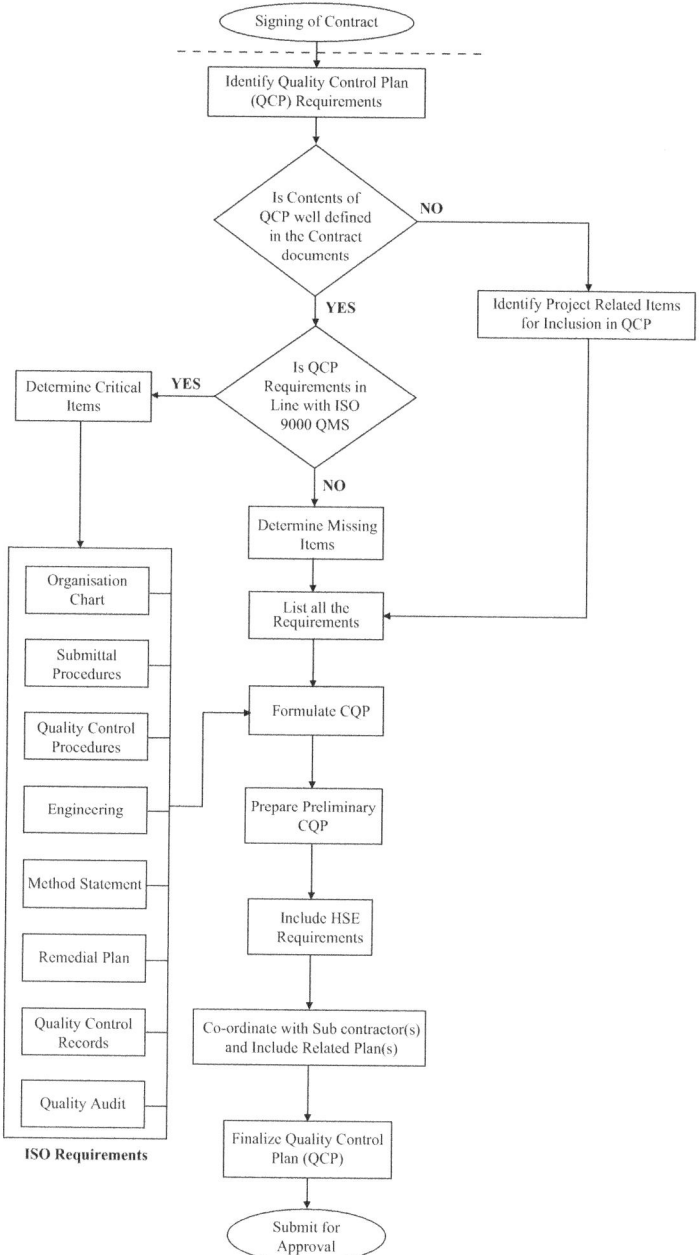

FIGURE 4.35 Logic flowchart for development of contractor's quality control plan.

TABLE 4.39
Contents of Contractor's Quality Control Plan

Section			Topic
1			Introduction
2			Description of project
3			Quality control organization
4			Qualification of QC staff
5			Responsibilities of QC personnel
6			Procedure for submittals
7			Quality control procedure
	7.1		Detailed engineering
		7.1.1	Accuracy and correctness of detailed engineering
		7.1.2	Quality of drawings
		7.1.3	Check specifications
		7.1.4	Check contract documents
		7.1.5	Number of drawings
	7.2		Procurement
		7.2.1	Vendor selection
		7.2.2	Cost of material, equipment, and system
		7.2.3	Off-site manufacturing, inspection, and testing
		7.2.4	Inspection of material received at site
		7.2.5	Material storage and handling
	7.3		Construction
		7.3.1	Inspection of site activities (checklists)
		7.3.2	Inspection of installed equipment (checklists)
		7.3.3	Inspection and testing procedure for systems
		7.3.4	Procedure for laboratory testing of material
		7.3.5	Corrective and preventive action
		7.3.6	Protection of works
8			Method statement for various installation activities
9			Project-specific quality procedures
10			Owner-specific processes
11			Company's quality manual and procedure.
12			Periodical testing of construction equipment
13			Subcontractor's quality plan
	13.1		Sub contractor's QA/QC personnel and resources
14			Vendor's quality plan
15			Quality control records
16			Resource management
17			Communication and meetings
18			Risk management
19			Health, Safety, and Environment
20			Testing. Commissioning and Handover
	20.1		Inspection and testing plan
	20.2		Inspection and testing of major equipment
	20.3		Performance test
21			Quality updating program
22			Quality auditing program

Management of Quality for Sustainability

4.6.8.1 Develop a Stakeholder Management Plan

In order to run a successful project, it is important to address the needs of the project stakeholders, effectively predicting how the project will be affected and how the stakeholders will be affected. Stakeholder management planning is a process to develop a stakeholder engagement plan depending on the roles and responsibilities of the stakeholders and their needs, expectations and influence on the project. Table 4.40 shows an example matrix for site administration of a construction project. A distribution matrix based on the interest and involvement of each stakeholder is prepared and the appropriate documents are sent for their action/information as per the agreed upon requirements.

4.6.8.2 Develop a Resource Management Plan

Resource management in construction is mainly related to management of the following processes:

1. Human resources (project teams)
   - Project owner team (project manager)
   - Supervision consultant
   - Designer (consultant)
2. Construction resources (contractor)
   - Contractor's core staff
   - Manpower for construction works
   - Material for project
     - Equipment
     - Materials
     - Systems

In most construction projects, the contractor is responsible for engaging subcontractors, specialist installers, and suppliers, to arrange for materials, equipment, construction tools and all types of human resources to complete the project as per the contract documents and to the satisfaction of the owner/owner's appointed supervision team. Workmanship is one of the most important factors to achieve quality in construction, therefore it is required that the construction workforce is fully trained and has full knowledge of all the related activities to be performed during the construction process.

Once the contract is awarded, the contractor prepares a detailed plan for all the resources needed to complete the project.

Contract documents normally specify a list of minimum number of core staff to be available at the site during the construction period. Absence of any of these staff may result in penalty to be imposed on the contractor by the owner.

The contractor's human resources mainly consist of two categories:

1. The contractor's own staff and workers
2. The subcontractor's staff and workers

TABLE 4.40
Matrix for Site Administration and Communication

Sr. No.	Description of Activities	Contractor	Consultant/PMC	Owner
1	**General**	-	-	P
	1.1 Notice to proceed	P	R	A
	1.2 Bonds and guarantees	-	P	A
	1.3 Consultant's staff approval	P	R/B	A
	1.4 Contractor's staff approval	P	R	A
	1.5 Payment guarantee	P	R	A
	1.6 Master schedule	-	P	A
	1.7 Stoppage of work	-	P	A
	1.8 Extension of time	P	R	A
	1.9 Deviation from contract Documents			
	a. Material			
	b. Cost			
	c. Time			
2	**Communication**			
	2.1 General correspondence	P	P	P
	2.2 Job site instruction	D	P	C
	2.3 Site works instruction	D	P/B	A
	2.4 Request for information	P	A	C
	2.5 Request for modification	P	B	A
3	**Submittals**			
	3.1 Subcontractor	P	B/R	A
	3.2 Materials	P	A	C
	3.3 Shop drawings	P	A	C
	3.4 Staff approval	P	B	A
	3.5 Pre-meeting submittals	P	D	C
4	**Plans and Programs**			
	4.1 Construction schedule	P	R	C
	4.2 Submittal logs	P	R	C
	4.3 Procurement logs	P	R	C
	4.4 Schedule update	P	R	C
5	**Quality**			
	6.1 Quality control plan	P	R	C
	6.2 Checklists	P	D	C

(*Continued*)

TABLE 4.40 CONTINUED
Matrix for Site Administration and Communication

Sr. No.	Description of Activities	Contractor	Consultant/PMC	Owner
	6.3 Method statements	P	A	C
	6.4 Mock-up	P	A	B
	6.5 Samples	P	A	B
	6.6 Remedial notes	D	P	C
	6.7 Non-conformance report	D	P	C
	6.8 Inspections	P	D	C
	6.9 Testing	P	A	B
6	**Site Safety**			
	7.1 Safety program	P	A	C
	7.2 Accident report	P	R	C
7	**Monitor and Control**			
	5.1 Progress	D	P	C
	5.2 Time	D	P	C
	5.3 Payments	P	R/B	A
	5.4 Variations	P	R/B	A
	5.5 Claims	P	R/B	A
8	**Meetings**			
	8.1 Progress	E	P	E
	8.2 Coordination	E	P	C
	8.3 Technical	E	P	C
	8.4 Quality	P	C	C
	8.5 Safety	P	C	C
	8.6 Close-out	–	P	
9	**Reports**			
	9.1 Daily report	P	R	C
	9.2 Monthly report	P	R	C
	9.3 Progress report	–	P	A
	9.4 Progress photographs	-	P	A
10	**Close-Out**			
	10.1 Snag list	P	P	C

(*Continued*)

TABLE 4.40 CONTINUED
Matrix for Site Administration and Communication

Sr. No.	Description of Activities	Contractor	Consultant/PMC	Owner
10.2	Authorities approvals	P	C	C
10.3	As-Built drawings	P	D/A	C
10.4	Spare parts	P	A	C
10.5	Manuals and documents	P	R/B	A
10.6	Warranties	P	R/B	A
10.7	Training	P	C	A
10.8	Hand-over	P	B	A
10.9	Substantial completion certificate	P	B/P	A

P	Prepare/Initiate
B	Advise/Assist
R	Review/Comment
A	Approve
D	Action
E	Attend
C	Information

Source: Abdul Razzak Rumane (2010). Quality Management in Construction Projects. Reprinted with permission of Taylor & Francis Group.

The main contractor has to manage all these personnel by:

1. Assigning the daily activities
2. Observing their performance and work output
3. Ensuring daily attendance
4. Highlighting safety during the construction process

Similarly, the contract documents specify that a minimum equipment set is to be available on site during the construction process to ensure smooth operation of all the construction activities. They are normally listed in the contract documents:

- Tower crane
- Mobile crane
- Normal concrete mixture machine
- Concrete mixing plant
- Dump trucks
- Compressors
- Vibrators
- Water pumps

Management of Quality for Sustainability

- Compactors
- Concrete pumps
- Trucks
- Concrete trucks
- Diesel generator sets

In most construction projects, the contractor is responsible for the procurement of materials, equipment, and systems to be installed on the project. The specifications are based on the approved contract specifications. The contractor also prepares a procurement log based on the project completion schedule. Figure 4.36 illustrates the Material Management Process for Construction Projects.

4.6.8.3 Develop the Communication Management Plan

A construction project has the involvement of many stakeholders. The project team must provide timely and accurate information to the identified stakeholders who

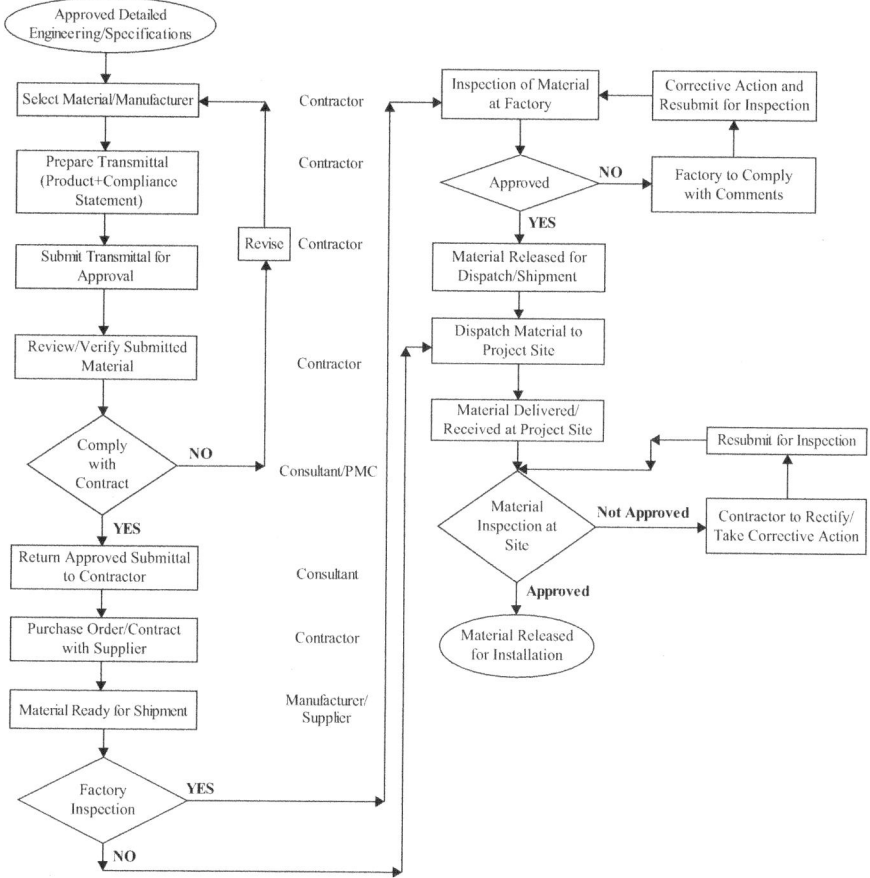

FIGURE 4.36 Material management process for construction project.

will receive communication. Effective communication is one of the most important factors contributing to the success of a project. For the smooth flow of construction process activities during the construction phase, proper communication and submittal procedure need to be established between all concerned parties at the beginning of the construction activities.

Table 4.41 illustrates an example of a guideline to prepare a communication matrix for site administration during the construction phase, and Table 4.42 illustrates the contents of a communication management plan.

The communication between contractor and supervisor takes place through a transmittal form. Table 4.43 lists the various types of project control documents used during a construction project. The number of copies, their distribution, and the communication method are specified in the contract documents. These documents are normally transmitted by the contractor using a transmittal form. Figure 4.37 illustrates a sample transmittal form.

Correspondence between the consultant and the contractor is normally through letters or job site instructions. Figure 4.38 is a site work instruction (SWI) form used by the consultant to communicate with the contractor.

4.6.4.4 Develop a Risk Management Plan

The probability of occurrence of risk during the construction phase is very high compared with design phases. During the construction phase, uncertainty comes from various sources as this phase involves various participants. Since the duration of the construction phase is longer than earlier phases, the contractor also has to consider the occurrence of financial, economic, commercial, political, and natural risks. The contractor has to develop a risk management plan. A risk management plan identifies how risks associated with the project will be identified, analyzed, managed, and controlled. Risk management is an integral part of project management as the risk is likely to occur at any stage of the project. Therefore, the risk has to be continually monitored and response actions taken immediately. The risk management plan outlines how risk activities will be performed, recorded, and monitored throughout the life cycle of the project. It is intended to maximize the positive impact for the benefit of the project and decrease, minimize, or eliminate the impact of events adverse to the project. The risk management must commence early in the project development stage (study stage) and proceed as the project evolves, when more and more information about the project is available. The project plan should:

- Define the risk management strategy/approach
- Define the project objectives, goals related to risk management
- Identify risk to the owner and team members
- Define the risk decisions
- Detail about risk resources
- Include the risk management process
  - Methods of risk identification

TABLE 4.41
Guidelines to Prepare a Communication Matrix

Project Name:

Name of Construction/Project Manager: Name of Consultant:

Contractor Name: Project Number:

Serial Number	Type of Document	Originator	Receiver(s)	Purpose	Frequency	Method	Responsible Person for Action	Comments

SAMPLE FORM

TABLE 4.42
Contents of Contractor's Communication Management Plan

Section	Topic
1	Introduction
	1.1 Project description
2	Stakeholders
	2.1 Project team directory
	2.2 Project organization chart
	2.3 Roles and responsibilities
	2.4 Stakeholders' requirement
3	Communication methods
4	Communication management constraints
5	Communication matrix
6	Distribution of communication documents
	5.1 General correspondence
	5.2 Submittals
	5.3 Status reports
	5.4 Meetings
	5.5 Management plans
	5.6 Change orders
	5.7 Payments
6	Regulatory requirements
7	Communication plan update

- Methods of risk assessment
- Level of risk
- Response to risk
- Management of risk
- Control of risk
- Process of integrating risk management activities into project scope, schedule, cost, and quality
- Documenting and recording of risks
- Communication procedure for risk reporting
- Update of risk management plan

4.6.8.5 Develop a Contract Management Plan

Contract management in construction projects is an organizational method, process, and procedure to obtain the required construction project/product. It includes the process to acquire the construction project complete with all the related material, equipment, system, and services from outside contractors/companies to the satisfaction of the owner/client/end-user.

TABLE 4.43
List of Project Control Documents

Serial Number		Document Name
I	**Administrative**	
	I-1	Material Entry Permit
	I-2	Material Removal Permit
	I-3	Vehicular Entry Permit
	I-4	Site Entry Permit
	I-5	Visitor Entry Permit
	I-6	Municipality Permit
	I-7	Request for Overtime
	I-8	Theft & Damage Report
	I-9	Performance Bonds
	I-10	Advance Payment Guarantee
	I-11	Insurance
	I-12	Accident Report
	I-13	Sample Tag
II	**Contracts Related**	
	II-1	Notice to Proceed
	II-2	Job Site Instruction
	II-3	Site Works Instruction
	II-4	Attachment to Site Works
	II-5	Request for Staff Approval
	II-6	Request for Subcontractor Approval
	II-7	Request for Vendor Approval
	II-8	Material Delivered at Site
	II-9	Variation Order
	II-10	Attachment to Variation Order
	II-11	Baseline Change Request Form
	II-12	Extension of Time
	II-13	Suspension of Work
	II-14	Attendees
	II-15	Minutes of Meeting
	II-16	Transmittal for Minutes of Meeting
	II-17	Submittal Form
III	**Engineering Submittal**	
	III-1	Master Schedule
	III-2	Cost-Loaded Schedule
	III-3	Engineering Drawings
	III-4	Material Approval
	III-5	Specification Comparison Statement
	III-6	Product Data
	III-7	Product Sample

(*Continued*)

TABLE 4.43 CONTINUED
List of Project Control Documents

Serial Number		Document Name
	III-8	Workshop Drawings
	III-8	Builders Drawings
	III-10	Composite Drawings
	III-11	Method Statement
	III-12	Request for Information
	III-13	Request for Modification
	III-14	Variation Order (Proposal)
	III-15	Request for Alternative or Substitution
IV	**PCS Reporting Forms**	
	IV-1	Contractor's Submittal Status Log E-1
	IV-2	Contractor's Procurement Log E-2
	IV-3	Contractor's Shop Drawing Status Log
	IV-4	Daily Progress Report
	IV-5	Weekly Progress Report
	IV-6	Look-Ahead Schedule
	IV-7	Monthly Progress Report
	IV-8	Progress Photographs
	IV-9	Daily Checklist Status
	IV-10	Progress Payment Request
	IV-11	Payment Certificate
	IV-12	Submittal Schedule
	IV-13	Schedule Update Report
V	**Management Plans**	
	V-1	Quality Control Plan
	V-2	Safety Management Plan
	V-3	Environmental Protection Plan
VI	**Quality Control Forms**	
	VI-1	Checklist (Request for Inspection)
	VI-2	Checklist for Form Work
	VI-3	Notice for Daily Concrete Casting
	VI-4	Check List for Concrete Casting
	VI-5	Quality Control of Concreting
	VI-6	Report on Concrete Casting
	VI-7	Notice for Testing at Lab
	VI-8	Concrete Quality Control Form
	VI-9	Checklist for Process Work
	VI-10	Checklist for Piping Work
	VI-11	Checklist for Instrumentation Work
	VI-12	Checklist for Utility Work
	VI-13	Checklist for Steel Fabrication Work

(*Continued*)

TABLE 4.43 CONTINUED
List of Project Control Documents

Serial Number		Document Name
	VI-14	Checklist for Detection and Protection Works
	VI-15	Checklist for Fire Fighting Work
	VI-16	Checklist for Architectural Work
	VI-17	Checklist for Civil Work
	VI-18	Checklist for Mechanical Work
	VI-19	Checklist for HVAC Work
	VI-20	Checklist for Electrical Work
	VI-21	Checklist for Elevator Work
	VI-22	Checklist for Low-Voltage System Work
	VI-23	Checklist for External Work
	VI-24	Checklist for Landscape
	VI-25	Remedial Note
	VI-26	Non-Conformance/Compliance Report
	VI-27	Material Inspection Report
	VI-28	Safety Violation Notice
	VI-29	Notice of Commencement of New Activity
	VI-30	Removal of Rejected Material
	VI-31	Testing and Commissioning
VII	**Closeout Forms**	
	VII-1	As-Built Drawings
	VII-2	Substantial Completion Certificate
	VII-3	Handing-Over Certificate
	VII-4	Taking-Over Certificate
	VII-5	Manuals
	VII-6	Handing-Over of Spare Parts
	VII-7	Defect Liability Certificate

Source: Modified from: Abdul Razzak Rumane (2013). Quality Tools for Managing Construction Projects. Reprinted with permission of Taylor & Francis Group.

Contract management in the construction project involves:

- Identification of
  - What services are available in-house
  - What services can be procured from outside agencies/organizations
  - How to procure (direct contract, competitive bidding)
  - How much to procure
  - How to select a supplier/contractor
  - How to arrive at an appropriate price, terms, and conditions

		Project Name			
		Consultant Name			
		SUBMITTAL TRANSMITTAL FORM			
Contractor Name:					
Contract No. :					
To.		Resident Engineer			
Transmittal No.:			Date :		

Submittal Type:		Action Requested:	
ED	Engineering Drawings	1	For Approval
DG	Shop Drawings	2	For Review and Comment
SK	Sketches	3	For Information
PR	Material/Product/System	4	For Construction
MD	Manufacturer's Data	5	For Incorporation Within the Design
SM	Sample	6	For Costing
MM	Minutes of Meeting	7	For Tendering
RP	Reports		
LG	Logs		
OT	Others (please specify)		

We are sending herewith the following:

			ENCLOSURES		
Item	Qty	Ref. No.	Description	Type	Action

Comments:
_____
_____
_____

Issued by:	Received by:
Signature:	Signature:
Date:	Date:

FIGURE 4.37 Transmittal form.

- Signing of contract
- Timely delivery
- Receiving the right type of material/system
- Timely execution of the work
- Inspection of work to maintain the quality of the project
- Completion of the project within the agreed-upon schedule
- Completion of the project within the agreed-upon budget
- Documenting the reports and plans

Management of Quality for Sustainability

| **Project Name** |
| Consultant Name |

SITE WORK INSTRUCTION (SWI)

CONTRACTOR: _____   JSI No. : _____

CONTRACT No.: _____   DATE : _____

The work shall be carried out in accordance with the Contract Documents without change in Contract Sum or Contract Time. Proceeding with the work in accordance with these instructions indicates your acknowledgement that there will be no change in the Contract Sum or Contract Time.

Subject:
_____

SAMPLE FORM

_____

ATTACHMENTS: (List attached documents that support description.)
_____

Signed: _____   Received by Contractor : _____
         Resident Engineer   Date:

Distribution: ☐ Owner   ☐ Consultant (Supervision)   ☐ Contractor

FIGURE 4.38 Job site instruction.

The administration of the contract is the process of formal governance of the contract and changes to the contract document. It is concerned with managing the contractual relationship between various participants to successfully complete the facility to meet the owner's objectives. It includes tasks such as:

- Administration of project requirement
- Administration of project team members

- Communication and management reporting
- Execution of contract
- Monitoring contract performance (scope, cost, schedule, quality, risk)
- Inspection and quality
- Variation order process
- Making changes to the contract documents by taking corrective action, as needed
- Payment procedures

It is required that the contract administration procedure is clearly defined for the success of the contract and that the parties to the contract understand who does what, when, and how. The following are some typical procedures that should be in place for management of the contract:

1. Contract document maintenance and variation
2. Performance review system
3. Resource management and planning
4. Management reporting
5. Change control procedure
6. Variation order procedure
7. Payment procedure

Table 4.44 lists the contents of a contract management plan.

4.6.8.6 Develop an HSE Management Plan

The construction industry has long been considered to be dangerous. The nature of work at the site always presents some dangers and hazards. There are a relatively high number of injuries and accidents at construction sites. Safety represents an important aspect of construction projects. In construction projects, the requirements to prepare a Safety Management Plan (SMP) by the contractor are specified in the contract documents. The contractor has to submit the plan for review and approval by the supervisor/consultant during the mobilization stage of the construction phase. The following are the guidelines normally specified in the contract documents which are to be considered by the contractor to establish an HSE management plan:

1. Project scope detailing the description of project and safety requirements
2. Safety policy statement documenting the contractor's/subcontractor's commitment and emphasis on safety
3. Regulatory requirements about safety
4. Roles and responsibilities of all individuals involved
5. Safety management of different activities
6. Site communication plan detailing how safety information will be shared
7. Hazard identification, risk assessment and control
8. Emergency evacuation plan
9. Accident reporting system

TABLE 4.44
Contents of Contract Management Plan

Serial Number	Topics
1	Contract summary, deliverables and scope of work
2	Type of contract
3	Contract schedule
4	Contract cost
5	Project team members with roles and responsibilities
6	Core staff approval procedure
7	Contract communication matrix/management reporting
8	Coordination process
9	Liaison with regulatory authorities
10	Engineering drawing submission/approval process
11	Vendor selection process
12	Material/product/system review/approval process
13	Shop drawing review/approval process
14	Project monitoring and control process
15	Contract change control process a) Scope b) Material c) Method d) Schedule e) Cost
16	Review of variation/change requests
17	Project hold-up areas
18	Quality of performance
19	Inspection and acceptance criteria
20	Risk identification and management
21	Progress payment process
22	Safety management
23	Claims, disputes, conflict, and litigation resolution
24	Contract documents and records
25	Post contract liabilities
26	Contract closeout and independent audit.

Source: Abdul Razzak Rumane (2013). Quality Tools for Managing Construction Projects. Reprinted with permission of Taylor & Francis Group.

10. Accident investigation to document root causes and determine corrective and preventive actions
11. Measures for emergency situations
12. Plant, equipment maintenance, and licensing
13. Routine inspections

14. Continuous monitoring and regular assessment
15. Health surveillance
16. System feedback and continuous improvements
17. Safety assurance measures
18. Evaluation of a subcontractor's safety capabilities
19. Site neighborhood characteristics and constraints
20. System education and training
21. Safety audit
22. Documentation
23. Records
24. Procedure for Project HSE Review
25. System update

The following are the main responsibilities of the safety engineer/officer:

1. Conducting safety meetings
2. Monitoring on-the-job safety
3. Inspecting the work and identifying hazardous areas
4. Initiating a safety awareness program
5. Ensuring the availability of first aid and emergency medical services per local codes and regulations
6. Ensuring that personnel are using protective equipment, such as hard hat, safety shoes, protective clothing, life belt, and protective eye covering
7. Ensuring that the temporary firefighting system is working
8. Ensuring that work areas are free from trash and hazardous material
9. Housekeeping

Table 4.45 illustrates a sample content of contractor's HSE management plan for a construction project detailing hazardous identification, risk assessment, and control for scaffolding work. However, the contractor has to take into consideration all the requirements listed under the contract documents, which depend on the complexity and nature of the project.

4.6.9 Construction/Execution of the Works

Construction projects are unique and nonrepetitive in nature. Construction projects consist of many activities aimed at the accomplishment of a desired objective. In order to achieve the quality objectives of the project, each activity has to be completed within the specified limit, using the specified product, and an approved method of installation. A construction project consists of a number of related activities that are dependent on other activities and cannot be started until others are completed, and some that can run in parallel.

For smooth implementation of a project, a proper communication system is established clearly identifying the submission process for correspondence and transmittals. Correspondence between consultant and contractor is normally done via job site

TABLE 4.45
Contents of Contractor's HSE Plan

Section	Topic
1	Introduction
	1.1 Project Description
2	Project General Requirements
3	HSE Policy
4	Regulatory Requirements
	4.1 Occupational Health, Health Surveillance
	4.2 Safety
	4.3 Environmental
5	Project HSE Issues
6	HSE Organization
	6.1 HSE Organization Chart
	6.2 Roles and Responsibilities
	6.3 Communication Plan
7	Safety Management Plan
	7.1 Design, Detailed Engineering
	7.2 Construction
	7.3 Process-related Activities
	7.4 Non-process-related Activities
	7.5 Project Start-up
8	Emergency Evacuation Plan
	8.1 Accident Reporting
	8.2 Action Plan
	8.3 Measures for Emergency Situations
	8.4 Disaster Control
9	Training
	9.1 Awareness
	9.2 Meetings
	9.3 Drills
10	Monitoring
	10.1 Routine Inspection
	10.2 Protective Equipment
	10.3 Plant, Equipment, Machinery
	10.4 Fire Prevention
	10.5 Gas and Chemical Leak Detection
	10.6 Safety Measures
	10.6.1 Preparatory Activities
	10.6.2 Execution/Installation/Construction Activities
	10.6.3 Start-up Activities
	10.7 Hygiene
	10.8 Housekeeping
	10.9 Analysis and Evaluation
	10.10 Subcontractor's Compliance

(*Continued*)

TABLE 4.45 CONTINUED
Contents of Contractor's HSE Plan

Section	Topic
11	Hazards Management
	11.1 Safety Hazards
12	Procedure for Project HSE Review (PHSER)
13	First Aid and Medical Services
14	Environmental Protection and Control Measures
15	Waste Management
16	Risk Identification, Management
17	Material Handling
18	Preventive Action
19	Documentation
20	Record
21	Internal Audit
22	Management Review
23	System Update/Continual Improvement

instructions whereas correspondence between the owner, consultant, and contractor is normally done through letters.

The contractor is responsible for executing the contracted works in accordance with the approved material, shop drawings, and specifications as specified in the contract documents.

Prior to starting the execution of construction/installation of works, the contractor has to submit material, systems, and shop drawings to the construction supervisor (consultant) for their review and approval.

The detailed procedure for submitting shop drawings, materials, and samples is specified under the section "SUBMITTAL" of contract specifications. The contractor has to submit the same to the owner/consultant for their review and approval. The consultant reviews the submittal and returns the transmittal to the contractor with an appropriate action.

4.6.9.1 Submittals

In construction projects, there are various types of documents which are to be sent to different stakeholders. Proper correspondence and reporting methods are important to distributing this information.

Construction projects involve many stakeholders. Large numbers of documents move forward and backward between these stakeholders for information or action.

During the construction phase, there are many documents being sent forth and back between the owner, construction supervisor (A/E, consultant), and contractor. The originator of these documents is mentioned in the communication matrix. The following are the types of documents exchanged among the owner, supervisor, and contractor:

Management of Quality for Sustainability

- Administrative
- Contract related
- Engineering submittals
- Project monitoring and control
- Quality

Table 4.43, discussed earlier, illustrates the list of forms (logs) normally used during the construction phase to communicate different types of documents.

The detailed procedure for submitting materials/products/systems, samples, and shop drawings is specified under the section "SUBMITTAL" of contract specifications. The contractor has to submit the same to the owner/consultant for their review and approval. The contract documents under Section 013300 Submittal Procedures (CSI-Format General Requirements) specifies administrative and procedural requirements for the submission of submittals and other documents. The contractor has to comply with the contractual requirements for submittal requirements.

The submittal process in construction projects is essential to ensure that contractor's understanding of product specifications, contract drawings, and installation methods matches with the designer's intent of product usage and installation method. The submittal process provides the owner with the assurance that the contractor is complying with the design concept, and that the installed material will function as required by the contract documents. Submittals are documents that are presented by the contractor for approval, review, decision, or consideration.

Generally, these submittals fall into three categories:

1. Approval submittals
2. Review submittals
3. Information submittals

Prior to the start of the execution/installation of the work, the contractor has to submit specified material/product/systems and shop drawings to the A/E /(consultant), construction/project manager as per project specification requirements for approval, review or information. The contractor, while preparing the submittal for shop drawing, has to consider the following:

1. Review contract specification
2. Review contract drawings
3. Determine and verify field/site measurements
4. Installation information about the material to be used
5. Installation details relating to the axis or grid of the project
6. Dimensions of the product or equipment to be installed
7. Roughing-in requirements
8. Coordination with other trade (discipline) requirements
9. Clearly marking the changes, deviations to the contract drawings

The consultant reviews the submittal to verify that the proposed product/sample/shop drawing comply with the contract specifications and returns the transmittal to the contractor, mentioning one of the following actions on the transmittal:

A-Approved
B-Approved As Noted
C-Revise and Resubmit OR Not Approved
D-For Information OR More Information Required

In the case of deviation from that of specified items, the contractor has to submit a schedule of such deviation(s), listing all the points that do not conform to the specifications.

4.6.9.1.1 Material Approval

In most construction projects, the contractor is responsible for procurement of material, equipment, and systems to be installed on the project. The contractors have their own procurement strategies. While submitting the bid, the contractor obtains the quotations from various vendors/subcontractors that are listed/registered with the companies. However, it is likely that these vendors may not be included under the approved list of manufacturers/ vendors in the contract specifications. Since the product specifications for oil and gas projects emphasize that certain items should be procured from specific manufacturers, it is required that the contractor follows proper vendor selection procedures to meet the contract requirements and also the organization's procurement strategy. Figure 4.39 illustrates a logic flowchart for vendor selection procedure and Figure 4.40 illustrates a logic flowchart for material approval and procurement procedure.

Figure 4.41 illustrates a Site Transmittal Form for material approval, whereas Figure 4.42 illustrates a Specification Comparison Form.

4.6.9.1.2 Shop Drawing Approval

The contract drawings and specifications prepared by the designer are indicative and are generally meant for determining the planning and pricing of the construction project. In many cases, they are not sufficient for the installation or execution of works at various stages. More details are required during the construction phase to ensure specified quality. These details are provided by the contractor on the shop drawings. Shop drawings are used by the contractor as reference documents to execute/install the works. Detailed shop drawings help the contractor to achieve zero defect in installation at the first stage itself, thus avoiding any rejection/rework.

The number of shop drawings to be produced is mutually agreed upon between the contractor and consultant, depending on the complexity of the work to be installed and to ensure that adequate details are available to execute the work.

The shop drawings are to be drawn accurately to scale and will have project-specific information on it. The shop drawings shall not be reproductions of the contract drawings.

Management of Quality for Sustainability

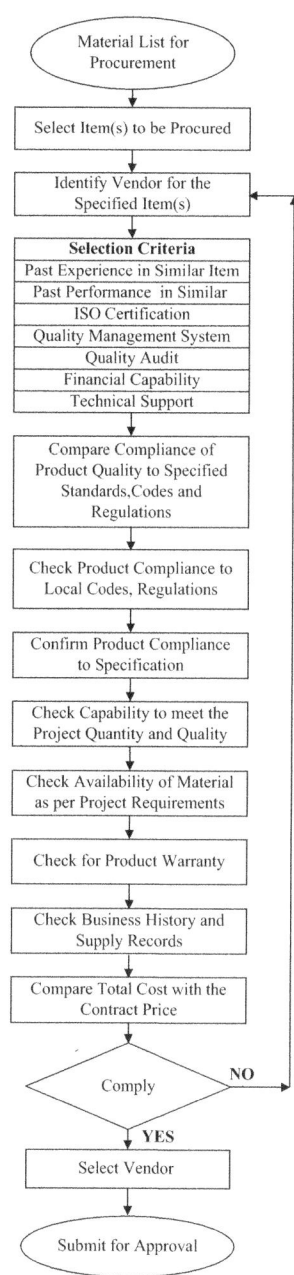

FIGURE 4.39 Logic flowchart for vendor selection procedure.

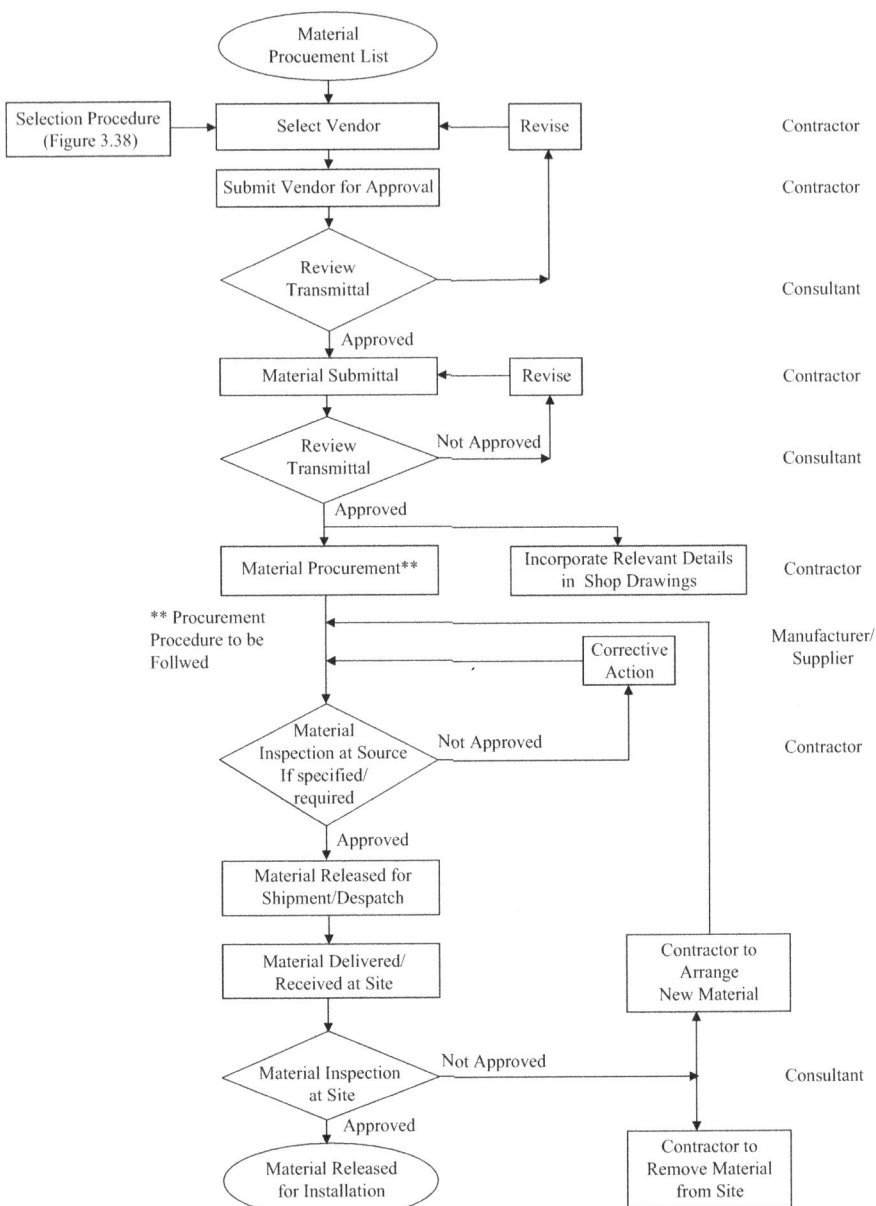

FIGURE 4.40 Logic flowchart for material approval and procurement procedure.

Management of Quality for Sustainability 321

Project Name
Consultant Name
SITE TRANSMITTAL
Request for Material Approval

CONTRACT No. :   TRANSMITTAL NO.   REV.

CONTRACTOR :

TO   :

WE REQUEST APPROVAL OF THE FOLLOWING MATERIALS/GOODS/PRODUCTS/EQUIPMENT

ITEM NO.	DWG., SPEC. OR BOQ. REF	DESCRIPTION	SUBMITTAL CODE *	ACTION CODE **
		SAMPLE FORM		

DETAILS OF INFORMATION, LITERATURE, CATALOG CUTS, AND THE LIKE ATTACHED ARE:

SAMPLES:
Enclosed [ ]   Submitted under separate cover [ ]   Not applicable [ ]

N.B: We certify that above items have been reviewed in detail & and are correct & in strict performance with the Contract Drawings & Specification except as otherwise stated.

CONTRACTOR'S REP. :   DATE :

RECEIVED BY CONSULTANT :   DATE :

cc: Owner Rep.

Resident Engineer to enter ACTION CODE and REMARKS

R.E.'s REMARKS :

                                            Initials   Date

Corrections or comments made relative to submittals during this review do not relieve Contractor from compliance with the requirements of the Drawings and Specifications. This check is only for review of general conformance with the design concept of the project and general compliance with the information given in the Contract Documents. Contractor is responsible for confirming and correlating all quantities and dimensions, selecting fabrication process and techniques of construction; coordinating his work with that of other trades, and performing his work in a safe and satisfactory manner.

Resident Engineer :   DATE :

Received by Contractor :   DATE :

cc: Owner Rep.

* SUBMITTAL CODE:	** ACTION CODE:		
1: Submitted for Approval	A: Approved	C:	Not Approved

FIGURE 4.41 Site transmittal for material approval.

	Project Name		
	Consultant Name		

SPECIFICATION COMPARISON STATEMENT (SCS)

Contractor: _____   Date: _____

Contract No. _____   A/S No.: _____

Submittal No. : _____	Revision: _____	Transmittal Ref.: _____
Submittal Title: _____		Specification Ref: _____

S No.	SPECIFICATION REQUIREMENTS	CONTRACTOR'S PROPOSAL	REMARKS
	SAMPLE FORM		

FIGURE 4.42 Specification comparison statement.

The contractor is required to prepare shop drawings taking into consideration the following as a minimum but not limited to:

1. Reference to contract drawings. This helps the consultant to compare and review the shop drawings relative to the contract drawings
2. Detailed plans and information based on the contract drawings
3. Notes of changes or alterations from the contract documents
4. Detailed information about fabrication or installation of works
5. All dimensions needed to verify at the jobsite
6. Identification of the product (material, equipment)
7. Installation information about the materials, equipment to be used in the construction
8. Type of finishes, color, and textures
9. Installation details relating to the axis or grid of the project
10. Roughing in and setting diagram
11. Coordination certification from all other related trades (subcontractors)

The contractor has to consider the following points at least, while developing shop drawings of different trades, to meet the design intents.

1. Review contract specification
2. Review contract drawings
3. Determine and verify field/site measurements
4. Installation information about the material to be used
5. Installation details relating to the axis or grid of the project

Management of Quality for Sustainability

6. Dimensions of the product, equipment to be installed in the project
7. Roughing in requirements
8. Coordination with other trade (discipline) requirements
9. Clearly marking the changes, deviations to the contract drawings

Figure 4.43 illustrates the shop drawing preparation and approval procedure.
Figure 4.44 illustrates the Site Transmittal Form for shop drawing approval.

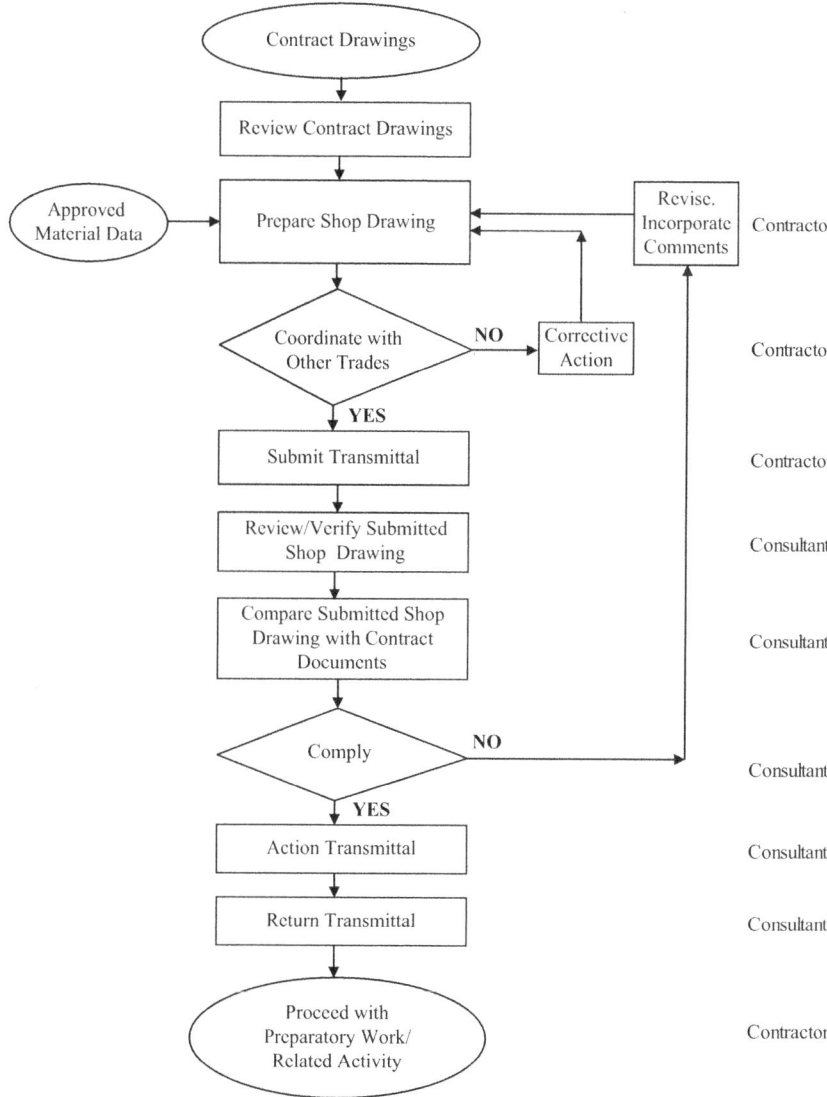

FIGURE 4.43 Shop drawing preparation and approval procedure.

Project Name
Consultant Name
SITE TRANSMITTAL
Request for Shop Drawings Approval

CONTRACT No.:　　　　　　　　　　TRANSMITTAL NO.　　　　REV.

CONTRACTOR :

TO :

WE REQUEST APPROVAL OF THE FOLLOWING ENCLOSED DRAWINGS

ITEM NO.	DWG., SPEC. OR BOQ. REF	DRAWING TITLE	DWG. NOS.	Rev.	SUBMITTAL CODE *	ACTION CODE **

N.B: We certify that above items have been reviewed in detail & and are correct & in strict performance with the Contract Drawings & Specification except as otherwise stated.

CONTRACTOR'S REP. :　　　　　　　　　　　　　　　DATE :

RECEIVED BY CONSULTANT :　　　　　　　　　　　　DATE :

cc: Owner Rep.

　　　　　Resident Engineer to enter ACTION CODE and REMARKS

R.E.'s REMARKS :

　　　　　　　　　　　　　　　　　　　　　　　Initials　　　　Date

Corrections or comments made relative to submittals during this review do not relieve Contractor from compliance with the requirements of the Drawings and Specifications. This check is only for review of general conformance with the design concept of the project and general compliance with the information given in the Contract Documents. Contractor is responsible for confirming and correlating all quantities and dimensions, selecting fabrication process and techniques of construction; coordinating his work with that of other trades, and performing his work in a safe and satisfactory manner.

Resident Engineer :　　　　　　　　　　　　　　　　DATE :

Received by Contractor :　　　　　　　　　　　　　　DATE :

cc: Owner Rep.

cc: Project Manager	** **ACTION CODE:**		
1: Submitted for Approval	A: Approved	C:	Not Approved
2: Submitted for your Information	B: Approved as Noted	D:	For Information
3:			

FIGURE 4.44 Site transmittal for work shop drawings.

Management of Quality for Sustainability

Immediately after approval of individual trade shop drawings, the contractor has to submit builders' workshop drawings and composite/coordinated shop drawings, taking into consideration the following drawings as a minimum.

4.6.9.1.2.1 Builders' Workshop Drawings

Builders' workshop drawings indicate the openings required in the civil or architectural work for services and other trades. These drawings indicate the size of openings, sleeves, level references with the help of detailed elevations and plans. Figure 4.45 illustrates the builders' workshop drawing preparation and approval procedure.

4.6.9.1.2.2 Composite /Coordination Shop Drawings

The composite drawings indicate the relationship among components shown on the related shop drawings and indicate the required installation sequence. Composite

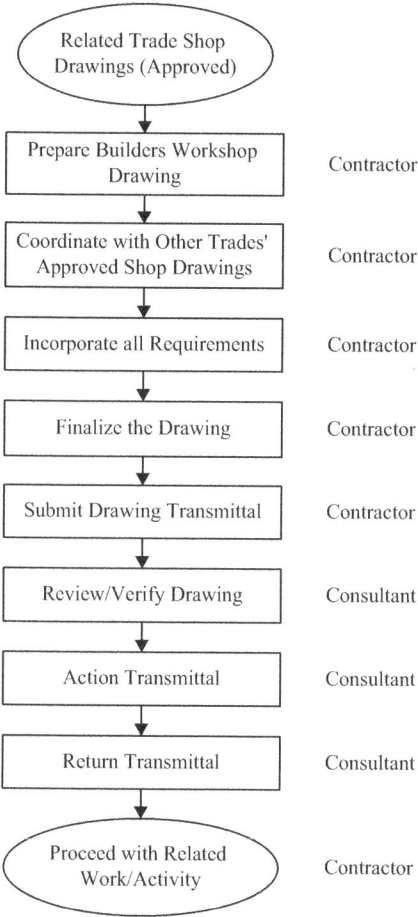

FIGURE 4.45 Builder's workshop drawing preparation and approval procedure.

drawings show the interrelationship among all services with each other and with the surrounding civil and architectural work. Composite drawings also show the detailed coordinated cross sections, elevations, reflected plans...etc., resolving all conflicts in levels, alignment, access, space, etc.... These drawings are to be prepared taking into consideration the actual physical dimensions required for installation within the available space. Figure 4.46 illustrates the composite drawing preparation and approval procedure.

4.6.9.1.2.3 Method Statement
The contractor has to execute the works as per the method statement specified in the contract. The contractor has to submit a method statement to the consultant for their approval as per the contract documents to ensure compliance with the contract

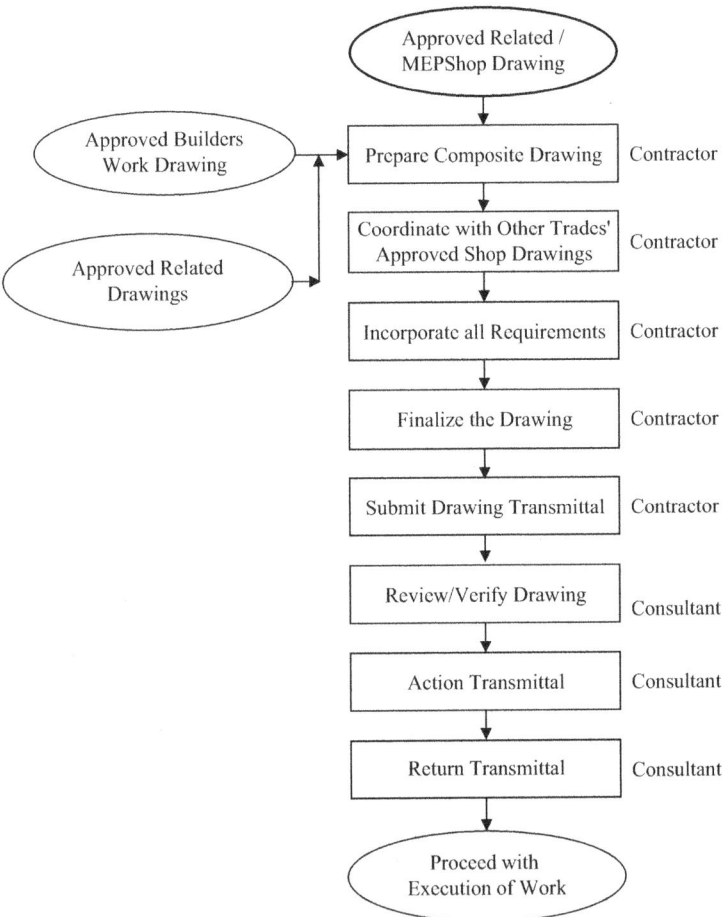

FIGURE 4.46 Composite/coordination shop drawing preparation and approval procedure.

Management of Quality for Sustainability

requirements. The method statement describes the steps involved in the execution/installation of work by ensuring safety at each stage. It has the following information:

1. Scope of work: Brief description of work/activity
2. Documentation: Relevant technical documents to undertake this work/activity
3. Personnel involved
4. Safety arrangement
5. Equipment and plant required
6. Personal Protective Equipment
7. Permits/authorities' approval to work
8. Possible hazards
9. Description of the work/activity: Detailed method of the sequence of each operation/key steps to complete the work/activity.

4.6.9.2 Execute the Project Works

The contractor is responsible for executing the contracted works in accordance with the contract drawings and specifications as specified in the contract documents. The contractor has to arrange necessary resources to complete the project within the schedule and the contracted amount. The contractor has to maintain the executed works until handing over the project to the owner/end-user and maintain it for an additional period, if contracted to do so. During the construction period, the contractor has to protect the executed/installed works to ensure that the works are not damaged. The contractor has to use new and approved material to construct the project/facility.

Construction activities mainly consist of the following:

1. Site work such as cleaning and excavation of project site
2. Site earth works
3. Dewatering and shoring
4. Excavation and backfilling
5. Construction of the foundation and platform for equipment, machinery
6. Construction of beams and columns
7. Forming, reinforcing, and placing the floor slab
8. Laying up masonry walls and partitions
9. Installation of roofing system
10. Finishes
11. Furnishings
12. Elevator and conveying system
13. Installation of fire suppression system
14. Installation of water supply, plumbing, and public health system
15. Drainage system
16. Installation of heating, ventilating, and air conditioning system
17. Equipment and machinery installation
18. Instrumentation works

19. Electrical works
20. Installation of electrical lighting, daylighting, sensor-controlled lighting
21. Electrical power system
22. Emergency power supply system
23. Integrated automation system
24. Fire alarm system works
25. Information and Communication Technology system
26. Electronic security and access control system
27. Low-voltage systems works
28. Landscape works
29. External works
30. Waste treatment works
31. Pollution control system

4.6.9.2.1 Manage the Construction Quality

The construction project quality control process is a part of the contract documents which provide details about specific quality practices, resources, and activities relevant to the project. The purpose of quality control during construction is to ensure that the work is accomplished in accordance with the requirements specified in the contract. Inspection of the construction works is carried out throughout the construction period either by the construction supervision team (consultant) or an appointed inspector agency. Quality is an important aspect of the construction project. The quality of the construction project must meet the requirements specified in the contract documents. Normally, the contractor provides onsite inspection and testing facilities at the construction site. On a construction site, inspection and testing are carried out at three stages during the construction period to ensure quality compliance.

1. During the construction process. This is carried out with the checklist request submitted by the contractor for the testing of ongoing works before proceeding to the next step.
2. Receipt of the subcontracted or purchased material, equipment, or services. This is performed by a Material Inspection Request (MIR) submitted by the contractor to the supervision consultant upon receipt of the material.
3. Before final delivery or commissioning and handover.

Quality management in construction is a management function. In general, quality assurance and control programs are used to monitor design and construction conformance to established requirements as determined by the contract specifications. Instituting quality management programs reduces costs while producing the specified facility.

In order to achieve quality in the construction projects, all the works have to be executed as per approved shop drawings, using approved materials and fully coordinating with different trades. Proper sequencing and method statement for the installation of work must be followed by the contractor to avoid rejection/rework. Figure 4.47 illustrates the sequence of execution of work.

Management of Quality for Sustainability

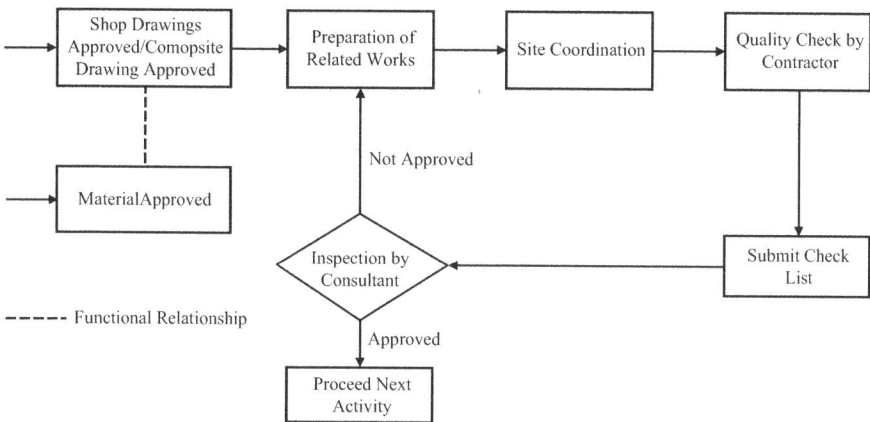

FIGURE 4.47 Sequence of execution of works. Source: Abdul Razzak Rumane. (2017). *Quality Management in Construction Projects*, Second Edition. Reprinted with permission from Taylor & Francis Group Company.

Following this sequence will help the contractor to avoid the rejection of works. Rejection of the checklist will result in rework, which will need time to redo the works and have cost implications for the contractor. Frequent rejection of works may delay the project, ultimately affecting the overall completion schedule.

In order to ensure that structural concrete works are executed without any defects or rejection and achieve the concrete strength specified, the proper sequencing of works is important. Figure 4.48 illustrates the work sequence for formwork and Figure 4.49 illustrates the process for concrete casting.

4.6.10 Monitoring and Controlling

Monitoring and controlling of project works depends on the type of project delivery system. In major projects, the owner engages a construction/project management firm that is responsible for supervising, monitoring, and controlling construction activities performed by the contractor. However, in most cases, supervision of the project during the construction phase is carried out by the consultant who is involved in designing the project. The resident engineer (RE), along with supervision team members, is responsible for supervising, monitoring, and controlling, implementing the procedure specified in the contract documents, and ensuring completion of the project within the specified time, budget, and per the defined scope of work. In order to ensure the smooth flow of supervision activities, the RE has to follow the organization's supervision manual and contractual requirements. Table 4.46 illustrates an example checklist, listing the items to be verified by the RE to ensure availability of all the necessary documents, information and to facilitate the smooth flow of the supervision work.

Monitoring and controlling the project is an ongoing process. It starts from the inception of the project and continues to handover of the project. Monitoring and

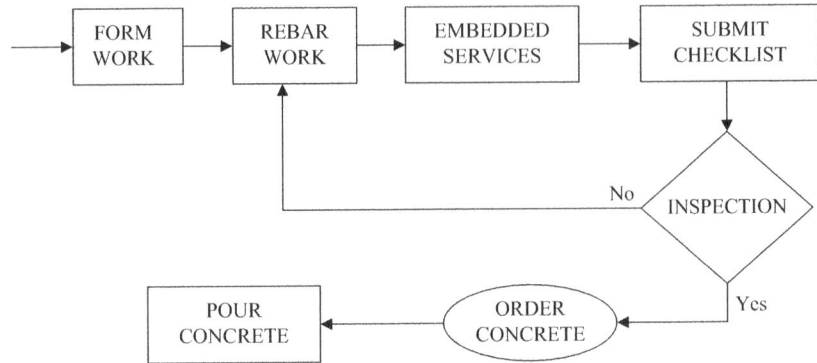

FIGURE 4.48 Flowchart for concrete casting. Source: Abdul Razzak Rumane. (2017). *Quality Management in Construction Projects*. Second Edition. Reprinted with permission of Taylor & Francis Group.

control of the construction project are operative during the execution of the project and its aim is to recognize any obstacles encountered during execution and to apply measures to mitigate these difficulties and to ensure that the goals and objectives of the project are being met.

Monitoring is collecting, recording, and reporting information concerning any and all aspects of project performance that the project manager or others in the organization need to know. Monitoring of the construction project is normally done by collecting and recording the status of various activities and compiling them in the form of progress reports. These are prepared by the consultant and contractor and distributed to the concerned members of project team.

Monitoring involves not only tracking time but also budget, quality, resources, and risk. Monitoring in construction projects is normally done by compiling the status of various activities in the form of progress reports. These are prepared by the contractor, the supervision team (consultant), and the construction/project management team. The objectives of project monitoring and control are:

1. To report the necessary information in detail and in an appropriate form which can be interpreted by management and other concerned personnel to provide information about how the resources are being used to achieve the project objectives.
2. To provide an organized and efficient means of measuring, collecting, verifying, and quantifying data reflecting the progress and status of execution of project activities, with respect to schedule, cost, resources, procurement, and quality.
3. To provide an organized, efficient, and accurate means of converting the data from the execution process into information.
4. To identify and isolate the most important and critical information about the project activities to enable decision-making personnel to take corrective action for the benefit of the project.
5. To forecast and predict the future progress of activities.

Management of Quality for Sustainability

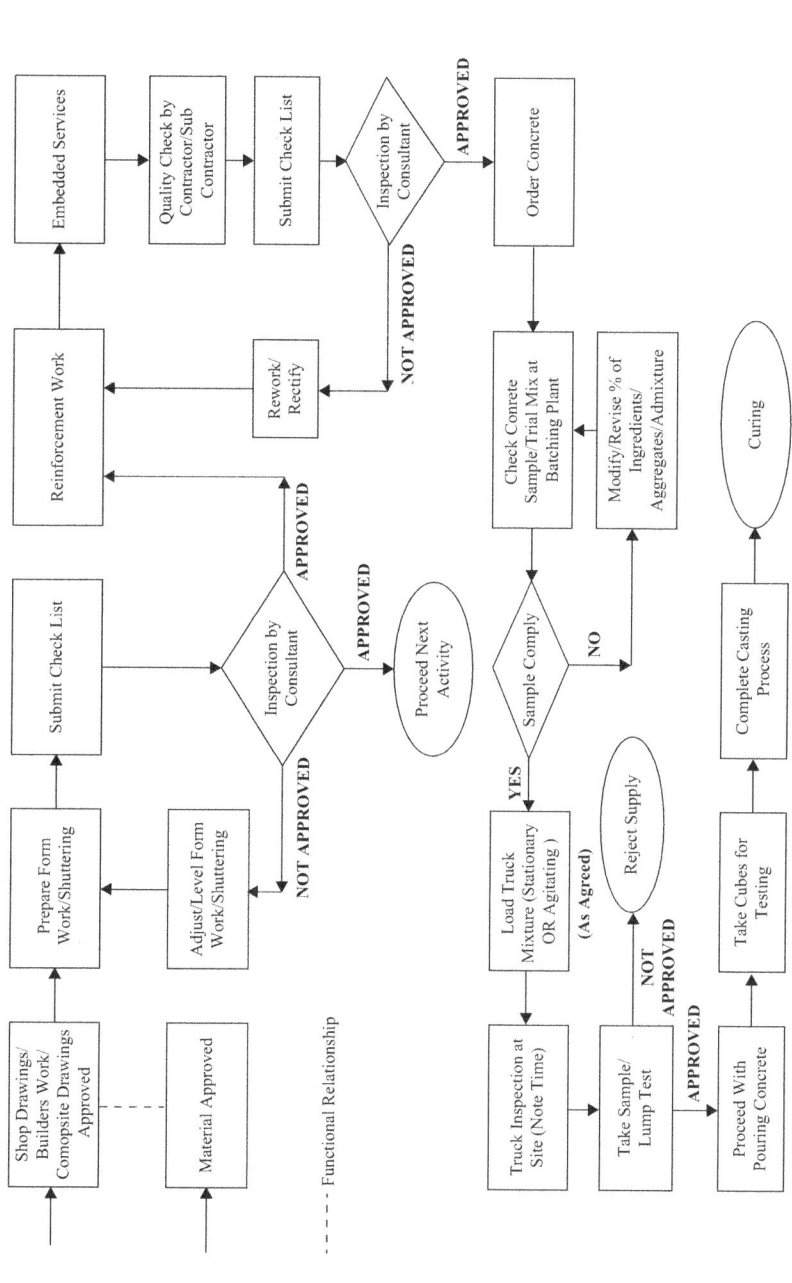

FIGURE 4.49 Process for structural concrete work. Source: Abdul Razzak Rumane. (2013). *Quality Tools for Managing Construction Projects*. Reprinted with permission of Taylor & Francis Group.

TABLE 4.46
Consultant's Checklist for Smooth Functioning of a Project

Serial Number		Items to be Checked/Verified
I	**Project Details**	
	I.1	Scope of work
	I.2	Project objectives
	I.3	Project deliverables
II	**Project Organization**	
	II.1	Organization chart and roles and responsibilities of defined supervision staff
	II.2	Supervision staff deployment matching with project requirements
	II.3	Contractor's staff deployment plan approved as per contract requirements
	II.4	Responsibility matrix prepared and approved by the client and distributed among all project parties
	II-5	Project directory
III	**Mobilization**	
	III.1	Site permit from authorities available
	III.2	Project plot boundaries are marked as per the permit
	III.3	Project commencement order issued
	III.4	Copy of permit issued to the contractor
	III.5	Temporary site offices drawings approved
	III.6	Temporary firefighting plan approved by respective authority
	III.7	Copies of contractor's performance bond, guarantees, insurance policies, and licences available at site
	III.8	Copies of consultant's performance bond, guarantees, insurance policies and licences available at site
	III.9	Preconstruction meeting conducted and submittal and approval procedures discussed and agreed
IV	**Project Administration**	
	III-1 Contract Documents	
	IV-1.1	Signed copy of contract between owner and contractor available at site
	IV-1.2	Copies of contract documents available at site
	IV-1.3	Contracted Bill of Quantity (BOQ) is available
	IV-1.4	All volumes of Particular Specifications available
	IV-1.5	Contracted drawings are available
	IV-1.6	Authority-approved drawings, duly stamped, are available
	IV-1.7	Addendum, if any, to the contract are available
	IV-1.8	Replies to tender queries are available
	IV-1.9	Copies of signed contract documents and drawings handed over to contractor are available, who has acknowledged the same
	IV-1.10	Log for Codes and Standards available
	IV-2 Document Management	
	IV-2.1	Document Control System is in place
	IV-2.2	Filing Index is available
	IV-2.3	Material Submittal Log is available
	IV-2.4	Shop Drawing Submittal Log is available

(Continued)

TABLE 4.46 CONTINUED
Consultant's Checklist for Smooth Functioning of a Project

Serial Number		Items to be Checked/Verified
	IV-2.5	Logs for Correspondence between various parties are available
	IV-2.6	Log for Checklist (Request for Inspection) is available
	IV-2.7	Log for JSI (Job Site Instruction) is available
	IV-2.8	Log for SWI (Site Work Instruction) is available
	IV-2.9	Log for RFI (Request for Information) is available
	IV-2.10	Log for VO (Variation Order) is available
	IV-2.11	Log for NCR (Non-Conformance Report) is available
	IV-2.12	Material sample log and place are identified
	IV-2.13	Log for Equipment test certificate is available
	IV-2.14	Log for Visitor's at site
	IV-2.15	Contractor's staff approval log is in place
	IV-2.16	Subcontractor's approval log is in place
	IV-2.17	Consultants staff approval is in place
	IV-2.18	Overtime request log available
V	**Communication**	
	V-1	Communication matrix established and agreed by all the parties
	V-2	Distribution system for transmittals/submittals agreed
VI	**Project Monitoring and Control**	
	VI-1	Daily report log in place
	VI-2	Weekly report log in place
	VI-3	Monthly report log in place
	VI-4	Progress meetings log in place
	VI-5	Minutes of meetings log in place
	VI-6	Progress payment log in place
	VI-7	Construction schedule log in place
VII	**Construction**	
	VII-1	Quality control plan log in place
	VII-2	Safety management plan log in place
	VII-3	Risk management plan log in place
	VII-4	Method statement submittal log in place
	VII-5	Accident and fire report
	VII-6	Off-site inspection visits
	VII-7	Location of gathering point established
VIII	**General**	
	VIII-1	Correspondence between site and head office
	VIII-2	Staff-related matters
	VIII-3	Copy of supervision manual available
	VIII-4	Emergency contact telephones and contact details displayed at site

Source: Abdul Razzak Rumane (2013). Quality Tools for Managing Construction Projects. Reprinted with permission of Taylor & Francis Group.

Figure 4.50 illustrates a logic flow diagram for monitoring and controlling process and Table 4.47 illustrates monitoring and controlling references for construction projects.

Project monitoring and controlling involves a regular comparison of performance against targets, a search for the cause of any deviation, and a commitment to check for adverse variance. It serves two major functions:

1. It ensures regular monitoring of performance
2. It motivates project stakeholders to strive to achieve project objectives

Construction project control is exercised through knowing where to put in the main effort at a given time and maintaining good communication. There are mainly three areas where project control is required:

1. Quality (scope)
2. Schedule
3. Budget.

All of these areas are to be kept in balance to achieve the project objectives. In order to accomplish the project objectives in construction projects and to achieve a successful project, the supervision consultant/contractor must monitor and control all of the major management processes and handle key attributes effectively and efficiently, tracking the progress of the work from inception to completion of the construction and handover of the project. For successful completion of a project, monitoring and controlling of the major management processes of a project are discussed below.

4.6.10.1 Integration Management

Integration management is coordination and implementation of five project management process groups (initiating, planning, executing, monitoring, and controlling) from the time the project is conceived right to the closeout stage. Integration management involves putting together all of the process groups.

Integration management of a construction project includes all the activities performed to effectively control the final output of project production (facility), and the input of the process is the owner's need for the construction project. To achieve the adequacy of the client brief which addresses the numerous complex client/user needs, it is necessary to monitor and control all of the related activities.

The supervision consultant/contractor has to monitor and control the following activities to ensure that these activities are performed in order to achieve a successful project that satisfies the owner's/end-user's needs.

1. Mobilization requirements
2. Development of project execution scope
    - The tender documents and all the requirements for execution of project are reviewed
    - Review of the contract documents

Management of Quality for Sustainability

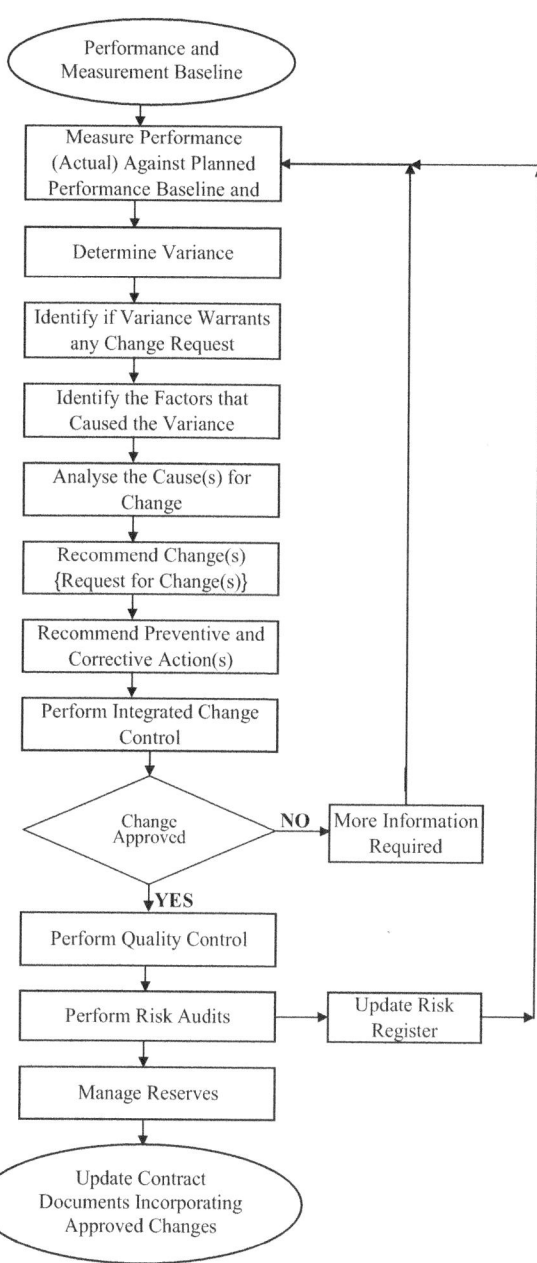

FIGURE 4.50 Logic flowchart for monitoring and controlling process. Source: Abdul Razzak Rumane. (2013). *Quality Tools for Managing Construction Projects*. Reprinted with permission of Taylor & Francis Group.

TABLE 4.47
Monitoring and Controlling Plan References for Construction Projects

Serial Number	Elements	Contract Reference	Contractor Reference
1	Performance Baseline	Specifications, drawings	1. Approved materials 2. Approved shop drawings 3. Approved composite drawings
		Schedule	1. Contractor's construction schedule
		Cost	1. Approved S-Curve
2	Data Collection Methods	Reports	1. Daily report 2. Weekly report 3. Monthly report 4. Safety report 8. Risk report 5. Accident report 6. Checklist 7. Risk report
3	Frequency of Data Collection	1. Daily 2. Weekly 3. Monthly	1. Daily report 2. Weekly report 3. Monthly report
4	Status Information Collection	1. Logs 2. Reports 3. Meetings 4. Checklists	1. Logs 2. Reports 3. Meetings 4. Checklists
5	Comparison Between Planned and Actual (Variance)	1. S-Curves 2. Milestones	1. Progress reports 2. Progress payment 3. Milestones
6	Analysis	1. Price analysis	1. Construction schedule attachment 2. Progress payment
7	Corrective Action	1. Comments by consultant	1. Incorporate comments
8	Change Order	1. Variation order	1. Request for information 2. Request for modification 3. Request for variation
9	Document Updates	1. On a regular basis	1. Reports

Source: Abdul Razzak Rumane (2013). Quality Tools for Managing Construction Projects. Reprinted with permission of Taylor & Francis Group.

Management of Quality for Sustainability

- Review of construction drawings
- Review of specifications

3. Sustainability requirements
4. Project execution plan is established
5. All the management plans are developed and approved
6. The construction/execution process and method statement is established
7. Change management – monitor and record all the changes during the execution of the project

4.6.10.2 Stakeholders Management

In order to run a successful project, it is important to address the needs of the project stakeholders, effectively predicting how the project will be affected and how the stakeholders will be affected.

In order to manage the stakeholder's expectations in construction projects, the following construction-related activities have to be monitored and controlled:

1. The responsibilities matrix is properly followed
2. To ensure that all the related information and reports are properly distributed to concerned stakeholders, the following are the major activities to be controlled and the information distributed and reported to the respective stakeholders:
   - Project status/performance
   - Changes in scope, schedule, and budget
   - Project status
   - Project updates
   - Project-related issues
   - Project payments
   - Project conflicts
   - Change orders
   - Conflicts
   - Variation orders
   - Anticipated/forecasted problems
   - Minutes of meetings

4.6.10.3 Scope Management

In construction projects, scope management is the process which includes the activities to formulate and define the client's need, by establishing project objectives and goals properly in order for the project to have clear direction, and controlling what is or is not involved in the project. The project scope documents explain the boundaries of the project, establish the project responsibilities for each team member and set up procedures for how completed works will be verified and approved. The scope describes the features and functions of the end-product or the services to be provided by the project. During the project, the scope documentation helps the project team remain focused and on-task. The scope statement also provides the project team with guidelines for making decisions about change requests during the project. It is essential that the scope statement should be unambiguous and clearly written to enable all the members of project team to understand the project scope to achieve the project objectives and goals.

4.6.10.3.1 Validate Scope

To validate scope is the process of formalizing acceptance of completed project deliverables. It is a method to ensure that:

- Project design conforms to the current applicable codes and standards
- Project design takes care of all the requirements listed under the Terms of Reference
- Value Engineering analysis has been performed and the recommendations are incorporated in the project baseline
- All the material and equipment installed comply with the specification requirements
- All the works on site are performed per the approved shop drawings of the approved material
- Regulatory approvals are obtained
- Installed/executed works are checked, inspected at every stage to confirm that they have been installed/executed as specified, using specified and approved materials and installation methods recommended by the manufacturer, to meet the intended use of the project
- Corrective actions or defect repairs to be completed
- Inspection and tests are carried out to ascertain operational requirement
- The work is documented and changes are recorded
- Records of inspections and tests are maintained to verify that approved construction methods and materials were used
- Outstanding defects, works (punch list) are listed and documented
- As built drawings, documents, manuals are ready for handover
- Start-up test plans are established to demonstrate that all the systems installed in the project meet the required operations and safety requirements
- Handover/takeover program is established
- All the requirements for facility management are documented, including all the information and knowledge that is required to strategically and physically manage the new facility

In a construction project, the project elements (intermediate deliverables) need to be verified, reviewed, approved, and accepted at different stages of the project life cycle to ensure that completed project deliverables meet the owner's needs and expectations (goals and objectives). It is the assessment of readiness of the construction or execution and to confirm the completeness and accuracy of the project as per the agreed-upon scope baseline.

It is performed mainly during the following phases of the construction project:

1. Design phase
2. Construction phase
3. Testing, commissioning, and handover phase

Management of Quality for Sustainability

The following are the four main steps needed to establish an assessment, review, approve, and accept the project:

1. Identify the elements, items, material, system, and products to be reviewed and checked in each of the trades (architectural, structural, fire suppression, plumbing, HVAC, electrical, low-voltage systems, landscape, external, etc.)
2. Identify the frequency of checking, inspection
3. Identify the agency/persons authorized to check and approve
4. Identify the indications/criteria for performance monitoring and control

Figure 4.51 illustrates the scope validation process for the construction project.

4.6.10.3.2 Control Scope

Control scope is the process of monitoring the project scope and managing any changes to the scope baseline. It is common that, despite all the efforts devoted to developing the contract documents (scope baseline), the contract documents cannot provide complete information about every possible condition or circumstance that the construction team may encounter. Figure 4.52 illustrates the scope control process in a construction project.

4.6.10.3.3 Manage Scope Changes

It is common that, during the construction process, there will be some changes to the original contract. Even under the most ideal circumstances, contract documents cannot provide complete information about every possible condition or circumstance that the construction team may encounter. The causes for changes in the construction project are listed in Table 4.48.

These changes help to construct the project to achieve the project objective. These changes are identified as the construction proceeds. Prompt identification of such requirements helps both the owner and contractor to avoid unnecessary disruption of work and its impact on cost and time.

The contractor uses the Request for Information (RFI) form to request technical information from the supervision team. These queries are normally resolved by the supervision engineer concerned. However, it is likely that the matter has to be referred to the designer as Request for Information (RFI) has many other considerations to be taken care of, which may be beyond the capacity of the supervision team member to resolve. Normally, there is a defined period to respond to RFI. Such queries may result in variations to the contract documents. It is in the interest of both the owner and contractor to resolve an RFI expeditiously to avoid its effect on the construction schedule.

Figure 4.53 illustrates the Request for Information (RFI) Form which the contractor submits to the consultant to clarify differences/errors observed in the contract documents, a change in the construction methodology, a change in the specified material, ... etc., and Figure 4.54 illustrates the Process to Resolve Request for Variation (Contractor Initiated).

Figure 4.55 illustrates the Process to Resolve Scope Change (Owner Initiated).

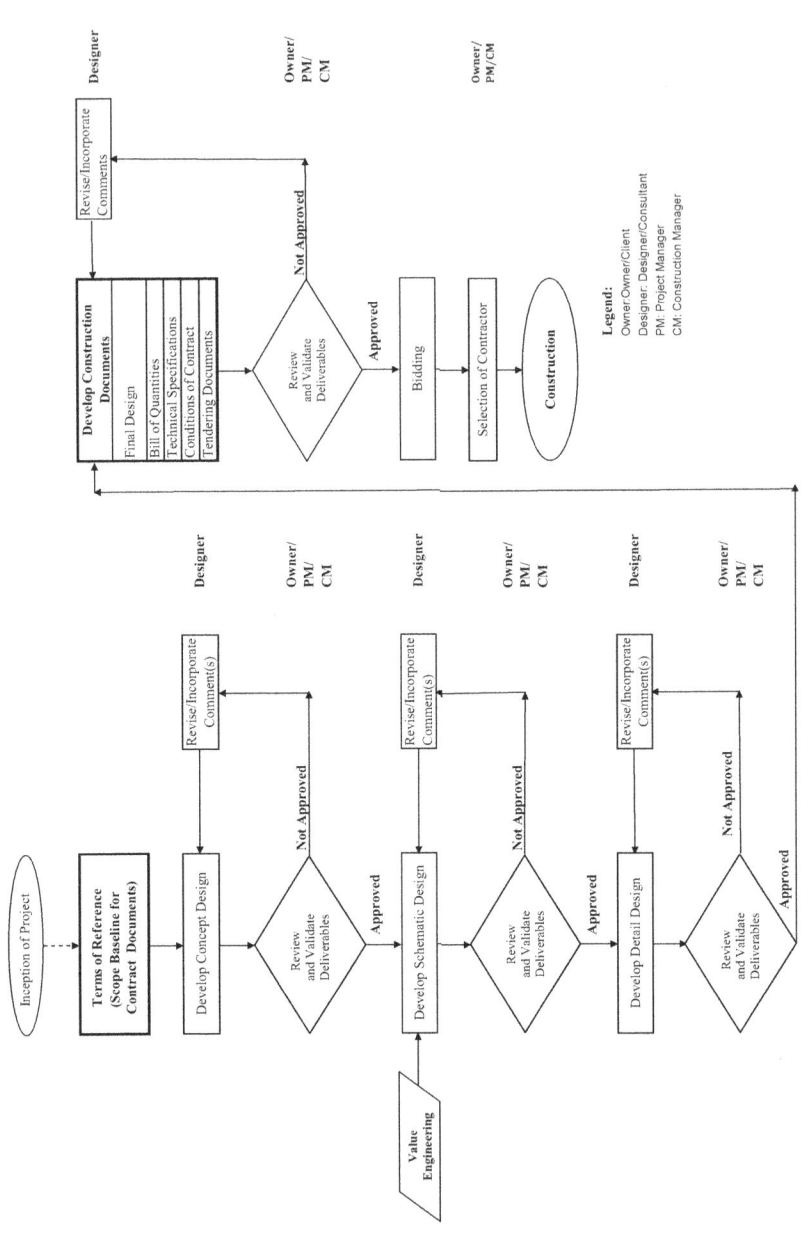

FIGURE 4.51 Scope validation process for construction project (design and bidding stage).

Management of Quality for Sustainability

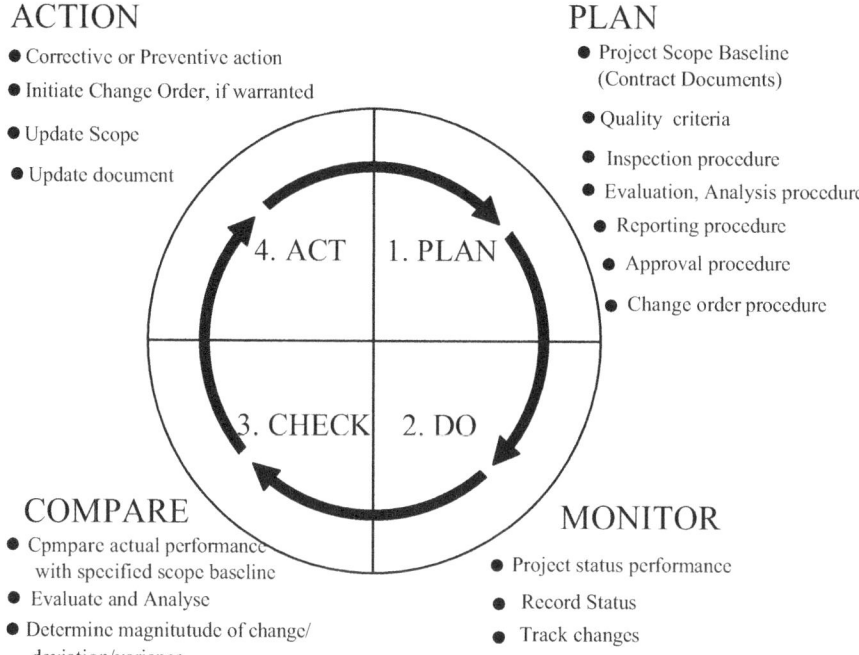

FIGURE 4.52 Scope control process.

Figure 4.38, discussed earlier, illustrates a Site Work Instruction (SWI) Form. It gives instructions to the contractor to proceed with the change(s). All the necessary documents are sent along with the SWI to the contractor. A SWI is also used to instruct the contractor for owner-initiated changes.

4.6.10.3.3.1 Resolve Conflict
It is essential that changes in the project are managed as quickly as possible in accordance with the conditions of the contract. However, disputes and conflicts in construction projects are inevitable due to the fact that, with all the precautionary steps, the discrepancies or errors do occur in the contract documents. The following methods are normally carried out to resolve any conflict:

- Negotiation
  - The economical method to resolve the conflict is negotiation. Negotiation involves compromise. Both parties should discuss the issue by arranging meetings of all the project team members involved and whose input to the issue will help resolve the conflict. The issue should be analyzed with the help of related documents, substantiation for claim, and justification for claim. If agreement is not reached, then senior representatives from both the parties should be involved.

TABLE 4.48
Causes of Changes in a Construction Project

Serial Number		Causes
I	**Owner**	
	I-1	Delay in making the site available on time
	I-2	Change of plans
	I-3	Financial problems/payment delays
	I-4	Change of schedule
	I-5	Addition of work
	I-6	Omission of work
	I-7	Project objectives are changed/modified
	I-8	Different site conditions
	I-9	Value engineering
II	**Contractor**	
	II-1	Process/methodology
	II-2	Substitution of material
	II-3	Non-availability of specified/approved material
	II-4	Non-availability of equipment, machinery
	II-5	Material not meeting the specifications
	II-6	Charges payable to an outside party due to cancellation of certain items/products
	II-7	Delay in approval
	II-8	Contractor's financial difficulties
	II-9	Non-availability of manpower
	II-10	Workmanship not up to the mark
IV	**Miscellaneous**	
	III-1	New regulations
	III-2	Safety considerations
	III-3	Weather conditions
	III-4	Unforeseen circumstances
	III-5	Inflation
	III-6	Fluctuation in exchange rate
	III-7	Government policies

- Mediation
  - Mediation is a process in which a neutral third-party group is involved to assist the parties to a dispute in reaching an amicable agreement that resolves the conflict.
- Arbitration
  - Arbitration is the voluntary submission of a dispute to one or more impartial persons for final and binding determination. There are certain agencies that certify arbitrators.
- Litigation
  - Litigation means to apply to the court to resolve the dispute.

Management of Quality for Sustainability

FIGURE 4.53 Request for information (RFI).

4.6.10.4 Schedule Management

Scheduling is the mechanical process of formalizing the planned functions, assigning the starting and completion dates to each part or activity of the work so that the whole work proceeds in a logical sequence and in an orderly and systematic manner. Scheduling is the process of determining the sequential order of the planned

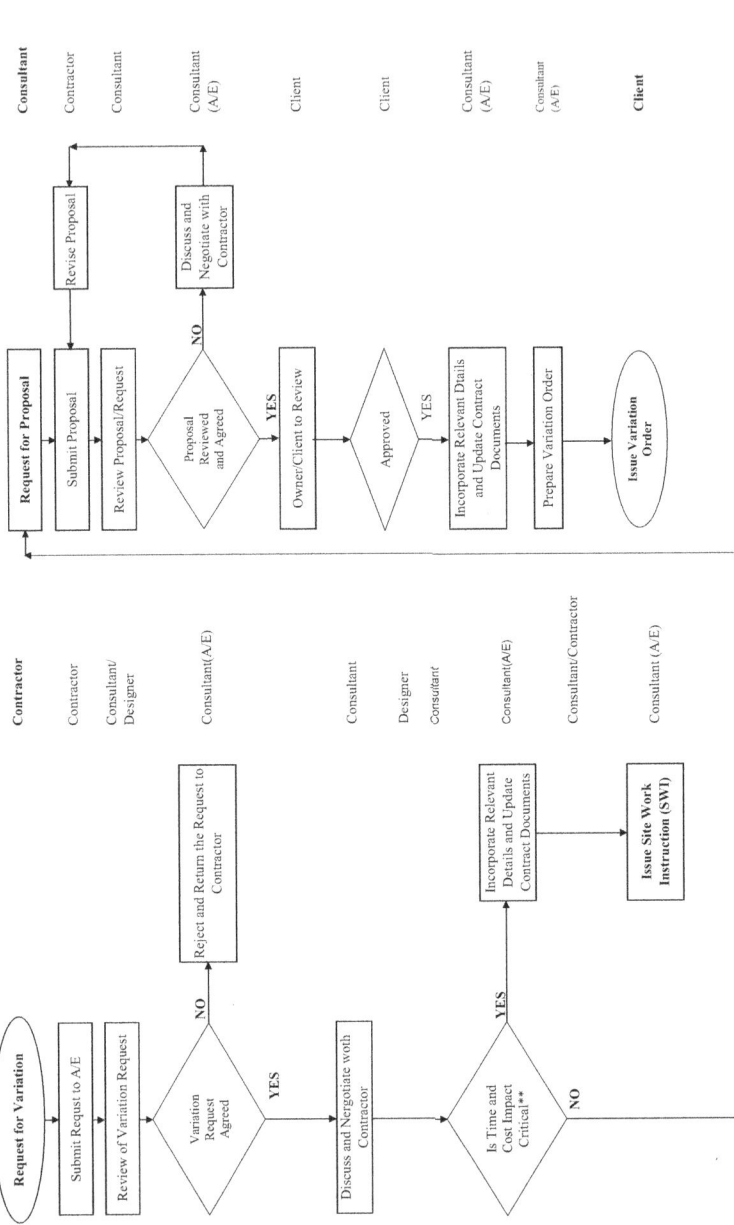

FIGURE 4.54 Process to resolve request for variation. Source: Abdul Razzak Rumane. (2016). *Handbook of Construction Management: Scope, Schedule, and Cost Control*. Reprinted with permission of Taylor & Francis Group **Critical: Delay decision will have negative Impact. No alternative available. Variations do not increase anticipated final cost.

Management of Quality for Sustainability 345

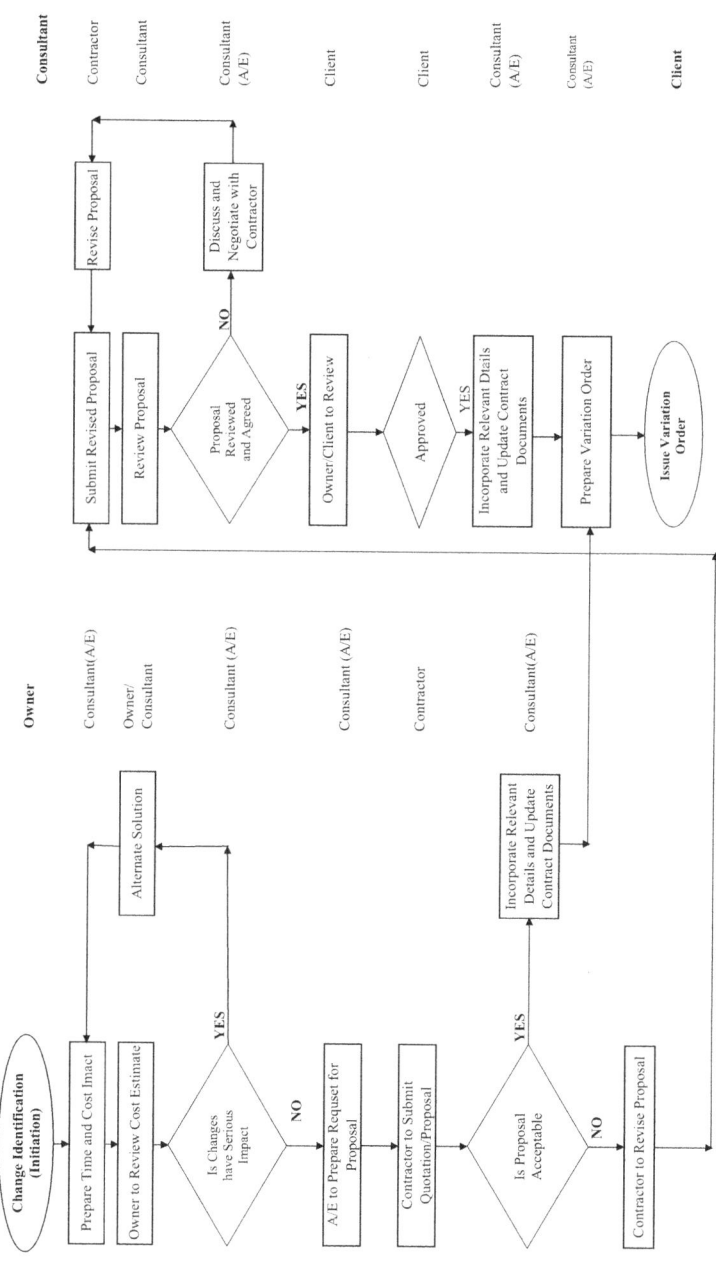

FIGURE 4.55 Process to resolve scope change (owner initiated). Source: Abdul Razzak Rumane. (2016). *Handbook of Construction Management: Scope, Schedule, and Cost Control*. Reprinted with permission of Taylor & Francis Group.

activities and the time required to complete each activity. Scheduling is a time-based graphical presentation of project activities/tasks, utilizing information on available resources and time constraints.

4.6.10.4.1 Monitoring and Controlling Schedule

Monitoring and Controlling Schedule is the process used to determine the current status of the schedule, identify the influencing factors that causes schedule changes, determine that the schedule has changed, and manage the changes in the approved project schedule baseline by updating and taking appropriate actions, if necessary, to minimize deviation from the approved schedule.

Monitoring is collecting, recording, and reporting information concerning project performance. Monitoring involves measurement of the current status of the project accomplishment and performance.

Monitoring in construction projects is normally done by compiling the status of various activities in the form of progress reports. These are prepared by the contractor, supervision team (consultant), and construction/project management team. The objectives of project monitoring and control are:

1. To report the necessary information in detail and in an appropriate form which can be interpreted by management and other concerned personnel to provide them with the information about how the resources are being used to achieve the project objectives.
2. To provide an organized and efficient means of measuring, collecting, verifying, and quantifying data reflecting the progress and status of execution of the project activities, with respect to schedule, cost, resources, procurement, and quality.
3. To provide an organized, efficient, and accurate means of converting the data from the execution process into information.
4. To identify and isolate the most important and critical information about the project activities to enable decision-making personnel to take corrective action for the benefit of the project.
5. To forecast and predict the future progress of activities to be performed.

Controlling is using the actual data collected through monitoring and comparing the same to the planned performance to bring the actual performance in line with the planned performance by correcting the variances or implementing approved changes. Analysis of variance between the baseline and the current schedule dates and duration provides necessary information for the decision of management and stakeholders. The control process is established for managing the current schedule.

The following information is required to prepare the current (as-built) schedule and compare it with the baseline schedule:

- Percentage completion of each activity based on the approved checklist
- Actual start/finish dates for the completed activities
- Activities scheduled to start but not yet started

Management of Quality for Sustainability

- Activities scheduled to complete but in progress
- Remaining duration to complete each activity
- Percentage completed activities
- Percentage partially completed activities
- Percentage not yet started
- Material, equipment yet to be received
- Available resources
- Regulatory approvals
- Milestones not yet reached
- Logic and duration revision to keep the schedule unchanged
- Problems and issues
- Risks
- Change orders

After analyzing the current status with the actual schedule, the schedule performance report is prepared consisting of the following information:

- Project status: where the schedule stands currently
- Project progress: plan versus actual
- Forecasting: prediction of future status and progress tend

Figure 4.56 illustrates the Schedule Monitoring and Controlling Process.

4.6.10.4.2 Monitoring of Work Progress

Work progress is normally monitored through daily and monthly progress reports. A monthly progress report consists of progress photographs to document the physical progress of work. These photographs are used to compare compliance with the planned activities and actual performance. Figure 4.57 illustrates the Traditional Monitoring System.

With the advent of technology, it is possible to monitor and evaluate construction activities using cameras and related software technologies. In this process, digital images are captured through the use of cameras. These photographs are processed

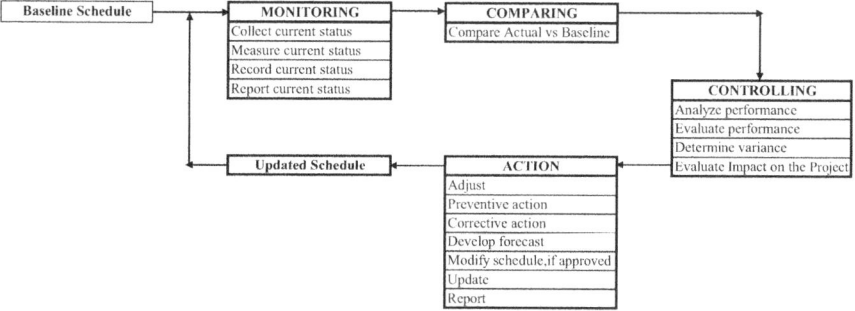

FIGURE 4.56 Schedule monitoring and controlling process.

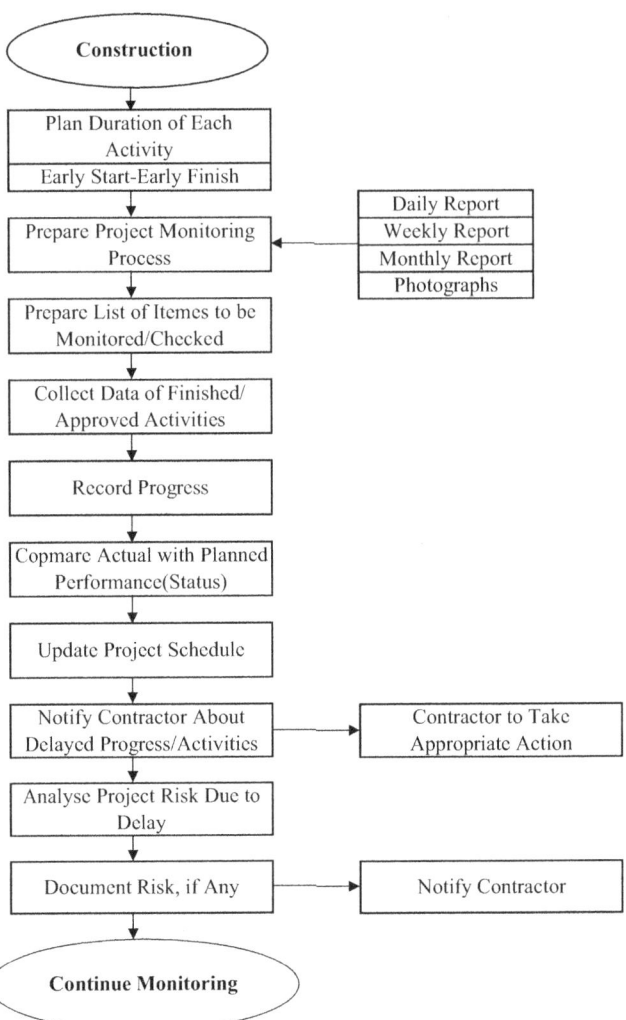

FIGURE 4.57 Traditional monitoring system.

using photo modeling software to develop a 3D model view of the digital picture captured from the site. The captured As-Built data is compared with the planned activities by interfacing through an Integrated Information Modeling System. The use of this system:

- Improves the accuracy of the information
- Avoids delays in getting the information
- Improves communication among all parties
- Improves effective control of the project
- Improves document recording
- Helps reduce claims

Management of Quality for Sustainability

Figure 4.58 illustrates a schematic for the Digitized Monitoring System.

The contractor's approved construction schedule is the performance baseline for construction projects, which is achieved by collecting information through different methods.

4.6.10.4.3 Logs

There are various types of logs used in construction projects to monitor and control construction activities. The main logs used in a construction project are as follows:

1. Subcontractors Submittal & Approval Log
2. Submittal Status Log
3. Shop Drawings and Materials Logs – E1
4. Procurement Log – E2
5. Manpower Log
6. Equipment Log

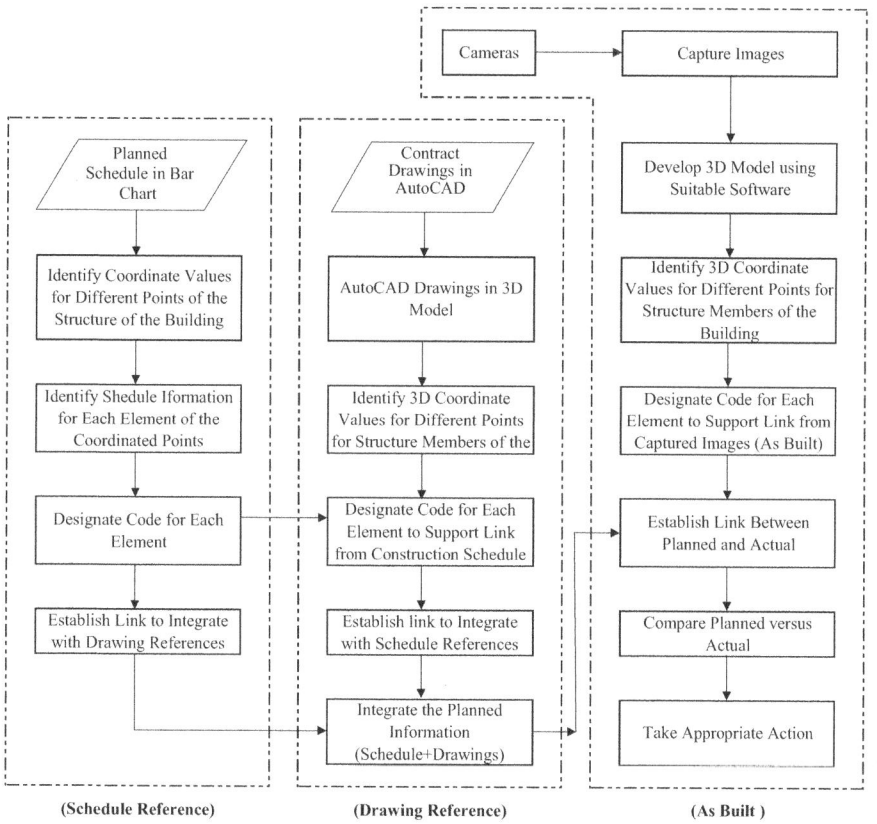

FIGURE 4.58 Digitized progress monitoring. Source: Abdul Razzak Rumane. (2013). *Quality Tools for Managing Construction Projects.* Reprinted with permission of Taylor & Francis Group.

These logs provide necessary information about the status of subcontractors, materials, shop drawings, procurement, and availability of contractor's resources, and help to determine its effects on the project schedule and project completion.

4.6.10.4.4 Reports

Apart from the different types of logs and submittals, progress curves, and time control charts, the contractor's progress is monitored through various types of reports and meetings. These are:

1. Daily reports
2. Weekly reports
3. Monthly reports

The contractor's daily progress is monitored through daily progress reports submitted by the contractor on the morning of the working day following the day to which the report relates. It gives the status of all the resources available at the site for that particular day. It shows the details of contractor's staff and manpower, contractor's plant and equipment, and material received at site. Details of the subcontractor's work and resources are also included in the report; along with the daily report, the contractor submits work in the progress report.

Monthly reports, giving details of all the site activities along with photographs, are submitted by the contractor to the consultant/owner for their information to know the progress of work during the month. Table 4.49 illustrates the contents of the contractor's monthly progress report and Table 4.50 illustrates the contents of the consultant's monthly progress report.

In addition to the reports listed above, the contractor submits the following reports to monitor the related activities:

1. Risk report
2. Safety report

4.6.10.4.5 Meetings

Progress meetings are conducted at an agreed-upon interval to review the progress of works and discuss any problems to achieve smooth progress of the construction activities. The contractor supplies a pre-meeting submittal to the project manager/consultant normally two days in advance of the scheduled meeting date. The submittal consists of:

1. List of completed activities
2. List of current activities
3. Two weeks look-ahead
4. Critical activities
5. Materials submittal log
6. Shop drawings submittal log
7. Procurement log

TABLE 4.49
Monthly Progress Report — EXAMPLE CONTENTS

Contents of Monthly Progress Report (Contractor) (Sample Contents)

Serial Number	Contents		Description
1	**Executive Summary (Tabular)**		
	1.1	Summary Status Report	Brief description of the project status up to date, i.e., manpower, cash, activities
	1.2	NOC's Report	No objection certificate report
	1.3	Project Manager Narratives	Narrative description of project status up to date
2	**Progress Layouts**		
	2.1	Updated Milestone Table	Comparison of planned v. actual for contractual Milestones per construction unit /design, You track the delays for major trades through the color theme
	2.2	Updated Major of Events Table	Comparison of planned v. actual for major trades per construction unit/design. Track the delays for major trades through the color theme
	2.3	Updated Layouts	Same as **2.1** above but presentation per milestone phase (drawing)
3	**Updated Execution Program.**		
	3.1	Updated Milestone Schedule – Roll-up "Update versus latest Target"	To indicate the status of the control and key milestones, comparing the current status with the baseline
	3.2	Updated Detailed Schedule	All activities in details
	3.3	One Month Look-Ahead Program	Same as **3.2**, but includes only the detailed activities for the coming month BUT ON EXCEL FORMAT
4	**Submittal Status Report (E1 Log).**		Updated status of submittals
	4.1	Submittal Status Report	Briefly, describe the project submittal status up to date
	4.2	Detailed E1 Log	Describe the project submittal status up to date in detail
5	**Procurement Status Report (E2 Log).**		Updated procurement status
6	**Status of Information Requested.**		
	6.1	RFI Status report	Request for Information (RFI)
	6.2	NCR Report	Non- conformance request (NCR)
	6.3	PCO and NOV Log	Potential change order and notice of variation summary

(*Continued*)

TABLE 4.49 CONTINUED
Monthly Progress Report EXAMPLE CONTENTS

Contents of Monthly Progress Report (Contractor) (Sample Contents)

Serial Number	Contents		Description
7	Updated Cost-Loaded Program.		
	7.1	T-7 Updated Status Report	Money progress = physical × budgeted cost for each running or completed activity
	7.2	Updated Cost-Loaded Schedule	Same as T-7 but money values not percentages
	7.3	Updated Work-In-Place (%) Report	Histogram and cumulative curve
	7.4	Updated Cash Flow Report	Histogram and cumulative curve
8	Updated Manpower Histogram.		Histogram and cumulative curve
9	Updated Schedule of Construction Equipment and Vehicles		Tabular report
10	Updated Critical Indicators.		
	10.1	Shop Drawings Status Report	Histogram and cumulative curve
	10.2	Material Status Report	Histogram and cumulative curve
	10.3	Construction. Leading Indicators	Each trade alone
	10.4	Line of Balance Diagram	It indicates the progress of all the major trades as a cumulative line chart
11	Updated Progress Photographs.		-
12	Updated Safety Inspection Checklist.		Tabular report
13	Contractor Information		Organization chart, tabular report

Source: Abdul Razzak Rumane (2013). Quality Tools for Managing Construction Projects Reprinted with permission of Taylor & Francis Group.

Apart from the issues related to the progress of works and programs, site safety- and quality-related matters are also discussed at these meetings. These meetings are normally attended by the owner's representative, the designer/consultant staff, the contractor's representative, and the subcontractor's responsible personnel.

Coordination meetings are held from time to time to resolve coordination matters among various trades.

TABLE 4.50
Contents of Progress Report (Consultant)

1.0 Contract Particulars
    1.1 Project description
    1.2 Project data

2.0 Construction Schedule

3.0 Progress of Works
    3.1 Temporary facilities and mobilization
    3.2 Summary of construction progress
        3.2.1 Status
        3.2.2 On-shore progress
        3.2.3 Off-shore progress

4.0 Time Control
    4.1 CPM Schedule-Level One (Target vs. Current) – Summary by Building/Marine.
    4.2 CPM Schedule-Level Two (Target vs. Current) – Summary by Building/Division.
    4.3 30-Days Look-Ahead Schedule
    4.4 Time Control Conclusion

5.0 Cost Control
    5.1 Financial progress
    5.2 Cash flow curve and Histogram
    5.3 Work-in-place S-Curve and histogram
    5.4 Cost control conclusion

6.0 Status of Contractor's Submittals
    6.1 Material status
    6.2 Shop drawing status

7.0 Subcontractors
8.0 Consultant's Staff
9.0 Quality Control
10.0 Meetings
11.0 Site Work Instructions
12.0 Variation Orders
13.0 Construction Photographs
14.0 Contractors Resources
15.0 Others Matters
    15.1 Safety
    15.2 Weather conditions
    15.3 Important developments/proposals/submissions

Quality meetings are conducted to discuss quality issues at the site and how to improve the construction process to avoid/reduce rejection and rework.

Safety meetings are also held to discuss related health, site safety, and environmental matters.

The frequency of meetings is agreed between all the parties at the beginning of the construction phase. Normally, the construction manager/resident engineer prepares the agenda for the meeting and circulates it to all the participants. The contractor informs the resident engineer in advance about the points the contractor would like to discuss, which are included in the agenda. The minutes of the meetings are recorded and circulated among all the attendees and others per the approved responsibility/site communication matrix.

4.6.10.5 Cost Management

Cost management is the process involving planning, cost estimating, budgeting, and cost controlling to ensure that the project is successfully completed within the approved budget.

Cost management in the construction project involves planning and managing the cost of the facility throughout the project life cycle.

4.6.10.5.1 Cost Control

Cost control is the process of monitoring the status of the project to update the project cost and managing changes to the baseline. This process provides the means to recognize variance from the approved plan, evaluate possible alternatives, and take corrective action to minimize the risk. In order to have successful cost control, it is essential to have the necessary information and data to take appropriate action. If the necessary information and regular updates are not available or if the action is inefficiently executed, then the risk to cost control on a project is raised considerably.

The purpose of cost control is to manage the project delivery within the approved budget. Regular cost reporting will facilitate:

- Establishing the project cost to date
- Anticipating the final budget of the project
- Predicting cash flow requirements
- Understanding any potential risk to the project

The cost control process focuses on:

- Identifying the factors that influence the changes to the cost baseline
- Determining whether the cost baseline has changed
- Ensuring that any changes are beneficial for the project
- Establishing the cost control structure and policy
- Managing the actual changes when and as they occur
- Monitoring the cost performance to detect cost variance from the actual budget
- Recording all appropriate changes
- Informing/reporting concerned stakeholders about the approved changes
- Preventing unauthorized changes to the cost baseline
- Working to bring cost overruns within acceptable limits

Monitoring and control of project payments is essential with the budgeted amount. This is done through monitoring cash flow with the help of S-curves and progress

Management of Quality for Sustainability

curves which give the exact status of payments and also identify whether they are exceeding the budget. Uninterrupted cash flow is one of the most important elements in the overall success of the project. Figure 4.59 illustrates the Planned and Actual Cost S-Curve.

4.6.10.6 Quality Management

Quality management is an organization-wide approach to understanding customer needs and delivering the solutions to fulfill the project and satisfy the customer. Quality management is managing and implementing a quality system to achieve customer satisfaction at the lowest overall cost to the organization while continuing to improve the process. A quality system is a framework for quality management. It embraces the organization structure, policies, procedures, and processes needed to implement the quality management system.

Quality management in construction projects is different to that of manufacturing projects.

Quality in construction projects is not only the quality of the products and equipment used in the construction; it is the total management approach to completing the facility as per the scope of works to the satisfaction of the customer/owner, to be completed within the specified schedule and within the budget to meet the owner's defined purpose. Quality management in construction addresses the management of the project, the product of the project, and all the components of the product. It also involves incorporation of changes or improvements, if needed. Construction project quality involves the fulfillment of the owner's needs as per the defined scope of works within the budget and specified schedule to satisfy the owner's/end-user's requirements.

The quality management system in construction projects consists mainly of:

- Quality management planning
- Quality assurance
- Quality control

4.6.10.6.1 Develop a Quality Management Plan

The quality management plan for construction projects is part of the overall project documentation, addressing and describing the procedures to manage construction quality and project deliverables. The quality management plan identifies the following key components:

- Details of the quality standards and codes to be complied with
- Project objectives, project scope of work
- Stakeholders' quality requirements
- Regulatory requirements
- Quality matrix for different stages
- Design criteria
- Design procedures
- Detailed construction drawings

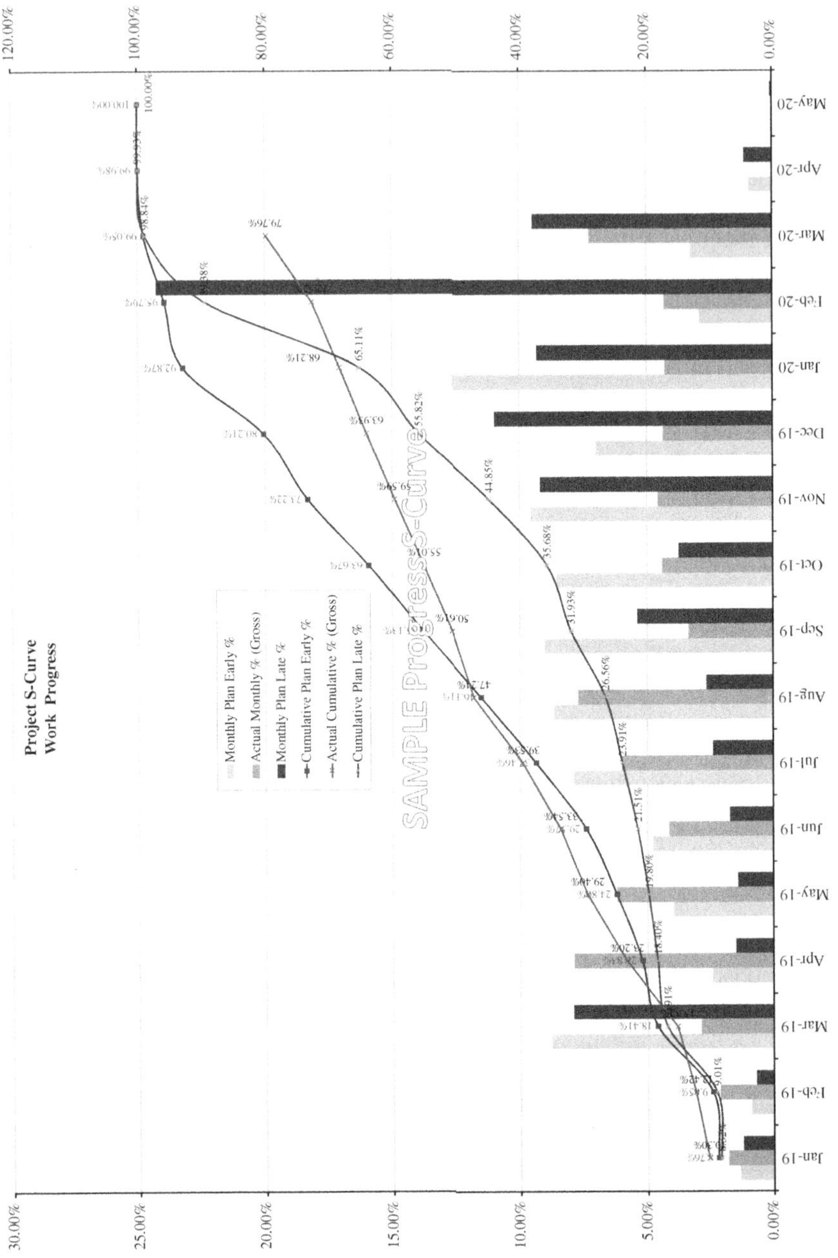

FIGURE 4.59 S-Curve (work progress).

- Detailed work procedure
- Well-defined specification for all the materials, products, components, and equipment to be used to construct the facility
- Manpower and other resources to be used for the project
- Inspection and testing procedures
- Quality assurance activities
- Quality control activities
- Defect prevention, corrective action, and rework procedures
- Project completion schedule
- Cost of the project
- Documentation and reporting procedure

4.6.10.6.2 Perform Quality Assurance

Quality assurance in construction projects covers all activities performed by the design team, contractor, and quality controller/auditor (supervision staff) to meet the owner's objectives as specified and to ensure and guarantee that the project/facility is fully functional to the satisfaction of the owner/end-user. Auditing is part of the quality assurance function.

Quality assurance is the activity for providing evidence to establish confidence among all concerned that quality-related activities are being performed effectively. All these planned or systematic actions are necessary to provide adequate confidence that a product or service will satisfy the given requirements for quality.

Quality assurance covers all activities from design, development, production/construction, installation, servicing to documentation, and also includes regulations of the quality of raw materials; assemblies, products, and components, services related to production, and management, production, and inspection processes. The following are the major activities to be performed for quality assurance of the construction project;

- Confirmation that the owner's needs and requirements are included in the scope of works (TOR)
- Review and confirm design compliance to Terms of Reference (TOR)
- Compliance of executed works with the specified standards and codes
- Conformance to regulatory requirements
- Works executed as per approved shop drawings
- Installation of approved material, equipment on the project
- Method of installation as per approved method statement or manufacturer's recommendation
- Coordination among all the trades
- Continuous inspection during the construction/installation process
- Identification and correction of the deficiencies
- Timely submission and review of transmittals

4.6.10.6.3 Control Quality

Quality control in construction projects is performed at every stage through the use of various control charts, diagrams, checklists, etc., and can be defined as:

- Checking and reviewing of the project design
- Checking and reviewing of bidding/tendering documents
- Analysis of contractor's bids
- Checking of executed/installed works to confirm that works have been performed/executed as specified, using specified/approved materials, installation methods, and specified references, codes, standards to meet the intended use
- Controlling budget
- Planning, monitoring, and controlling project schedule

The construction project quality control process is a part of the contract documents which provide details about specific quality practices, resources, and activities relevant to the project. The purpose of quality control during construction is to ensure that the work is accomplished in accordance with the requirements specified in the contract. Inspection of construction works is carried out throughout the construction period, either by the construction supervision team (consultant) or appointed inspector agency. Quality is an important aspect of a construction project. The quality of a construction project must meet the requirements specified in the contract documents. Normally, the contractor provides on-site inspection and testing facilities at the construction site. On a construction site, inspection and testing are carried out at three stages during the construction period to ensure quality compliance:

1. During construction process. This is carried out with the checklist request submitted by the contractor for testing of ongoing works before proceeding to the next step.
2. Receipt of subcontractor-purchased material or services. This is performed by a material Inspection Request submitted by the contractor to the consultant upon the receipt of material.
3. Before final delivery or commissioning and handover.

Quality management in construction is a management function. In general, quality assurance and control programs are used to monitor design and construction conformance to established requirements as determined by the contract specifications. Instituting Quality Management programs reduces costs while producing the specified facility. The contractor's Quality Control plan developed as per the contents, discussed in Table 4.39 under Sections 4.6.7.3, is to be followed throughout a construction project. Table 4.51 illustrates the responsibilities for site quality control.

4.6.10.7 Resource Management

Resource management in construction mainly relates to management of the following processes:

TABLE 4.51
Responsibility for Site Quality Control

Sr. No.	Description	Owner / Owner/Project Manager	Supervisor/Consultant / Consultant/Designer	Contractor Manager	Quality In charge	Quality Engineers	Site Engineers	Safety Officer	Head Office
				(Contractor — Linear Responsibility Chart)					
1	Specify quality standards	□	■					□	□
2	Prepare quality control plan			□	■	□			
3	Control distribution of plans and specifications			□	■	□			■
4	Submittals		■	■ □	□				
5	Prepare procurement documents								
6	Prepare construction method procedures			□	□		■ ■		
7	Inspect work in progress		■	■		□	■ ■		
8	Accept work in progress		■				■		
9	Stop work in progress	■	□						
10	Inspect materials upon receipt		■	□ □	■		■ ■		
11	Monitor and evaluate quality of works		□		■	■	■ ■		
12	Maintain quality records	□		□	■				
13	Determine disposition of non-conforming items			■					

(Continued)

TABLE 4.51 CONTINUED
Responsibility for Site Quality Control

Sr. No.	Description	Owner	Supervisor/Consultant	Contractor					
		Owner/Project Manager	Consultant/Designer	Contractor Manager	Quality In charge	Quality Engineers	Site Engineers	Safety Officer	Head Office
14	Investigate failures	□	□	■	■	□	■		
15	Site Safety	□		□				■	
16	Testing and commissioning	■	■	□			■		
17	Acceptance of completed works	■	□	□					

■ Primary Responsibility
□ Advise/Assist

Source: Abdul Razzak Rumane (2017). Quality Management in Construction Projects, Second Edition. Reprinted with permission from Taylor & Francis Group Company.

Management of Quality for Sustainability

1. Human resources (project teams)
   - Project owner team (project manager, construction manager)
   - Designer (consultant)
2. Construction resources (contractor)
   - Manpower
   - Material
   - Equipment

4.6.10.7.1 Manage the Project Team

This is a process to keep track of team member performance, provide feedback, resolve issues, and manage changes to ensure project performance optimization.

Construction projects are of a temporary nature, and project team members are collected from different backgrounds and disciplines, therefore it is inevitable that issues/conflicts may arise among the members. There are mainly three project teams involved in a construction project. Each project manager has to resolve any issues as they arise and has to manage the team effectively by maintaining cohesion among all the team members. This can be achieved by:

- Establishing ground rules
- Coordinating with all team members to understand their issues
- Creating shared vision among team members
- Tracking team performance
- Training, recognition, and rewards
- Conducting meetings, exchanging relevant information, and resolving issues
- Problem identification and providing quick solutions
- Conflict management

4.6.10.7.1.1 Managing Conflict

There exist several types of conflict. Each conflict can assume a different intensity at different stages of the project. The causes of disagreement vary in different phases of the project as different members are involved at different phases of the project. The following are typical source of conflict that may arise during the project;

- Priorities
- Scope change
- Project schedule
- Cost
- Technical opinions
- Resources
- Personality conflict
- Lack of coordination by team members
- Communication problem
- Administrative procedures
- Claims

The project manager has to resolve conflict by searching for an alternative solution. The following methods are normally used to resolve conflicts:

1. Withdrawing/Avoiding
   - It means both parties retreat from the conflict issue
2. Smoothing/Accommodating
   - It emphasizes friendly relationships and agreements rather than differences of opinions
3. Collaborating
   - In this method, parties try to incorporate multiple viewpoints in order to lead to a consensus
4. Confronting/Problem-Solving
   - It is a fact-based approach where both parties solve their problems by focusing on the issues, looking at alternatives, and selecting the best alternative
5. Compromising
   - Compromising is finding a solution that brings some degree of satisfaction to both parties.
6. Forcing
   - Forcing is the use of authority and power in resolving a conflict by exerting one's viewpoint over another at the expense of the other party.

4.6.10.7.2 Managing Construction Resources

The success of a construction project depends largely on the availability, performance, and utilization of resources. In a construction project, the resources are linked to the duration of the project and each activity is allocated a specific resource to be available at the specific time. The construction resource mainly consists of:

1. Construction workforce
   - Contractor's own staff and workers
   - Subcontractor's staff and workers
2. Construction material, equipment, and systems to be installed on the project.
3. Construction equipment, machinery.

In most construction projects, the contractor is responsible for engaging with all types of human resources to complete the project, namely subcontractors, specialist installers and suppliers, as well as arranging for equipment, construction tools, and materials as per the contract documents and to the satisfaction of the owner/owner's appointed supervision team. Workmanship is one of the most important factors in achieving quality in construction, therefore it is required that the construction workforce is fully trained and has full knowledge of all the related activities to be performed during the construction process. The contractor has to give preferences to locally available resources.

Management of Quality for Sustainability

4.6.10.7.3 Construction Workforce

Once the contract is awarded, the contractor prepares a detailed plan for all the resources he needs to complete the project. Contract documents normally specify a list of a minimum number of core staff to be available at site during the construction period. Absence of any of these staff may result in a penalty to be imposed on the contractor by the owner.

A typical list of contractor's minimum core staff needed during the construction period for execution of the work on an oil and gas project is specified in the contract documents.

Contractor's human resources mainly consist of two categories:

1. Contractor's own staff and workers
2. Subcontractor's staff and workers

The human resources required to complete the projects are based on a resource-loading program. It is necessary that all the construction resources are coordinated and brought together at the right time in order to complete the project on time and within budget.

The main contractor has to manage all these personnel by:

1. Assigning the daily activities
2. Observing their performance and work output (productivity)
3. Ensuring daily attendance
4. Ensuring safety during the construction process

4.6.10.7.4 Subcontractor

In most construction projects, the main contractor engages subcontractors and specialist contractors to execute certain portions of the contracted project work. The main contractor has to monitor the performance of the subcontractor throughout the project to ensure project success. In order to achieve project objectives, it is essential that the main contractor and subcontractor maintain a partnering relationship. In order for the smooth operation of the project, the two parties should have cooperative, collaborative, and a joint problem-solving attitude. Their aim should be to achieve a successful project and maintain a long-term business relationship.

The main contractor/subcontractor relationship starts the moment the subcontractor is selected and a contract/agreement is signed to execute the project.

It is important that necessary precautions and care are taken to prequalify and select the subcontractor. Table 4.38, discussed earlier under Section 4.6.1.4, lists questionnaires to prequalify the subcontractor.

In certain cases, the subcontractor is involved with the main contractor from the tendering stage. The main contractor takes into consideration the prices quoted by the subcontractor when submitting the proposal. The contractor price between main contractor and the subcontractor can be either:

- Back-to-back on the main contractor prices, keeping an agreed-upon margin for the main contractor, OR
- Negotiated prices with the subcontractor, in which case the main contractor's contract-awarded prices are not known to the subcontractor.

In order for the successful execution of the project, it is necessary to have a subcontractor management plan in place to ensure that each of the subcontractors executes the project as specified without affecting the project quality and schedule.

The following are the contents of a typical Subcontractor Management Plan:

1. Introduction
2. Organization
    i. Organization chart
    ii. Roles
    iii. Responsibilities
3. Scope of work
4. Quality management plan
5. Project coordination method
6. Submittals
    i. Material
    ii. Shop drawings
7. Resource management
    i. Training
8. Change management
9. Communication
    i. Meetings
10. Risk management
11. HSE management
12. Documentation
13. Invoicing, payments
14. Closeout contract

4.6.10.7.5 Construction Material

The contractor also prepares a procurement log based on the project completion schedule. The contractor has to ensure that the construction material is available at the project site on time to avoid any delay. The delivery of long lead items has to be initiated at an early stage of the project and monitored closely. Late order placements for materials result in delayed delivery of material, which, in turn, affect the timely completion of the project and are a common scenario in construction projects. Hence these logs have to be updated regularly and prompt actions have to be taken to avoid delays. The contractor is required to provide twice a month, or at any time requested by the owner/consultant, full and complete details of all products/systems procurement data relating to all the approved products, systems which have been ordered and/or procured by the contractor for use in the construction project. The contractor maintains contractors' procurement log E-2 to keep track of the material status. Figure 4.60 is a sample procurement log. The log E-2 is normally submitted by the contractor along with the monthly progress report.

Project Name

Consultant Name

CONTRACTOR'S PROCUREMENT LOG (REPORT E-2)

CONTRACT NO. : ______________________ NO. : ______

CONTRACTOR : ______________________ DATE : ______

Activity No.	Description	B.O.Q./ Specification. No.	Estimated Quantity	Ordered Quantity	Reqd. Order Date	Date Pur. Ord. Issued	Purchase Order No.	Supplier	Method Of Shipping	Shipping Date to Kuwait	Required On Job Date	E.D.A. To Site	A.D.A. To Site	Lead Time	Remarks

FIGURE 4.60 Contractor's procurement log.

4.6.10.7.6 Construction Equipment

Likewise, the contract documents specify that the minimum number of items of equipment are to be available at site during the construction process to ensure smooth operation of all the construction activities.

4.6.10.7.7 Supply Chain Management

Supply chain management in construction projects involves managing and optimizing the flow of construction materials, systems, equipment, and resources to ensure timely availability of all the construction resources without affecting the progress of works at the site. Figure 4.61 illustrates the supply chain management process in a construction project.

In construction projects, supply chain management starts from the inception of the project. The designer has to consider the following while specifying the products (materials, systems, equipment) for use/installation in the project:

- Quality management system followed by the manufacturer/supplier
- Quality of the product
- Reliability of the product
- Reliability of the manufacturer/supplier
- Durability of the product
- Availability of the product requirement for the entire project
- Price economy/cost-efficient
- Sustainability
- Conformance to applicable codes and standards
- Manufacturing time
- Location of the manufacturer/supplier relative to the project site

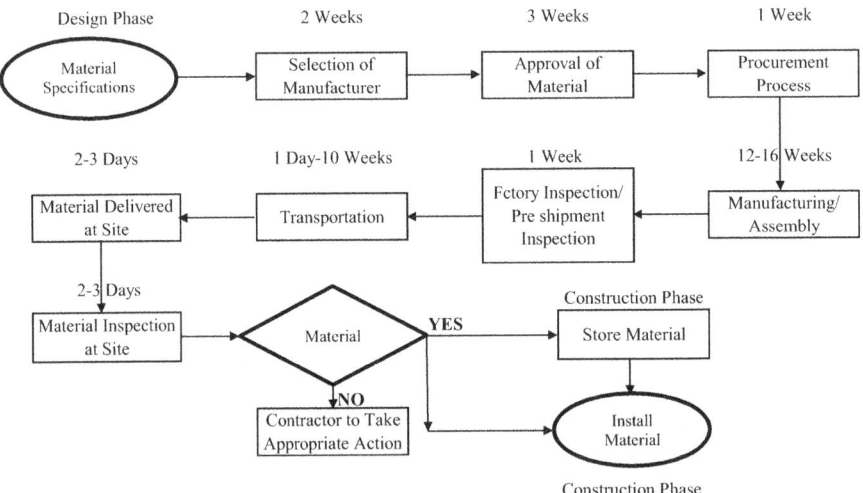

FIGURE 4.61 Supply chain process in construction project.

Management of Quality for Sustainability

- Interchangeability
- Avoid a monopolistic product

Product specifications are documented in the construction documents (particular specifications). In certain projects, the documents lists the names of recommended manufacturers/suppliers. However, with the aim of achieving a continuous and uninterrupted supply of the specified product, the contractor has to consider the following:

- Quality management system followed by the manufacturer/supplier
- Historical rejection/acceptance record
- Reliability of the manufacturer/supplier
- Product certification
- Financial stability
- Proximity to the project site
- Manufacturing/lead time
- Availability of the product as per the activity installation/execution schedule
- Manufacturing capacity
- Availability of the quantity to meet the project requirements
- Timeliness of delivery
- Location of manufacturer/supplier
- Product cost
- Transportation cost
- Product certification
- Risks in delivery of product
- Responsiveness
- Cooperative and collaborative nature to resolve problem

In order to ensure a supply chain management system, payments are made promptly, and the cash flow system should be projected accordingly as the supply chain may be affected due to interruption in payments for supply of products.

4.6.10.8 Communication Management

Communication is the process by which information is transmitted by a sender to a receiver via a channel/medium. A construction project involves many stakeholders. The project team must provide timely and accurate information to identified stakeholders who will receive communications. Effective communication is one of the most important factors contributing to the success of a project. Project communication is the responsibility of everyone on the project team. In order to manage communication, it is necessary to plan communication. A communication plan helps project team members to identify internal and external stakeholders involved in the project.

A comprehensive communication plan for a construction project can be developed by analyzing the questions listed in Table 4.52.

For the smooth flow of communication in a construction project, an appropriate communication matrix among all the stakeholders needs to be established at the

TABLE 4.52
Analysis for Communication Matrix

Serial Number	What and How	Related Analyzing Question
1	What?	What is the purpose of communication?
2	What?	What type of information needs to be communicated?
3	Who?	Who will initiate (send) the communication?
4	Who?	Who are the stakeholders to receive the information?
5	When?	When is the information to be sent (frequency)?
6	What?	What method of communication is to be used?
7	How many?	How many copies are to be distributed?
8	How much?	How much time should one wait to receive the feedback?
9	How?	How should the documents be archived?

start of each stage of the project. A communication matrix is used as a guideline as to what information to communicate, who is the team member to initiate, who will receive and take appropriate action, when to communicate, and the method of communication. Table 4.53 presents an example guideline to prepare a communication matrix for site administration during the construction phase

Contract documents specify a number of original (paper print) and copies to be transmitted to the A/E (consultant) for review and approval. Figure 4.62 A illustrates an example of a submittal process (paper based) and Figure 4.62 B illustrates an example of a submittal process (electronic).

4.6.10.8.1 Manage Submittals

There are different types of logs used in construction projects to monitor the submission of material, shop drawings, and other submittals. Figure 4.63 illustrates a Contractor's Submittal Status Log (Report E-1), normally called Log E-1, and Figure 4.64 illustrates a Contractor's Shop Drawing Status Log.

4.6.10.8.2 Meetings

A project meeting refers to a face-to-face communication method to exchange information among team members and stakeholders.

In a construction project, there are many types of meetings held during all the stages/phases of the project. Each meeting has its own purpose and structure. The meetings are used to distribute information, discuss issues, make proposals, suggest solutions, share information, and contribute to ideas for the improvement and successful completion of the project. The agenda and minutes of meetings are to be distributed to the stakeholders concerned.

4.6.10.3 Manage Documents

In construction projects, all the related documents are sent and received (exchanged) through the use of a transmittal form. These forms are generally issued to the

TABLE 4.53
Communication Matrix

SAMPLE FORM

Project Name:

Name of Construction/Project Manager: Name of Consultant:

Contractor Name: Project Number:

Serial Number	Type of Document	Originator	Receiver(s)	Purpose	Frequency	Method	Responsible Person for Action	Comments

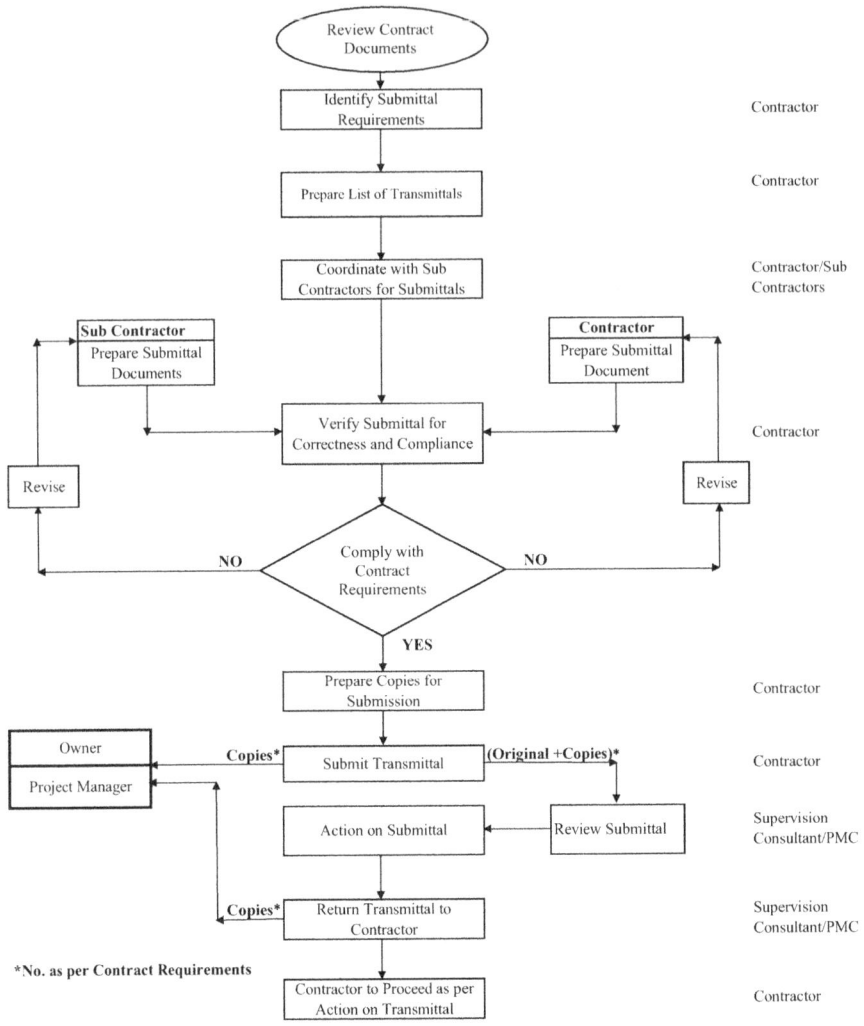

FIGURE 4.62A Submittal process (paper copy).

contractor along with other contract documents. Project team members are required to follow the procedures specified in the contract.

For any communication with external stakeholders, such as regulatory bodies, company letterhead is used to communicate the matter.

4.6.10.8.4 Control Documents

Contract conditions specify the time allowed to process the transmittal and other contract-related communication. The contractor, consultant, and construction/project manager maintain logs for all incoming and outgoing documents. Follow-up is also done among internal project team members to expedite the required action to

Management of Quality for Sustainability

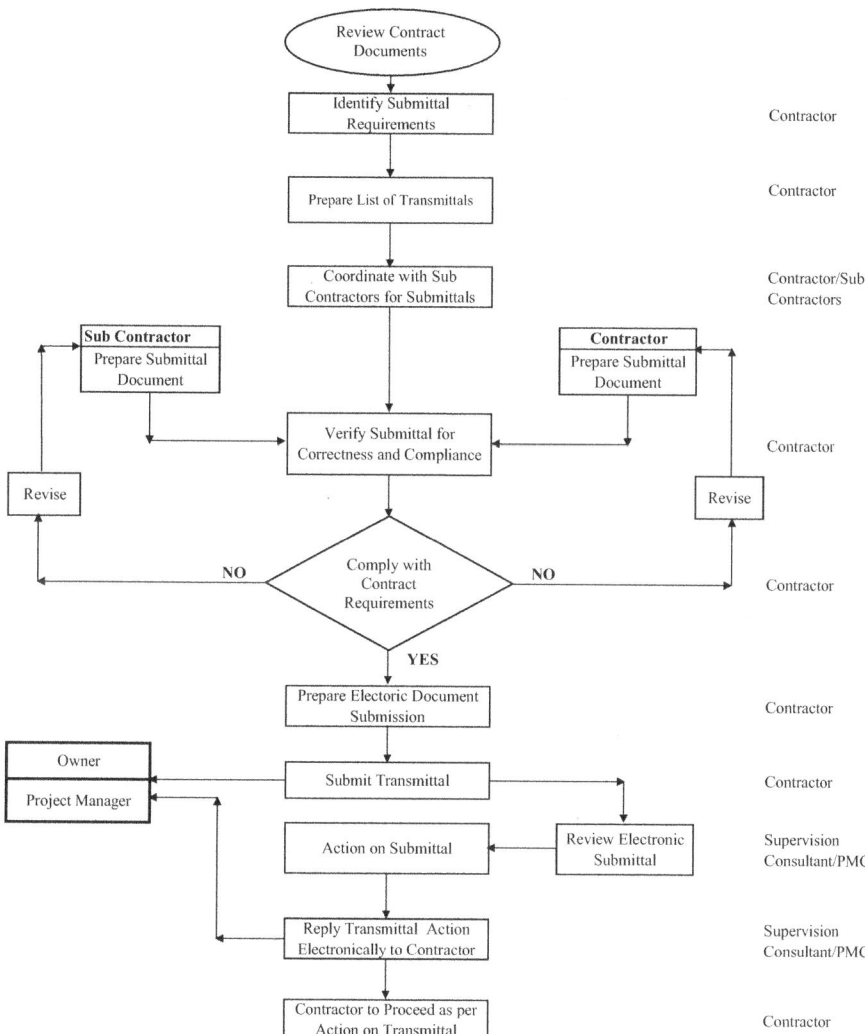

FIGURE 4.62B Submittal process (electronic).

be taken in response to the transmittal. The List of Forms (logs) and control documents normally used during the construction phase to communicate different types of documents was discussed earlier (see Table 4.43 under Section 4.6.8.3)

4.6.10.9 Risk Management

Risk management is the process of identifying, assessing, and prioritizing different kinds of risks, planning risk mitigation, implementing mitigation plan, and controlling the risks. It is a process of thinking systematically about the possible risks, problems, or disasters before they happen and setting up the procedure that will avoid

Project Name
Consultant Name

CONTRACTOR'S SUBMITTAL STATUS LOG (REPORT E-1)

CONTRACT NO. : ____________________ NO. : ________

CONTRACTOR : ____________________ DATE : ________

Document Specification or BOQ Page	Submittal Sequence No.	Rev	Description	Type of Submittal: Shop Drawing SD	Samples SM	Guarantee GT	Mf's Dats MF	Mt's Dats MT	Certificates CT	Test TT	Otr OT	Planned Sub. date	Submitted By Contrac.	Returned By A/E	Action	Approval Needed By	Required On Job	Notes

Action Code :
A = Approved as submitted
B = Approved as noted
C = Not Approved
D = For Information

SAMPLE FORM

FIGURE 4.63 Contractor's submittal status log.

Project Name
Consultant Name

CONTRACTOR'S SHOP DRAWINGS STATUS LOG

CONTRACT NO. : ____________________ NO. : __________
CONTRACTOR : ____________________ DATE : __________

Document Specification or BOQ Page	Item No.	Submittal Sequence No.						Drawing No.	Description	Submittal Type	Action Code : A = Approved as submitted B = Approved as noted BR = Approved as noted, Resubmit C = Not Approved D = For Information						Remarks
		Div	Sec	Typ	Sr.	Rev	Ser			SD	Planned Sub. date	Submitted By Contrac.	Returned By A/E	Action	Approval Needed By	Required On Job	

SAMPLE FORM

FIGURE 4.64 Contractor's shop drawing submittal log.

the risk, or minimize the impact, or cope with the impact. The objectives of project risk management are to increase the probability and impacts of positive events and decrease the probability and impacts of events adverse to the project's objectives. Risk is the probability that the occurrence of an event may turn into an undesirable outcome (loss, disaster). It is virtually anything that threatens or limits the ability of an organization to achieve its objectives. It can be unexpected and unpredictable events which have the potential to damage the functioning of organization in terms of money or, in the worst-case scenario, it may cause the business to close.

Construction projects have many varying risks. Risk management throughout the life cycle of the project is important and essential to prevent unwanted consequences and effects on the project. Construction projects involve many stakeholders, such as project owners, developers, design firms (consultants), contractors, and the banks and financial institutions funding the project who are affected by the risk. Each of these parties has an involvement with certain portions of the overall construction project risk, although the owner has a greater share of the risks as the owner is involved from the inception until the completion of the project and beyond. The owner must take initiatives to develop risk consciousness and awareness among all the parties emphasizing the importance of explicit consideration of risk at each stage of the project as the owner is ultimately responsible for the overall project construction. Traditionally:

1. The owner/client is responsible for the investment/finance risk
2. The designer (consultant) is responsible for design risk
3. The contractors, subcontractors are responsible for construction risk.

Table 4.54 lists typical categories of risks in construction projects

Figure 4.65 illustrates a typical flowchart for risk management procedure developed based on ISO 31000 risk management clauses.

4.6.10.9.1 Monitor and Control Risk

It is a systematic process of tracking identified risks, monitoring residual risks, identifying new risks, execution of risk response plan and evaluating the effectiveness of implementation of actions against established levels of risk in the area of scope, time, cost, and quality throughout the life cycle of the project. It involves timely implementation of risk response to identified risk to ensure the best outcome for a risk to a project. Figure 4.66 illustrates a typical flowchart for risk monitoring process.

4.6.10.10 Contract Management

Contract management/procurement management in construction projects is an organizational method, process, and procedure to obtain the required construction products. It includes the process to acquire a construction facility complete with all the related products/materials, equipment, and services from outside contractors/companies to the satisfaction of the owner/client/end-user.

Conventional notions of the procurement/purchasing cycle, which is normally applied in batch production, mass production, or in merchandising, are less

TABLE 4.54
Typical Categories of Risks in Construction Projects

Sr. No.	Category	Types
1	Management	Selection of project delivery system
		Selection of project/construction manager
		Selection of designer
		Selection of contractor
2	Contract (project)	Scope/design changes
		Schedule
		Cost
		Conflict resolution
		Delay in changer order negotiations
3	Statutory	Statutory/regulatory delay
4	Technical	Incomplete design
		Incomplete scope of work
		Design changes
		Design mistakes
		Errors and omissions in contract documents
		Incomplete specifications
		Ambiguity in contract documents
		Inconsistency in contract documents
		Inappropriate schedule/plan
		Inappropriate construction method
		Conflict with different trades
		Improper coordination with regulatory authorities
		Inadequate site investigation data
5	Technology	New technology
6	Construction	Delay in mobilization
		Delay in transfer of site
		Different site conditions to the information provided
		Changes in scope of work
		Resource (labor) low productivity
		Equipment/plant productivity
		Insufficient skilled workforce
		Union and labor unrest
		Failure/delay of machinery and equipment
		Quality of material
		Failure/Delay of material delivery
		Delay in approval of submittals
		Extensive subcontracting
		Subcontractor's subcontractor
		Failure of project team members to perform as expected
		Information flow breaks
7	Physical	Damage to equipment
		Structure collapse

(*Continued*)

TABLE 4.54 CONTINUED
Typical Categories of Risks in Construction Projects

Sr. No.	Category	Types
		Damage to stored material
		Leakage of hazardous material
		Theft at site
		Fire at site
8	Logistic	Resource availability
		Spare part availability
		Consistent fuel supply
		Transportation facility
		Access to worksite
		Unfamiliarity with local conditions
9	Health, Safety, and Environment	Injuries
		Health and Safety rules
		Environmental protection rules
		Pollution rules
		Disposal of waste
10	Financial	Inflation
		Recession
		Fluctuations in exchange rate
		Availability of foreign exchange (certain countries)
		Availability of funds
		Delays in payment
		Local taxes
11	Economic	Variation of construction material price
		Sanctions
12	Commercial	Import restrictions
		Custom duties
13	Legal	Permits and licences
		Professional liability
		Litigation
14	Political	Change in laws and regulations
		Constraints on employment of expatriate workforce
		Use of local agent and firms
		Civil unrest
		War
15	Natural	Flood
		Earthquake
		Cyclone
		Sandstorm
		Landslide
		Heavy rains
		High humidity
		Fire

Source: Abdul Razzak Rumane (2013). Quality Tools for Managing Construction Projects. Reprinted with permission of Taylor & Francis Group.

Management of Quality for Sustainability

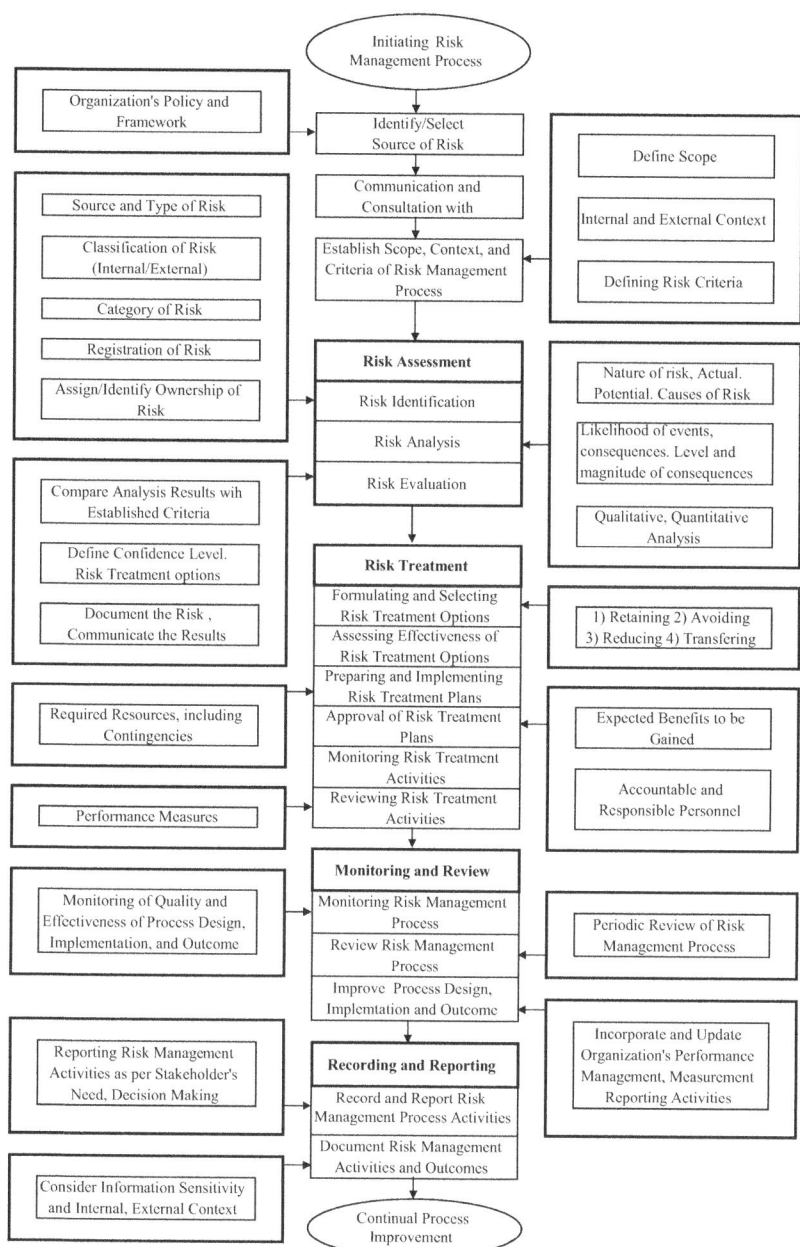

FIGURE 4.65 Typical flowchart for risk management procedure. Source: Abdul Razzak Rumane. (2022). *Risk Management Applications Used to Sustain Quality in Projects*, CRC Press, Florida. Reprinted with permission of Taylor & Francis Group.

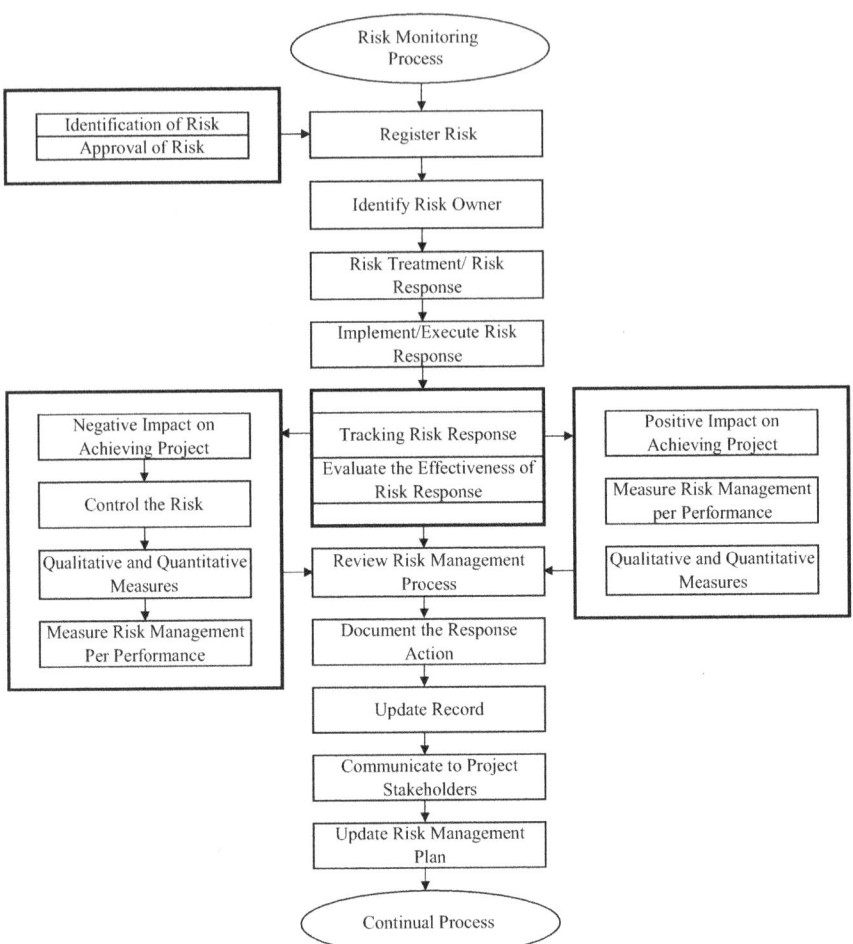

FIGURE 4.66 Typical flowchart for risk monitoring process. Source: Abdul Razzak Rumane. (2022). *Risk Management Applications Used to Sustain Quality in Projects*, CRC Press, Florida. Reprinted with permission of Taylor & Francis Group.

appropriate to the realm of construction projects. The procurement of construction projects also involves commissioning professional services and creating a specific solution. The process is complex, involving the interaction of the owner/client, project/construction manager, designer/consultant, contractor(s), suppliers, and various regulatory bodies. Generally, a construction project comprises of building materials (civil), electromechanical items, finishing items, and equipment. Construction involves installation and integration of various types of materials/products, equipment, systems, or other components to complete the project/facility to ensure that the facility is fully functional to the satisfaction of the owner/end-user. Contract management involves:

Management of Quality for Sustainability 379

1. Identification of:
   - What services are available in-house
   - What services are to be procured from outside agencies/organizations
   - How to procure (direct contract, competitive bidding)
   - How much to procure
   - How to select a supplier/contractor
   - How to arrive at an appropriate price, terms and conditions
2. Signing of contract
3. Timely delivery
4. Inspection of incoming material (on-site or factory inspection)
5. Receiving the right type of material/system
6. Timely execution of the work
7. Inspection of the work to maintain quality of the project
8. Completion of the project within the agreed-upon schedule
9. Completion of the within the agreed-upon budget
10. Documenting reports and plans

4.6.10.10.1 Procurement Management Process

In construction projects, the involvement of outside companies/parties starts at the early stage of the project development process. The owner/client has to decide which work is to be procured, and which is to be constructed by others. Every organization has their procurement system to procure services, contracts, products from others. Figure 4.67 illustrates a procurement management process, the stages at which an outside agency (contractor) is selected as per the procurement strategy for a particular type of project delivery system. At each of these stages, the bidding and tendering process takes place to select the agency (contractor). Figure 4.68 illustrates the contract management process.

4.6.10.10.2 Administer Contract

Contract administration is the process of the formal governance of a contract and changes to the contract document. It is concerned with managing the contractual relationship between various participants to successfully complete the facility to meet the owner's objectives. It includes tasks such as:

- Administration of the project requirement
- Administration of the project team members
- Administration of the contract management process/plan
- Execution of the contract
- Communication and management reporting
- Monitoring contract performance (scope, cost, schedule, quality, risk)
- Inspection and quality
- Variation order process
- Making changes to the contract documents by taking corrective action as needed
- Payment procedures

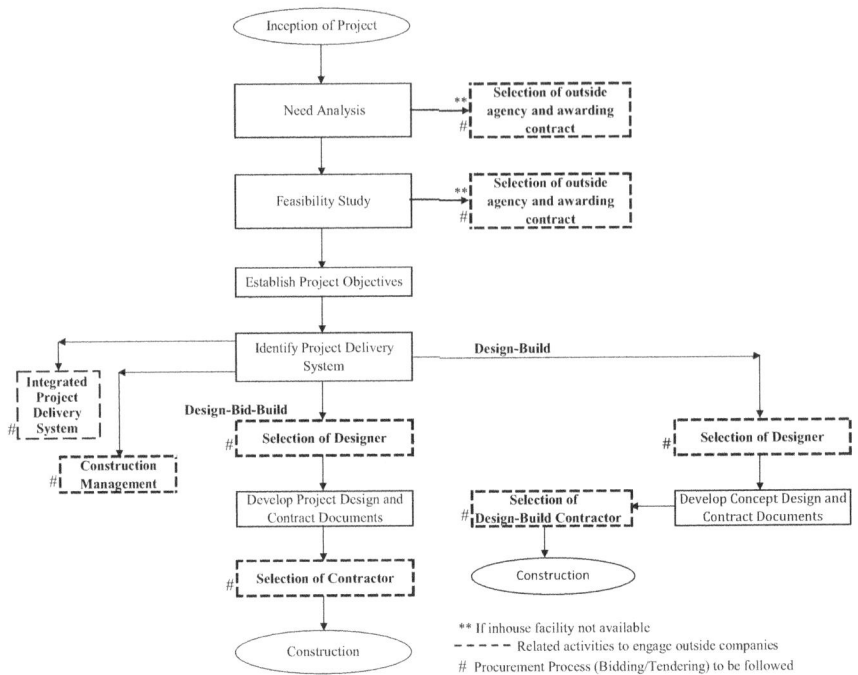

FIGURE 4.67 Procurement management process stages for construction projects. Source: Abdul Razzak Rumane. (2013). *Quality Tools for Managing Construction Projects*. Reprinted with permission of Taylor & Francis Group.

It is necessary that the contract administration procedure is clearly defined to achieve the success of the contract and that the parties to the contract understand who does what, when, and how. Table 4.44 discussed earlier (Section 4.6.8.5) lists the contents of a contract management plan.

4.6.10.10.3 Manage Contracts

Contract management during the construction phase is an organizational method, process, and procedure to manage all contract agreements involved between the owner, contractor, subcontractor, manufacturers, and suppliers. During the construction phase, contracts are managed mainly by the following parties who are directly involved in the execution of the project:

- Consultant/construction (project) manager
- Contractor

Apart from these two parties, subcontractors and vendors also have their contract management system.

The contract management process starts once the contract is signed. The consultant is responsible for managing the contract on behalf of the owner. The consultant

Management of Quality for Sustainability 381

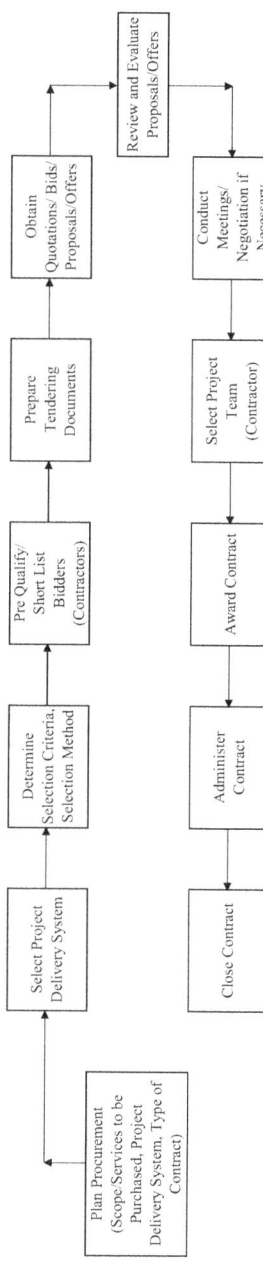

FIGURE 4.68 Contract management process.

monitors the scope, schedule, cost, and quality of the construction to ensure that the contract conditions are met. The contractor is responsible for ensuring that all project works are executed within the agreed-upon time and cost in accordance with the contract conditions and specification.

For successful contract management, the contractor as well as the consultant/CM/PM has to consider the following points while executing the project:

1. Use of Request for Information (RFI) to get clarification on some aspects of the project. There are two parts to the RF:
   i. "Question" by the contractor
   ii. "Answer" by the owner (consultant)
2. Cooperating with all team members to fulfill their contractual obligations
3. Developing a project execution plan with a realistic duration for each activity
4. Execution of contracted works in a timely manner in accordance with an agreed-upon schedule
5. Execution of project works using specified and approved materials, equipment, and systems
6. Installation of equipment, machinery, and system using the manufacturer's recommended method of installation
7. Providing resources to ensure timely availability of a competent workforce as per the resource schedule
8. Conducting meetings to monitor progress and clarify prevailing project issues
9. Dealing with variations to the specified product or method, work in accordance with related specification, contract clauses, and by providing substantiation and justifications, that has resulted in proposing alternative or substitute materials
10. Resolving disputes in an amicable way by adopting a cooperative approach
11. Taking action on all the transmittals within the agreed-upon period
12. Replying in a timely fashion to all correspondences, queries
13. Communicating issues and problems well in advance
14. Maintaining proper logs and records
15. Not ignoring problems/issues with the hope that they might go away
16. Resolving risk by taking proper action to eliminate/mitigate the risk
17. Managing errors, omissions, and additions strictly in accordance with the contract terms and avoiding any delays to the project
18. Arranging payment of monthly progress payments as per the contractual entitlement within the stipulated time
19. Maintaining list of claims on a monthly basis
20. Settling claims in accordance with the contract terms

4.6.10.11 HSE Management

The construction industry has been considered to be dangerous for a long time. The nature of works on-site always presents some dangers and hazards. There is a

Management of Quality for Sustainability

relatively high number of injuries and accidents at construction sites. Safety represents an important aspect of construction projects. Every project manager tries to ensure that the project is completed without major accident on the site.

The construction site should be a safe place for those who are working at sites. Necessary measures are always required to ensure the safety of all those working at construction site. Effective risk control strategies are necessary to reduce and prevent accidents.

Contract documents normally stipulate that the contractor, upon signing of the contract, has to submit a safety and accident prevention program. It emphasizes that all the personnel have to put efforts to prevent injuries and accidents. In the program, the contractor has to incorporate safety and health requirements of local authorities, manuals of accident prevention in construction, and all other local codes and regulations. The contractor also has to prepare an Emergency Evacuation Plan (EEP). The EEP is required to protect personnel and to reduce the number of fatalities in case of major accidents at site. The evacuation routes have to be displayed at various locations in a required manner. Transfer points and gathering points have to be designated and sign boards have to be displayed all the time. Evacuation sirens are to be sounded on a regular basis in order to ensure the smooth functioning of the evacuation plan.

A safety violation notice is issued to the contractor/employee if the contractor or any of his employees are not complying with safety requirements. Figure 4.69 illustrates a safety violation notice, which is to be actioned by the contractor.

Penalties are also imposed on contractors for noncompliance with the site safety program. Figure 4.70 illustrates a sample disciplinary notice form for a breach of safety rules. Different colors of cards may be issued along with the notice. Figure 4.71 illustrates concepts of issuance of different colors of cards.

Penalties are also imposed on the contractor for noncompliance with the site safety program. The safety program embodies the prevention of accidents, injury, occupational illness, and property damage. The contract specifies that a safety officer is engaged by the contractor to follow safety measures. The safety officer is normally responsible for:

1. Conducting safety meetings
2. Monitor on-the-job safety
3. Inspect the works and identify any hazardous area
4. Initiate a safety awareness program
5. Ensure availability of first aid and emergency medical services as per local code and regulations
6. To ensure that the personnel are using protective equipment, such as a hard hat, safety shoes, protective clothing, life belt, and protective eye coverings
7. To ensure that the temporary firefighting system is working
8. To ensure that work areas and access are free from trash and hazardous material
9. Housekeeping

Project Name
Consultant Name

Contract No.: SVN No.
Contractor: Date :
  Time:

SAFETY VIOLATION REPORT

SAFETY RELATED ITEMS			
Sr.No.	Description	Sr.No.	Description
1	Access Facilities	13	Hygienine
2	Barricade/Railing	14	Lifting Gears
3	Construction Equipment	15	Poor lighting
4	Crane	16	Protective Equipment
5	Earthwork/Excavation	17	Safety Gears
6	Electrical	18	Scaffolding
7	Fire Fighting/Protection	19	Site Fencing
8	First Aid	19	Storage Facilities
9	Formwork	20	Unsafe access
10	Hand and Power Tools	21	Vehicles
11	Hazars/Imflamable Material	22	Welding/Hot Work
12	Hoist	23	Others
12	House Keeping	24	

VIOLATION DESCRIPTION Action code: ☐

Item No.	Location	Description

ORIGINATOR: RESIDENT ENGINEER:

CONTRACTOR'S ACTION

Item No.	Location	Action	Date	Time

SAFETY OFFICER: CONTRACTOR'S PROJECT MANAGER:

Action Code: [A] For immediate action /( )hours   [B] Within ( ) days

FIGURE 4.69 Safety violation report.

4.6.10.12 Finance Management

In construction projects, the maximum amount of expenditure occurs during the construction phase. During this phase:

1. The owner has to make payments to:
    a. Main contractor
    b. Supervisor (consultant)
    c. Construction/project manager, if applicable

	PROJECT NAME
	CONSULTANT NAME

Safety Disciplinary Notice

Notice No.:		Date:
Name of Employee:		
Contractor Name:		
Area/Floor:		
Date & Time of Observance:		

Type of Notice

☐	**First/Verbal Warning** (White Card)	
☐	**Second/Written Warning** (Yellow Card)	
☐	**Suspension from Site** (Red Card)	

Reason for Issuance of Notice:

Action Required by Recipient:

Date by Which Action is Required:

Issued by:

Signature: Date:

Revewed by: Date:

CC: Owner ☐   Resident Engineer ☐   Project Manager ☐

SAMPLE FORM

FIGURE 4.70 Safety disciplinary notice.

    d. Specialist consultant
    e. Specialist contractor
    f. Any other party, such as direct appointments
    g. Owner-supplied items, if any
2. The main contractor has to make payments to:
    a. Subcontractors
    b. Suppliers (material procurement)
    c. Designer, if any design work is involved
    d. Workforce
    e. Rent (equipment rent, rental vehicles)

Serial Number	Card Color	Type of Disciplinary Action	Warning Validity	Reasons for Disciplinary Action
1	White	Verbal followed by Safety Discipline Notice	One month to Three months	1. Failure to use personal protective equipment 2. Failure to used define access 3. Working on plant, crane, vehicle without license 4. Working with unsafe scaffolding 5. Working on unsafe platform 6. Using unsafe sling or ropes for lifting 7. Working on unsafe ladders
2	Yellow	Issuance of Safety Discipline Notice and suspension from the work for rest of the day	Six months	1. Repetition of activities listed under "White Card" within one month of issuance of first notice 2. Failure to observe HSE related instructions 3. Failure to work as per instructed method of work, as per given task 4. Failure to follow storage principles about hazardous materials
3	Red	Issuance of Safety Disciplinary Notice and suspension from site for one month	One Year	1. Breach of safety rules where there is risk to life 2. Removal of safety devices, interlocks, guard rails, barriers without any authority 3. Deliberately exposing public to danger by not complying with agreed safe methods of work 4. Disposal of hazardous material in unsafe area

FIGURE 4.71 Concept of safety disciplinary action.

    3. The subcontractor has to make payment to:
        a. Suppliers
        b. Specialists

The owner's payment is mainly related to progress payment claimed and approved by the consultant, advance payments (if any, as per contract) to the contractor, and monthly fees to the consultant and construction/project manager. The contractor's and subcontractor's payments are linked to the approved executed works. The contractor submits the payment on a monthly basis. Figure 4.72 illustrates the progress payment submission format generally used by the contractor, and Figure 4.73 illustrates the process for the progress payment approval process.

Contractor Name:
Project Number

Project Name

Serial Number	BOQ Reference	Description	Original Reference				Work Completed (Quantity/Percentage)							Cummulative Work			Approved Amount
							Previous			Current				Previous+Current			
			Unit	BOQ Quantity	Unit Price	Total Amount	Quantity	Percentage (%)	Total Amount	Quantity	Percentage (%)	Check List/ Approval Refrence	Total Amount	Quantity	Percentage (%)	Total Amount	

FIGURE 4.72 Progress payment submission format.

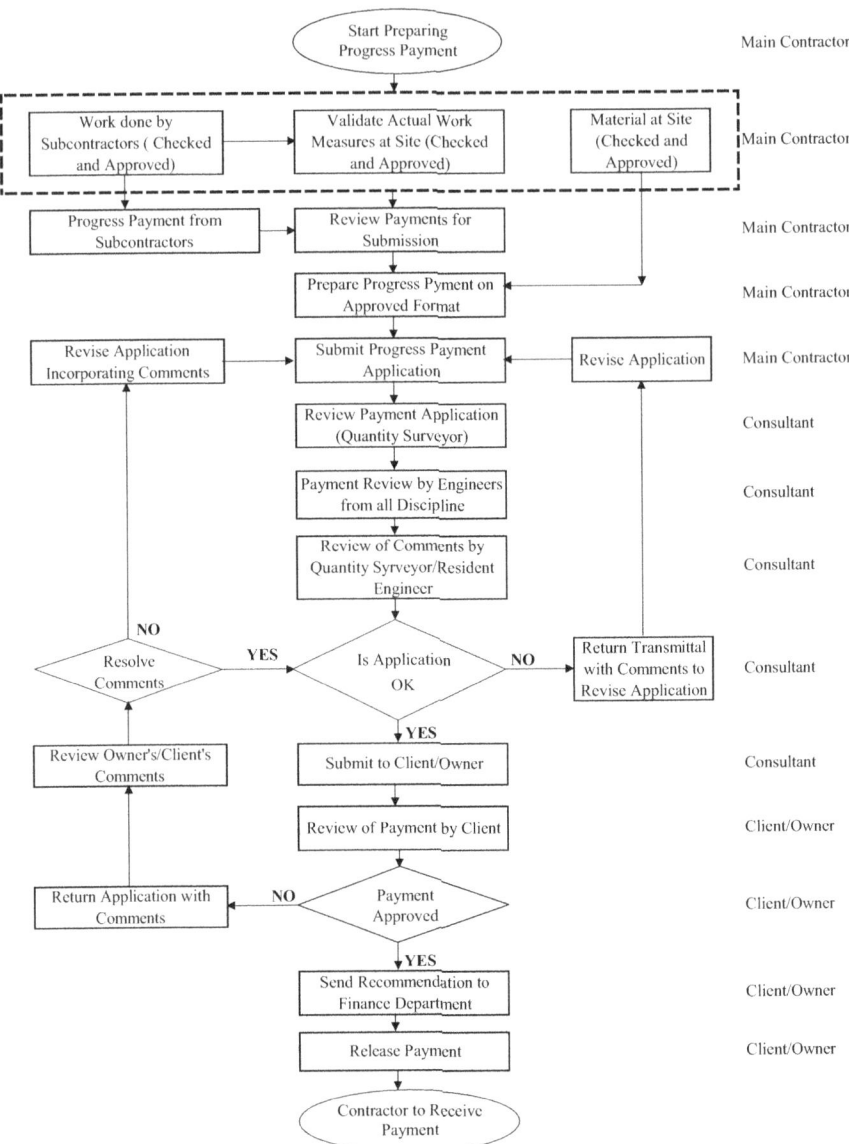

FIGURE 4.73 Progress payment approval process.

4.6.10.13 Claim Management

Most claims in construction projects are due to:

- Errors and omissions in the contract documents
- Incomplete design
- Design changes
- Delay in payment by the owner

Management of Quality for Sustainability

- Delay in transfer of site
- Delay in approval of submittals

During the construction phase, the contractor has to identify the errors or omissions in the contract, if any, and follow the contractual procedure to resolve the issues.

Any claim submitted by the contractor has to be resolved by the project team members amicably as per the condition of contract.

Figure 4.74 illustrates the claim resolution process.

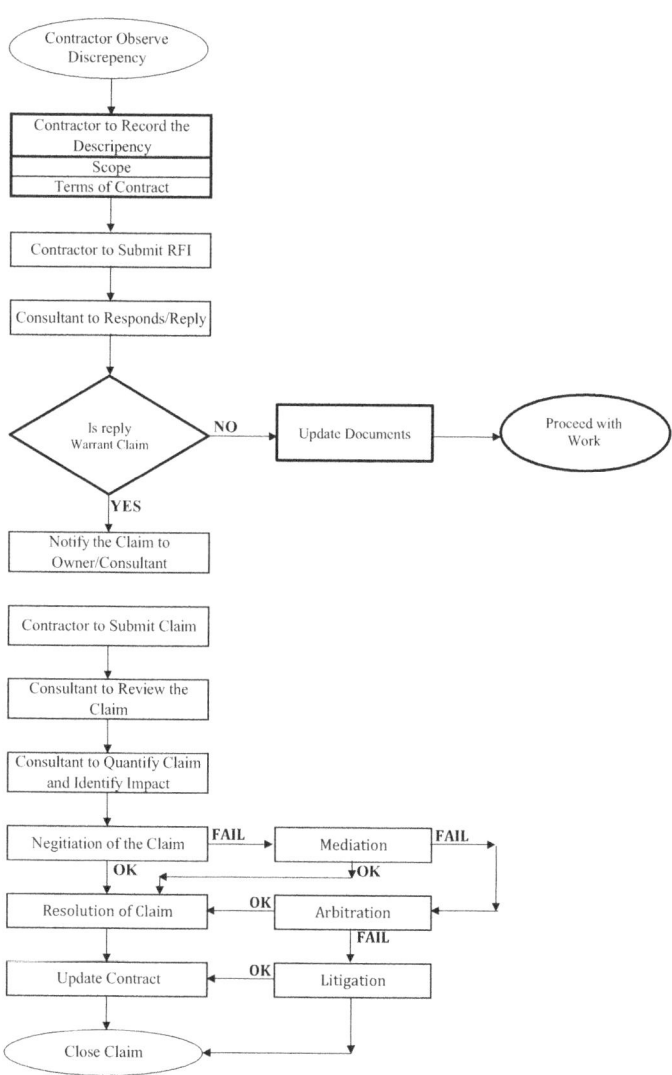

FIGURE 4.74 Claim resolution process.

4.6.11 Inspection of the Executed Works/Systems

The inspection of construction works is performed throughout the execution of the project. Inspection is an ongoing activity to physically check the installed works. Checklists (Inspection Notice) are submitted by the contractor to the consultant, who inspects the works/installations executed. If the work is not carried out as specified, then it is rejected, and the contractor has to rework or rectify the same to ensure compliance with the specifications. During construction, all the physical and mechanical activities are accomplished on-site. The contractor carries out the final inspection of the works to ensure full compliance with the contract documents.

During the construction process, the contractor has to submit the checklists to the consultant to inspect the works. Submission of checklists or request for inspection is an ongoing activity during the construction process to ensure proper quality control of the construction. Concrete work is one of the most important components of building construction. The concrete work has to be inspected and checked at all the stages to avoid rejection or rework. Necessary care has to be taken right from the control of the design mix of the concrete till the casting is complete and cured. Contractor has to submit the checklist at different stages of concrete work and has to make certain tests, specified in the contract, during the casting of concrete. In order to ensure that structural concrete works are executed without any defects or rejection and achieve concrete strength as specified, proper sequencing of works is important.

Figure 4.75 illustrates a checklist for general works to be inspected by the consultant.

If the consultant finds that an item has been executed at site not as per the contract documents, specification, or general code of practice, then a remedial note is issued by the consultant. Figure 4.76 illustrates a remedial note.

The contractor is required to reply on the same form after taking the necessary action. On finalization of the issue, the withdrawal notice is issued by the consultant/supervision staff.

The consultant's supervision staff always make a routine inspection during the construction process. A non-conformance report is prepared and sent to the contractor to take corrective/preventive action toward this activity. Figure 4.77 illustrates a non-conformance report used for a building construction project.

Upon receipt of the material at site, the contractor submits the material inspection report.

Figure 4.78 shows a material inspection report.

In addition to the above-mentioned checklists, the contractor submits the following checklist at different stages of the concreting process:

1. Checklist for quality control of Form Work
2. Notice for Daily Concrete Casting
3. Check list for Concrete Casting
4. Check list for Quality Control of Concreting
5. Report on Concrete Casting
6. Notice for Testing at Lab
7. Concrete Quality Control Form

Management of Quality for Sustainability

```
+------------------------------------------------------------------+
|                        Owner Name                                |
|                        Project Name                              |
|                       Consultant Name                            |
|                         CHECK LIST                               |
| CONTRACTOR :  _____       CHECK LIST No. : [    ]      |
| CONTRACT No.: _____                QC Code No. [    ]  |
| TO          : Resident Engineer                                  |
+------------------------------------------------------------------+
| CCS ACTIVITY NO : _____  SPECIFICATION DIVISION : ___  SECTION : ___ |
|           [ ] Process Work       [ ] Civil Works    [ ] Information and Communicati |
|  AREA :   [ ] Piping Work        [ ] Electrical Works [ ] Low Voltage System |
|           [ ] Mechanical Works   [ ] Instrumentation Works  [ ] Other (Specify) |
+------------------------------------------------------------------+
| Please inspect the following :-                                  |
|                                                                  |
| Location : _____                            |
| Work :     _____                            |
|                                                                  |
|                     Sketch(es) attached { } No.                  |
|                                                                  |
| The work to be inspected has been coordinated with all related subcontractors. |
| Estimated Quantity of Work : _____  Date & Time Inspection Required : _____ |
| Contractor Signature : _____     Date & Time _____ |
| Received By : _____      Date & Time _____ |
| C.C.: Owner Rep: _____     Date & Time _____ |
|    (All request must be submitted at least 24 hours prior to the required inspection) |
+------------------------------------------------------------------+
| Reply:  The above is Approved/Not approved for the following :-  |
| _____ |
| _____ |
| _____ |
| _____ |
| _____ |
|                                                                  |
| Inspected by _____ Date & Time ____ Resident Engineer _____ Date & Time ____ |
+------------------------------------------------------------------+
| Received by Contractor _____     Date & Time _____  |
| C.C.: Owner Rep:       _____     Date & Time _____  |
+------------------------------------------------------------------+
```

FIGURE 4.75 Check list.

Project Name
Consultant Name

REMEDIAL NOTE (RN)

Contractor: _____   R.N. No.: _____

Contract No.: _____   DATE: _____

> Your attention is drawn to the following works which have not been carried out in accordance with the Contract and are therefore not acceptable. Failure to carry out remedial works within a reasonable period of time may result either in additional work at your expense, or the Employer may elect to invoke Clause ---of the General Conditions of Contract.

LOCATION:

DEFECTS:

SAMPLE FORM

Signed: _____   Received by: _____
        Resident Engineer   Contractor/Date

Distribution: ☐ Owner   ☐ Consultant   ☐ Contractor   ☐

FIGURE 4.76 Remedial note.

4.6.12 VALIDATE THE EXECUTED WORKS

The inspection of construction works is performed throughout the execution of the project. Inspection is an ongoing activity to physically check the installed works. Checklists are submitted by the contractor to the consultant to inspect the executed works/installations.

If the work is not carried out as specified, then it is rejected, and the contractor has to rework or rectify the same to ensure compliance with the specifications. During construction, all the physical and mechanical activities are accomplished on the site. The contractor carries out the final inspection of the works to ensure full compliance with the contract documents.

Management of Quality for Sustainability

FIGURE 4.77 Non-conformance report.

4.7 TESTING, COMMISSIONING, AND HANDOVER

Testing, commissioning, and handover is the last phase of the construction project life cycle. This phase involves testing of the electro-mechanical systems, commissioning of the project, obtaining authorities' approval, training of user's personnel, handing over of technical manuals, documents, and as-built drawings to the owner/owner's representative. During this period, the project is transferred handed over to the owner/end-user for their use and substantial completion certificate is issued to the contractor. Figure 4.79 illustrates the testing and commissioning Process.

Project Name

Consultant Name

CONTRACT No.:

CONTRACTOR :

MATERIAL INSPECTION REPORT

Description of material for inspecition: -			MIR No.	:
			Date	:
			Contract No.	:
			Transmittal No.:	
			Spec/Drg.ref.	:

Qty. required	Qty. delivered	Total delivered	Attachments

SAMPLE FORM

Inspection Location :   Date of Material Delivery :

Contractor's Comments: -

Contractor's Signature :   Date :

Inspection Comments :

Comply with Approved Transmittal: YES ☐   NO ☐

Signature of Inspection Engineer :   Date :

Signature of R.E. :   Date :

FIGURE 4.78 Material inspection report

Management of Quality for Sustainability 395

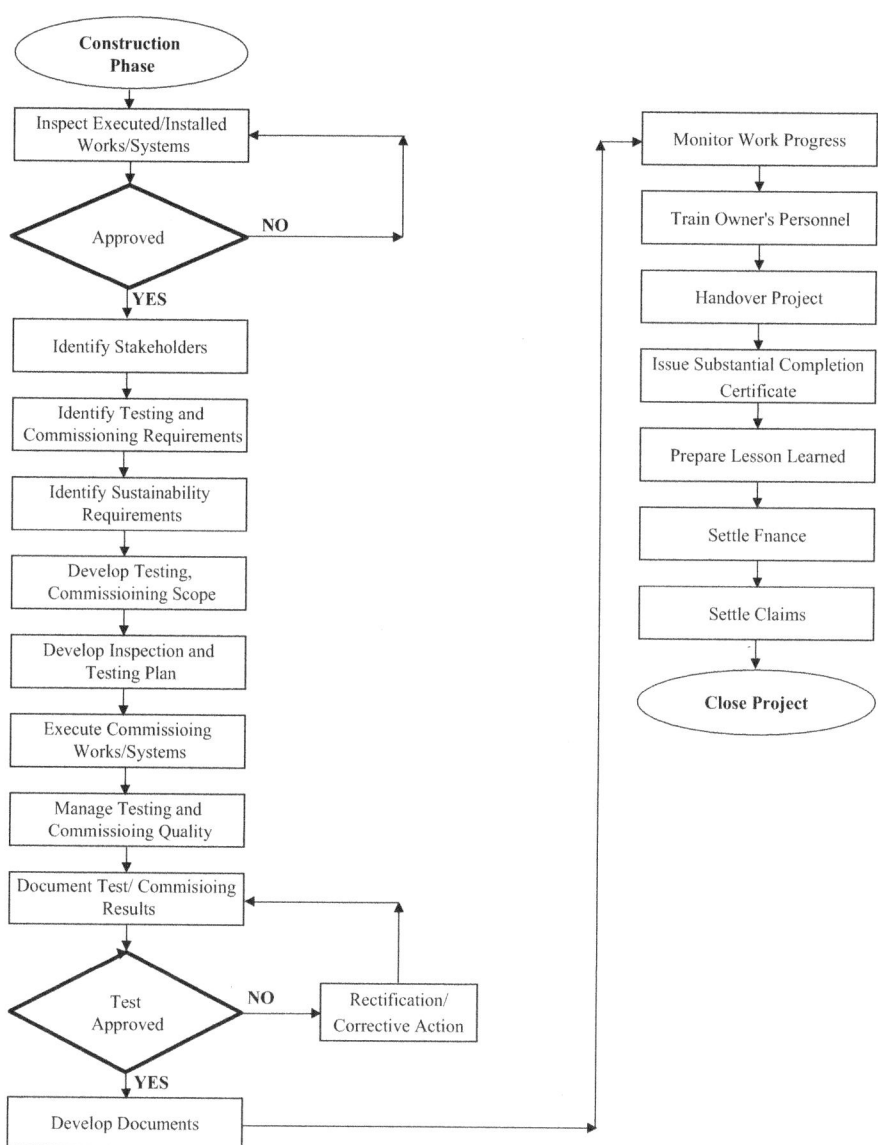

FIGURE 4.79 Logic flowchart for testing, commissioning, and handover phase.

Figure 4.80 illustrates the major activities relating to the testing, commissioning, and handover Phase developed based on the Project Management Process Groups methodology.

4.7.1 Identify Stakeholders

The following stakeholders are involved during the testing, commissioning, and handover phase:

1. Owner
   - Owner's representative/project manager
2. Construction supervisor
   - Construction supervisor (consultant)
   - Specialist contractor
3. Contractor
   - Main contractor
   - Subcontractor
   - Testing and commissioning specialist
4. Regulatory authorities
5. End-user
6. Third-party inspecting agency

4.7.1.1 Select Team Members

It is essential to select team members who have experience in testing and commissioning of major projects. The team members can be from the same supervision team which was involved during the execution of the project, if they have experience in carrying out testing and commissioning. In most cases, the manufacturer's representative is involved in testing the supplied equipment/systems. The owner may engage specialist firm(s) to perform start-up activities and commission the project.

4.7.1.2 Develop a Responsibility Matrix

Table 4.55 illustrates the contribution of various participants during the testing, commissioning, and handover phase and Table 4.56 lists the responsibilities of the consultant during the closeout phase.

4.7.2 Identify Testing, Commissioning, and Handover Requirements

Testing and start-up requirements are specified in the contract documents. It is essential to inspect and test all the installed/executed works prior to handover of the project to the owner/end-user. Generally, all works are checked and inspected on a regular basis while the construction is in progress; however, there are certain inspection and tests to be carried out by the contractor in the presence of the owner/ consultant. These are especially for rotating equipment, systems, conveying systems, electrical works, low-voltage systems, information- and technology-related products, emergency power supply system, and electrically operated equipment which are

Management of Quality for Sustainability

Testing, Commissioning and Handover Phase

Project Management Process Groups

Management Processes	Initiating Process	Planning Process	Execution Process	Monitoring & Controlling Process	Closing Process
Integration Management	Executed Project/Facility Testing and Commissioning Program Contract Documents	Testing, Commissioning, and Handover Plan	Testing and Commissioning Authorities Approval Punch List/Snag List As Built Drawings Manuals Spare Parts Move In Plan	Punch List/Snag List Testing and Commissioning Requirements	Handover of Project/Facility
Stakeholder Management	Identify Stakeholders	Stakeholders Requirements			Project Acceptance/Takeover
Scope Management		Contract Documents Sustainability Requirements		Authorities Approval Stakeholders Approval	Lesson Learned
Schedule Management		Testing Schedule Commissioning Schedule			
Cost Management					
Quality Management				Project Quality	
Resource Management		Demobilization Plan			Assign New Project/Termination
Communication Management		Test Results			
Risk Management		Plan Start up Risk		Control Risk	
Contract Management		Prepare Contract Closeout Documents	Finalize Closeout Documents	Check Documents for Compliance to Contract Requirements	Close Contracts
HSE Management					
Financial Management		Financial Administration and Records	Payment to All Contractors and Sub contractors		Payment to All Contractors and Sub contractors Payment towards all Purchases
Claim Management		Claim Resolution		Check for Claims	Settlement of Claims

FIGURE 4.80 Major activities relating to testing, commissioning, and handover. Note: These activities may not be strictly sequential; however, the breakdown allows implementation of project management function to be more effective and easily manageable at different stages of the project phase.

TABLE 4.55
Responsibilities of Various Participants during Testing, Commissioning, and Handover Phase

Phase	Responsibilities		
	Owner	Consultant/Supervisor	Contractor
Testing, Commissioning, and Handover	• Acceptance of project • Take-over • Substantial completion certificate • Training • Payments	• Witness tests • Check close-out requirements • Recommend take-over • Recommend issuance of substantial completion certificate	• Testing • Commissioning • Authorities' approvals • Documents • Training • Handover

TABLE 4.56
Typical Responsibilities of Supervision Consultant during Project Testing, Commissioning, and Handover Phase

Serial Number	Responsibilities
1	Ensure that all the equipment, systems are functioning and operative
2	Ensure that performance tests are carried out on all the equipment, systems, and equipment, systems are performing as per intended/design requirements
3	Ensure that Job Site Instruction (JSI), Non-Conformance Report (NCR) are closed
4	Ensure that the site is cleaned and all the temporary facilities and utilities are removed
5	Ensure as-built drawings handed over to the client/end-user
6	Ensure that operation and maintenance manuals are handed over to the client
7	Ensure that the record books are handed over to the client
8	Ensure that guarantees, warrantees, bonds are handed over to the client
9	Ensure that Test Reports, Test Certificates, Inspection Reports are handed over to the client
10	Ensure that spare parts are handed over to the client
11	Ensure that the snag (punch) list prepared and handed over to the client
12	Ensure that training for client/end-user personnel is completed
13	Ensure that the substantial completion certificate is issued and the maintenance period is commissioned
14	Ensure that all the dues of suppliers, subcontractors, contractors are paid
15	Ensure that the retention money is released
16	Ensure that the supervision completion certificate from the owner is obtained
17	"Lessons learned" is documented

Source: Modified from Abdul Razzak Rumane (2013). Quality Tools for Managing Construction Projects. Reprinted with permission of Taylor & Francis Group.

Management of Quality for Sustainability

energized after connection to a permanent power supply. Testing of all this equipment and systems starts after the completion of installation works. By this time, the facility is connected to a permanent electrical power supply and all the equipment is energized.

Testing and commissioning is to be carried out on installed equipment, machinery, and systems to ensure that they are safe and meet the intended requirements of the project to the satisfaction of the owner/end-user. The testing is normally undertaken to prove the quality and workmanship of the installation. It is also known as static testing. Upon completion of static testing, dynamic testing can be undertaken: this is "commissioning." Commissioning is carried out to prove that the systems operate and perform to their designed intent and specification.

Commissioning is the orderly sequence of testing, adjusting, and balancing the equipment and systems and bringing the equipment, systems, and subsystems into operation, and starts when the construction and installation of works are complete. Commissioning is normally carried out by the contractor or specialist in the presence of the consultant and owner/owner's representative and the user's operation and maintenance personnel to ascertain proper functioning of the systems to the specified standards.

4.7.3 Identify the Sustainability Requirements

The following items are to be checked for compliance with the requirements:

1. Energy-efficient equipment, systems
2. Daylighting system
3. Sensor-controlled lighting system
4. Performance testing to meet the owner's requirements and usage

4.7.4 Develop Testing, Commissioning Scope

The contract documents specify the testing and commission works to be performed by the contractor, subcontractor, and specialist supplier of equipment/systems. The following is the main scope of work to be carried out during this phase:

1. Testing of all equipment
2. Commissioning of all equipment
3. Testing of all systems
4. Commissioning of all systems
5. Performance testing of all equipment, systems
6. Conformance of all systems meeting sustainability requirements
7. Obtaining authorities' approvals
8. Submission of as-built drawings
9. Submission of technical manuals and documents
10. Submission of record books
11. Submission of warranties and guarantees

12. Training of owner's/user's personnel
13. Handover of spare parts
14. Handover of the project to the owner/end-user
15. Preparation of the punch list
16. Issuance of the substantial certificate
17. Lessons learned

Table 4.57 lists the items to be tested and commissioned prior to handing-over of the project.

4.7.5 Develop the Inspection and Testing Plan

Figure 4.81 illustrates a logic flowchart for the development of the inspection and testing plan.

4.7.6 Execute the Commission Works/Systems

The testing is mainly carried out mainly electromechanical works/systems and electrically operated equipment/systems, and rotating equipment which is energized after connection of a permanent power supply to the facility. These include the following:

1. Pumps (all types)
2. Piping works (pressure test)
3. Compressors
4. Generators
5. Coolers
6. Conveying System
7. Supervisory control system
8. Water supply, plumbing, and public health system
9. Fire suppression system
10. HVAC system
11. Integrated automation system (Building Automation System)
12. Electrical switchgear
13. Electrical lighting and power system
14. Instrumentation and control system
15. Grounding (earthing) and lightning protection system
16. Fire alarm System
17. Information and Communication System
18. Electronic security and access control system
19. Public address system
20. Parking control system
21. Material handling equipment
22. Emergency power supply system
23. Electrically operated equipment

TABLE 4.57
Major Items for Testing and Commissioning of Equipment

Serial Number	Discipline	Items
1.0	Elevator	1. Power supply 2. Speed 3. Capacity to carry design load 4. Number of stops 5. Emergency landing 6. Emergency call system 7. Elevator management system 8. Interface with fire alarm system
2.1	Mechanical (Fire suppression)	1. Sprinklers 2. Piping 3. Fire pumps 4. Power supply and controls 5. Emergency power supply for fire pumps 6. Hydrants 7. Hose reels 8. Water storage facility 9. Gaseous protection system for communication rooms 10. Fire protection system for diesel generator room 11. Interface with fire alarm system 12. Interface with BMS
2.2	Mechanical (Public Health)	1. Piping 2. Pipe flushing and cleaning 3. Pumps 4. Boilers 5. Hot water system 6. Water supply and purity 7. Fixtures 8. Power supply for equipment 9. Controls 10. Drainage system 11. Irrigation system
3.0	HVAC	1. Pipe cleaning and flushing 2. Chemical treatment 3. Pumps 4. Duct work 5. Air handling unit 6. Heat recovery unit 7. Split nuits 8. Chillers 9. Cooling towers 10. Heating system (controls, piping, pumps) 11. Fans (ventilation, exhaust)

(Continued)

TABLE 4.57 CONTINUED
Major Items for Testing and Commissioning of Equipment

Serial Number	Discipline	Items
		12. Humidifiers
		13. Starters
		14. Variable frequency drive
		15. Motor control centers (MCC panels)
		16. Chiller control panels
		17. Building Management System (BMS)
		18. Interface with fire alarm system
		19. Thermostat
		20. Air balancing
4.1	Electrical (Power)	1. Lighting illumination levels
		2. Working of photo cells and controls
		3. Wiring devices (sockets)
		4. Lighting control panels
		5. Electrical distribution boards
		6. Electrical bus duct system
		7. Main switch boards/sub main switch boards
		8. Min low-tension panels
		9. Isolators
		10. Emergency switch boards
		11. Motor control centers (MCC panels)
		12. Audiovisual alarm panel
		13. Diesel generator
		14. Automatic transfer switch (ATS)
		15. UPS (uninterrupted power supply)
		16. Earthing (grounding) system
		17. Lightning protection system
		18. Surge protection system
		19. Power Supply to Equipment (HVAC, mechanical, elevators, others)
		20. IP rating of outdoor switches, isolators, switch boards
		21. Interface with BMS
		22. Emergency power system
		23. Exhaust emission of generator
4.2	Electrical (low- voltage)	1. Fire alarm system
		2. Communication system
		3. CCTV system
		4. Access control system
		5. Public address system
		6. Audiovisual system
		7. Master satellite antenna system

(*Continued*)

TABLE 4.57 CONTINUED
Major Items for Testing and Commissioning of Equipment

Serial Number	Discipline	Items
5	External works	1. Lighting poles
		2. Boundary wall lighting
		3. Lighting bollards
		4. Irrigation system
		5. Electrical distribution boards
6	General	1. Power supply for gate barriers
		2. Automatic gates
		3. Rolling shutters
		4. Window cleaning system
		5. Gas detection system
		6. Water/fluid leak detection system
		7. Waste treatment system

Source: Abdul Razzak Rumane (2013). Quality Tools for Managing Construction Projects. Reprinted with permission of Taylor & Francis Group.

The testing of these works/systems is essential to ensure that each individual work/system is fully functional and operates as specified. The tests are normally coordinated and scheduled with specialist contractors, local inspection authorities, third-party inspection authorities, and manufacturer's representatives. Sometimes, the owner's representative may accompany the consultant to witness these tests.

Test procedures are submitted by the contractor along with the request for final inspection. Standard forms, charts, and checklists are used to record the testing results.

4.7.7 MANAGE TESTING, COMMISSIONING QUALITY

The contract document identifies the testing and commissioning procedure to be followed on the installed equipment.

In order to follow the specified procedure and smooth functioning of testing and commissioning, the contractor/consultant has to plan quality (planning of design work), perform quality assurance, and control quality for managing testing and commissioning activities. This will mainly consist of the following:

1. **Plan Quality:**
    - Identify testing and commissioning requirements
    - Identify codes and standards to be followed for testing and commissioning
    - Identify contract specifications to be followed for testing and commissioning
    - Identify regulatory requirements to be followed for testing and commissioning

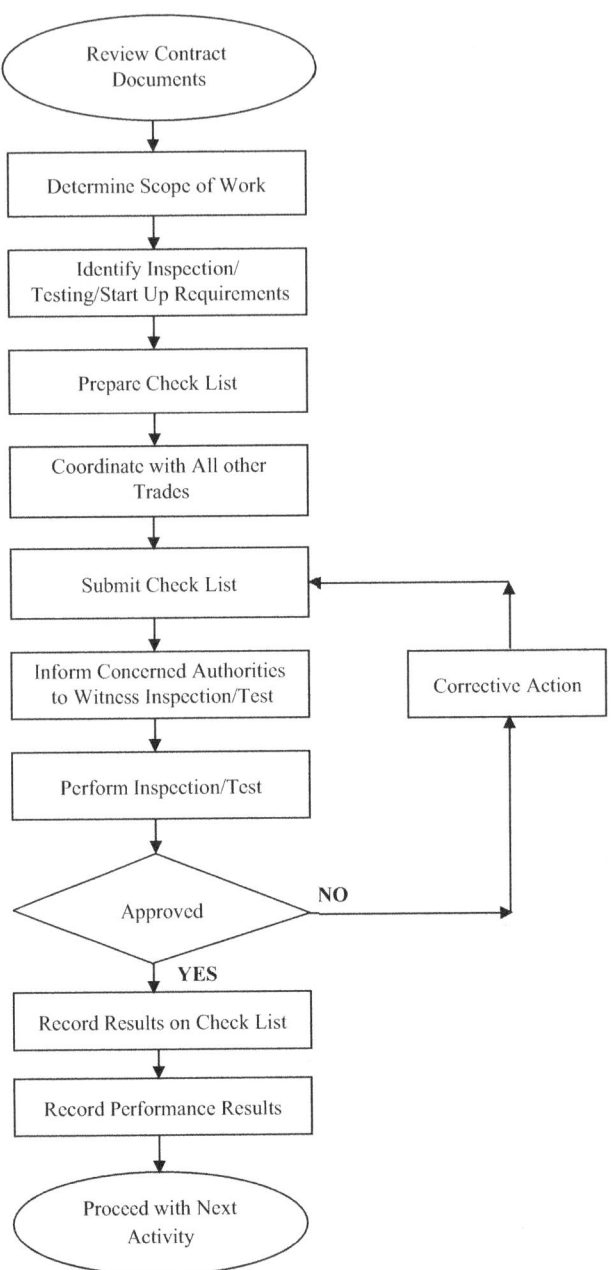

FIGURE 4.81 Logic flowchart for development of inspection and test plan. Source: Abdul Razzak Rumane. (2016). *Handbook of Construction Management*. Reprinted with permission of Taylor & Francis Group.

Management of Quality for Sustainability

- Identify (list) the items to be tested
- Identify (list) the items to be commissioned
- Establish the testing and commissioning criteria
- Identify a manufacturer recommendation procedure
- Establish a detailed procedure for the testing of equipment
- Ensure the mechanical completion of the plant/equipment
- Identify the pre-commissioning requirement
- Establish the schedule for testing and commissioning
- Establish the sequence of the commissioning of equipment
- Identify and prequalify the testing and commissioning agency, if mandated by the contract
- Identify the stakeholders to witness the testing and commissioning
- Calibration/certification of testing equipment to be used for testing and commissioning
- Acceptance criteria
- Establish corrective action plan

2. **Perform Quality Assurance**
   - Witnessing of tests by the stakeholders
   - Ensure inspection of all installed equipment is finished
   - Manufacturer's recommended method statement
   - Ensure operating and maintenance manual requirements are followed
   - Sequence of testing and commissioning
   - Testing test points
   - Approval of third-party inspection agency
   - Approval of specialist commissioning agency
   - Records of testing and commissioning
   - Calibration certificate of testing equipment used for Testing and Commissioning
   - Ensure availability of "Punch List."

3. **Quality Control**
   - Check line by line as per flowchart
   - Test points (location)
   - Flow test (water)
   - Control system
   - Storage capacity of tank
   - Installation safety
   - Environmental compatibility

4.7.8 DEVELOP DOCUMENTS

Table 4.58 illustrates the list of project closeout documents to be submitted for project closeout.

4.7.8.1 As-Built Drawings

Most contracts require the contractor to maintain a set of record drawings. These drawings are marked to indicate the actual installation wherever the installation

TABLE 4.58
Project Closeout Documents

Description	Testing and Commissioning	As-Built Drawings	Operation and Maintenance Manuals	Guarantees	Warranties	Government Authorities Approvals	Record Documents	Test Certificates	Samples	Spare Parts	Punch Lists	Final Cleaning	Training	Taking Over Certificate	Remarks
Architectural Works															
Civil Works															
Mechanical Works															
HVAC Works															
Electrical Works (Light & Power)															

SAMPLE FORM

(Continued)

Management of Quality for Sustainability

TABLE 4.58 CONTINUED
Project Closeout Documents

Description	Testing and Commissioning	As-Built Drawings	Operation and Maintenance Manuals	Guarantees	Warranties	Government Authorities Approvals	Record Documents	Test Certificates	Samples	Spare Parts	Punch Lists	Final Cleaning	Training	Taking Over Certificate	Remarks
Electrical Works (Low Voltage)															
Finishes															
External Works															

varies appreciably from the installation shown in the original contract. Revisions and changes to the original contract drawings are almost certain for any construction project. All such revisions and changes are required to be shown on the record drawings. As-built drawings are prepared by incorporating all the modifications, revisions, and changes made during the construction. These drawings are used by the user/operator after taking over the project for their reference purpose. It is the contractual requirements that the contractor handover as-built drawings along with record drawings, record specifications, and record product data to the owner/user before handing-over of the project and issuance of substantial completion certificate. In certain projects, the contractor has to submit field records on excavation and underground utility services, detailing their location and levels.

4.7.8.2 Technical Manuals and Documents

Technical manuals, design and performance specifications, test certificates, and warranties and guarantees of the installed equipment are required to be handed over to the owner as part of the contractual conditions.

Systems and equipment manuals submitted by the contractor to the owner/end-user generally consist of:

- Source Information
- Operating Procedures
- Manufacturer's Maintenance Documentation
- Maintenance Procedure
- Maintenance and Service Schedules
- Spare Parts List and Source Information
- Maintenance Service Contract
- Warranties and Guarantees

The procedure for submission of all these documents is specified in the contract document.

4.7.8.3 Record Books
- Manufacturing record book
- Project record book
- Engineering record book
- Construction record book

4.7.8.4 Warranties and Guarantees

The contractor has to submit warranties and guarantees in accordance with the contract documents. Normally, the guarantee for waterproofing works varies from 15 to 20 years. Similarly, the warranty for diesel generators is set at five years.

4.7.9 MONITOR THE WORK PROGRESS

The schedule for testing, commissioning, and handover phase is to be prepared and all the activities are to be performed as per the agreed-upon plan.

Management of Quality for Sustainability

4.7.10 Train the Owner's/End-User's Personnel

Normally, training of the user's personnel is part of the contract terms. The owner's/user's commissioning, operating, and maintenance personnel are trained and briefed before commissioning starts in order to familiarize the owner's/user's personnel with the installation works and also to ensure that the project is put into operation rapidly, safely, and effectively without any interruption. Timings and details of training vary widely from project to project. Training must be completed well in advance of the requirement to make the operating teams fully competent to be deployed at the right time during commissioning. This needs to be planned from project inception, so that the roles and activities of the commissioning and operating staff are integrated into a coherent team to maximize their effectiveness.

4.7.11 Handover of Project

Once the contractor considers that the construction and installation of works have been completed as per the scope of contract, and final tests have been performed and all the necessary obligations have been fulfilled, the contractor submits a written request to the owner/consultant for handing-over of the project and for issuance of the substantial completion certificate. This is done after testing and commissioning is carried out and it is established that project can be put in operation or the owner can occupy same. In most construction projects, there is a provision for a partial hand-over of the project.

4.7.11.1 Obtain the Authorities' Approval

Necessary regulatory approvals from the respective authorities concerned are obtained so that the owner can occupy the facility and start using/operating it. In certain countries, all such approvals are needed before the electrical power supply is connected to the facility. It is also required that the building/facility is certified by the relevant fire department authority/agency that it is safe for occupancy.

4.7.11.2 Handover of Spare Parts

Most contract documents include the list of spare parts, tools, and extra materials to be delivered to the owner/end-user during the close-out stage of the project. The contractor has to properly label these spare parts and tools, clearly indicating the manufacturer's name, and model number if applicable. Figure 4.82 illustrates the spare parts handing-over form used by the contractor.

4.7.11.3 Accept/Takeover of the Project

Normally, a final walk-through inspection of the project is carried out by the committee which consists of the owner's representative, design and supervision personnel, and contractor to decide on the acceptance of the works and that the project is complete enough to be put into use and made operational. If there are any minor items remaining to be finished, then such a list is attached with the certificate of substantial completion for conditional acceptance of the project. Issuance of substantial completion certifies acceptance of the works. If the works remaining are of minor nature then the contractor has to submit

Project Name
Project Name
HANDING OVER OF SPARE PARTS

CONTRACTOR: _____ CERTIFICATE No. : ☐

SUBCONTRACTOR: _____ DATE ☐

SPECIFICATION NO : _____ DIVISION _____ SECTION : _____

DRAWING No. _____ BOQ Ref. _____

AREA : ☐ Process Works ☐ Electrical Works ☐ Mechanical Works

☐ Piping Works ☐ Instrumentation Works ☐ Low Voltage System

Following Spare Parts have been handed over to the owner/end user

Description of Spare Parts

Sr.No.	Description	BOQ Reference	Spec. Ref.	Manufacturer	Specified Qty	Delivered Qty

(Attach additional sheet, if required)

SIGNED BY:

OWNER/END USER: _____ CONTRACTOR: _____

CONSULTANT: _____ SUBCONTRACTOR: _____

FIGURE 4.82 Handing over of spare parts.

a written commitment that he shall complete said works within the agreed-upon period. A memorandum of understanding is signed between the owner and the contractor that the remaining works will be completed within an agreed-upon period.

The contractor starts the handing-over of all the completed works/systems which are fully functional, and the owner agrees to take them over. A handing-over certificate is prepared and signed by all the concerned parties. Figure 4.83 illustrates a sample handing-over certificate.

4.7.11.4 Prepare the Punch List

The owner/consultant inspects the works and informs the contractor of any unfulfilled contract requirements. A punch list (snag list) is prepared by the consultant, listing all the items still requiring completion or correction. The list is handed over to the contractor for rework/correction of the works mentioned in the punch list. The contractor resubmits the inspection request after completing or correcting the previously notified works. A final snag list is prepared if there are still some items which need corrective action/completion by the contractor; any such remaining works are to be completed within the agreed period to the satisfaction of the owner/consultant. Table 4.59 is a sample form for preparation of the punch list

4.7.12 Issue the Substantial Completion Certificate

A substantial completion certificate is issued to the contractor once it is established that the contractor has completed works in accordance with the contract documents and to the satisfaction of the owner. The contractor has to submit all the required certificates and other documents to the owner before issuance of the certificate.

The certificate of substantial completion is issued to the contractor and the facility is taken over by the owner/end-user. By this stage, the owner/end-user has already taken possession of the facility, and operation and maintenance of the facility have commenced. The project is declared complete and is considered as the end of the construction project life cycle.

The defect liability period starts after issuance of the substantial completion certificate.

During this period, the contractor has to complete the punch list items and also to rectify any defects identified in the project/facility.

4.7.13 Lessons Learned

The construction/project manager, consultant, and contractor have to prepare "Lessons Learned" and document the same for future reference to improve the processes and organizational performance. This includes:

- Reasons for delay
- Reasons for cost overrun
- Reasons for rejection/rework
- Reasons for preventive/corrective actions
- Causes for claims

Project Name
Project Name
HANDING OVER CERTIFICATE

CONTRACTOR : _____ CERTIFICATE No. : ☐

SUBCONTRACTOR: _____ DATE ☐

SPECIFICATION NO : _____ DIVISION : _____ SECTION : _____

DRAWING No. _____ BOQ REF: _____

AREA : ☐ Process Works ☐ Electrical Works ☐ Mechanical Works

☐ Piping Works ☐ Instrumentation Works ☐ Low Voltage Systems

Description of Work/System: *SAMPLE FORM*

The work/system mentioned above is completed by the contractor as specified and has been inspected and tested as per contract documents. The work/system is fully functional to the satisfation of owner/end user. The contractor hand over the said work/system to the owner/end user as on --------------. The guarentee/warranty of work/system shall start as of ------------ and shall be valid for a period of ----------------- years(duration) from the date of issuance of substantial completion certificate. The contractor shall be liable contractually till the end of warranty/guarentee period.

SIGNED BY:

OWNER/END USER: _____ CONTRACTOR: _____

CONSULTANT: _____ SUBCONTRACTOR: _____

FIGURE 4.83 Handing over certificate.

TABLE 4.59
Punch List

Owner's Name
Project List
Punch List
Punch List Number: Date:
Area   Type of Work
Zone

Serial Number	Item	Remark
1	Piping	
2	Storage tank	
3	Pumps	
4	Vessel	
5	Fire alarm detectors	
6	Sprinklers	
7	Communication system devices	
8		
9		
10		
11		
12		
13		
14		
15		
16		
17		
18		
19	Any other item	

Sample List

4.7.14 Settle Payments

The owner has to settle all the payments due to the consultant, contractor, and other parties involved. Similarly, the contractor has to settle due payments to their subcontractors and suppliers.

4.7.15 Settle Claims

The entire project-related claims are to be amicably settled as per contract conditions to close the project.

4.7.16 Close the Contract

Table 4.60 illustrates a list of activities that need to be considered for project closeout.

TABLE 4.60
Project Close-out Checklist

Sr. No.	Description	Yes/ No
Project Execution		
1	Contracted works completed	
2	Site work instructions completed	
3	Job site instructions completed	
4	Remedial notes completed	
5	Non-Compliance reports completed	
6	All services connected	
7	All the contracted works inspected and approved	
8	Testing and Commissioning carried out and approved	
9	Any snags?	
10	Is project fully functional?	
11	Are all other deliverable completed?	
12	Have spare parts been delivered?	
13	Is waste material disposed of?	
14	Are safety measures for the use of hazardous material established?	
15	Is the project safe for use/occupation?	
Project Documentation		
16	Authorities' approval obtained	
17	Record drawings submitted	
18	Record documents submitted	
19	As-built drawings submitted	
20	Technical manuals submitted	
21	Operation and maintenance manuals submitted	
22	Equipment/material warranties/guarantees submitted	
23	Test results/test certificates submitted	
Training		
24	Training to owner's/end-user's personnel carried out	
Payments		
25	All payments to subcontractors/specialist suppliers released	
26	Bank guarantees received	
27	Final payment released to main contractor	
Handing over/Taking over		
28	Project handed over/taken over	
29	Operation/maintenance team taken over	
30	Excess project material handed over/taken over	
31	Facility manager in action	

Bibliography

AACE International Recommended Practice No. 18R–97 (2010)
AACE International Recommended Practice No. 27R-03 (2010)
AACE International Recommended Practice No. 37R-06 (2010)
Abdul Razzak Rumane (2021), *Quality Management in Oil and Gas Projects*, CRC Press, Florida (A Taylor & Francis Group Company)
Abdul Razzak Rumane (2017), *Quality Management in Construction Projects*, Second Edition, CRC Press, Florida (A Taylor & Francis Group Company)
Abdul Razzak Rumane (2016), *Handbook of Construction Management: Scope, Schedule, and Cost Control*, CRC Press, Florida (A Taylor & Francis Group Company)
Abdul Razzak Rumane (2013), *Quality Tools for Managing Construction Projects*, CRC Press, Florida (A Taylor & Francis Group Company)
Adedeji Badiru (2021), *Sustainability A Systems Engineering Approach to Global Grand Challenge*, CRC Press, Florida (A Taylor & Francis Group Company)
https://sdgs.un.org/2030agenda
https://www.iso.org/iso26000
https://www.cdp.net/
https://www.globalreporting.org/
http://integratedreporting.org/
https://www.sasb.org/
https://www.unglobalcompact.org/
http://www.accountability.org/standards/

Index

A

Accessibility, 148, 183, 191, 205, 214, 259
Accident, 16, 74, 136, 267, 273, 284, 301, 307, 312, 313, 315, 333, 336, 383
Activity network, 45–47, 49
Addendum, 261, 262, 265, 267–269, 271, 332
Aesthetic, 10, 148, 149, 157, 162, 165, 191, 214, 219, 250
Affordability, 4, 122
Agency, 103, 106, 139, 143, 145, 147, 151, 191, 213, 233, 261, 265, 328, 339, 358, 379, 380, 396, 405, 409
Agenda, 5, 6, 11, 12, 274, 275, 354, 368
Air quality, 10, 148, 163, 168, 191, 192, 197, 214, 223
Alternate energy, 10, 174
Alternative, 149, 151, 154, 162, 172, 186, 191, 192, 198, 204, 214, 215, 224, 291, 308, 344, 354, 362, 382
Analyse, 149, 335, 341, 348
Analysis, 22, 30, 37, 43, 53, 59–64, 66–69, 73, 88, 92, 95, 98–106, 109, 111, 115, 118, 120, 124, 126, 128, 130, 134, 135, 139–144, 147–149, 151, 154, 171, 172, 184, 197, 203, 204, 206, 209, 218, 231, 237, 248, 252, 255, 260, 264, 271, 279, 294, 315, 336, 338, 341, 346, 358, 368, 377, 380
Analyze, 37, 39, 45, 46, 53, 59–61, 66, 74, 94–97, 99, 100, 105, 109, 119, 141, 144, 147, 148, 151, 158, 165, 183, 205, 225, 304, 341, 347
Arbitration, 159, 342, 389
Architect/engineer (A/E), 14–16, 31, 103, 104, 107, 158, 186, 204, 316, 343–345, 368, 372, 373
As Built drawings, 137, 302, 309, 338, 393, 397–399, 405–408, 414
Assumptions, 3, 128, 145, 152, 175, 178, 236, 250, 258, 294
Audit, 28, 83, 102, 122, 132, 135, 136, 181, 273, 283, 297, 313, 314, 316, 319, 335
Auditing, 29, 35, 102, 119, 180, 273, 282, 298, 357
Auditor, 29, 102, 284, 357
Authority, 31, 103, 107, 112, 117, 130, 197, 261, 272, 286, 287, 332, 362, 386, 409
Avoidance, 136
Award, 17, 104, 114, 118, 128, 260–262, 267, 269–271, 286, 287, 299, 363, 380, 381

B

Bar Chart, 42, 186, 294, 349
Baseline, 97, 99, 100, 119, 127, 128, 178, 203, 294, 307, 335, 336, 338–341, 346, 347, 349, 351, 354
Benchmarking, 59, 60, 162
Benefits, 1, 2, 9, 26, 30, 90, 95, 98, 107, 109, 122–124, 139, 142, 144, 145, 203, 204, 304, 330, 346, 377
Bidder, 17, 104, 118, 251, 252, 254, 260–263, 265, 267–271, 287, 381
Bidding and Tendering, 22, 29, 102, 104, 105, 116, 118, 124, 128, 130, 150, 153, 159, 178, 203, 209, 211, 231, 237, 246, 248, 249, 254, 256, 259–262, 267–270, 272, 309, 340, 358, 379, 380
Bill of material, 22
Bill of quantities (BOQ), 18, 22, 33, 34, 127, 128, 153, 209, 212, 215, 216, 231, 234, 237, 248, 250–256, 260, 270, 287, 295, 340
Bill of quantity, 58, 205
Biodiversity, 2, 4, 6, 12, 122
Bond(s), 104, 110, 131, 136, 231, 252, 255, 265, 266, 268, 269, 271, 275, 283, 286, 300, 307, 332, 398
Bottom-up planning, 177, 233, 255
Brainstorming, 49, 69–71, 101, 147, 204
Budget, 3, 13, 18–21, 27, 29, 31, 58, 83, 84, 103, 106, 110, 116, 117, 125, 128, 139, 142, 146, 152, 154, 158, 162, 165, 175, 177, 178, 188, 191–193, 197, 200, 203, 209–211, 214, 230, 247–250, 261, 269–271, 273, 276, 277, 287, 289, 290, 292, 295, 310, 329, 330, 334, 337, 352, 354, 355, 358, 363, 379
Budgetary cost, 175, 201, 203
Builder, 68, 99, 138, 273, 308, 325, 326, 331
Builder's workshop drawings, 325
Building construction, 13–15, 152, 153, 167, 186, 230, 252, 278, 390
Business case, 9, 106, 107, 138, 139, 142, 145

C

Capital, 7, 17, 20, 131, 175, 351
Cash flow, 75, 77, 116, 131, 136, 200, 236, 273, 295, 352–354, 367

417

Categories, 15, 37–39, 46, 82, 83, 150, 151, 158, 293, 299, 317, 363, 374–376
Certification ISO, 266, 283, 319
Change management, 98, 134, 273, 337, 364
Characteristic(s), 2, 14, 17, 59, 68, 86, 122, 152, 177, 217, 218, 220, 314
Charter, project, 31, 95, 104, 107, 137, 139–141, 147–151, 156, 159, 162, 164, 165, 175, 179–181, 183
Checklist, 35, 36, 70, 84, 85, 121, 135, 136, 166, 167, 233, 246, 255, 265, 298, 300, 308, 309, 328, 329, 332, 333, 336, 346, 352, 358, 390, 392, 403, 414
Civil construction projects, 15
Claim, 118, 126, 131, 133, 136, 137, 139, 141, 157, 186, 197, 209, 217, 248, 262, 267, 272, 273, 288, 301, 313, 341, 348, 361, 382, 386, 388, 389, 395, 397, 411, 413
Clause(s), 22, 24, 25, 268, 270, 288, 374, 382, 392
Climate change, 4–6, 143
CLIPSCFM, 21, 25
Close out, 102, 104, 295, 301, 398, 409, 414
Closing process, 5, 125, 131, 137, 141, 157, 186, 209, 248, 262, 273, 297
Codes, 3, 20, 22, 29, 32, 80, 83, 90, 112, 117, 129, 136, 141, 148, 149, 152, 157, 162, 163, 166, 167, 171–175, 179, 181–183, 185, 186, 188, 190–192, 194, 197, 200, 201, 203, 206, 209, 211–214, 222–224, 233, 235, 236, 248, 250, 251, 256, 288, 289, 314, 319, 321, 324, 332, 338, 349, 355, 357, 358, 366, 372, 373, 383, 384, 390, 391, 403
Commercial Projects A/E, 15
Commissioning, 8, 29, 35, 36, 104, 118, 127, 128, 130, 137, 177, 201, 209, 223, 248, 272, 278, 295, 298, 309, 328, 338, 358, 360, 378, 393, 395–399, 401–403, 405–409, 414
Communication, 15, 22, 26, 41, 85, 86, 91, 111, 126, 129, 133, 135, 141, 156, 157, 161, 170, 174–176, 180, 181, 186, 196, 197, 207, 209, 217, 222, 225, 248, 253, 262, 272, 273, 275, 278, 279, 287, 290–292, 295, 296, 298, 300–306, 312–316, 328, 333, 334, 348, 354, 361, 364, 367–370, 377, 379, 397, 400–402, 413
Community development, 4
Compliance, 8, 16, 18, 19, 21, 22, 25, 34, 36, 83, 84, 113, 119, 132, 134, 135, 140, 141, 143, 149, 157, 159, 168, 172, 173, 181, 185, 186, 190, 199–201, 209, 212, 213, 230, 234, 238, 248, 256, 258, 269, 273, 296, 303, 309, 315, 319, 321, 324, 326, 328, 347, 357, 358, 370, 371, 383, 390, 392, 397, 399, 414

Components, 14, 18, 20, 27, 82, 85, 91, 122, 125, 203, 217, 218, 223, 229, 232, 325, 355, 357, 378, 390
Composite drawing, 99, 273, 308, 325, 326, 336
Concept Design, 33, 76, 77, 104, 105, 108, 112–115, 127, 128, 132, 148, 152, 155–159, 161, 162, 165–167, 169–171, 174, 175, 177–189, 192, 193, 202, 205, 236, 340, 380
Conflict, 26, 67, 73, 117, 132, 133, 135, 211, 235, 237, 246, 258, 273, 287, 288, 313, 326, 337, 341, 342, 361, 362, 375
Conformance, 67, 82, 83, 117, 134, 141, 157, 175, 321, 324, 328, 357, 358, 366, 399
Constraints, 46, 103, 106, 110, 128, 143, 145, 152, 154, 183, 186, 197, 204, 205, 227, 288, 292, 294, 306, 314, 346, 376
Constructability, 113, 115, 117, 157, 162, 181, 183, 186, 187, 198, 205, 238, 246, 288
Construction management, 17, 126, 131, 133, 136, 137, 151, 271, 273, 275, 279, 280, 294, 344, 345, 380, 404
Construction manager, 16, 17, 20, 132, 151, 153, 158, 187, 210, 246, 247, 261, 272, 286, 340, 353, 361, 375, 378
Construction Project life cycle, 77, 138, 150, 199, 270, 293, 411
Construction quality, 27, 296, 328, 355
Constructor, 18, 138
Consultant, 16–22, 29, 36, 41, 55, 65, 80, 83, 89, 92, 108, 131, 138, 141, 147, 151–153, 157–159, 166, 175, 178, 180, 181, 186, 187, 200, 208–211, 230, 236, 246–248, 257, 261–263, 267, 269, 271–273, 276–278, 286, 287, 292, 296, 299–305, 310–312, 314, 316–318, 320–326, 328–334, 336, 339, 340, 343–346, 350, 352, 353, 358–361, 364, 365, 368–374, 378, 380, 382, 384–386, 388–394, 396, 398, 399, 403, 409–411
Context, 24, 377
Contract documents, 13, 16–19, 21, 22, 25, 26, 29, 32–34, 36, 57, 58, 78, 102, 104, 111, 112, 114, 116, 117, 126, 128, 130, 132–134, 136, 153, 155, 159, 179, 185, 186, 190, 199–201, 203, 207–210, 212, 213, 230–232, 234, 237, 247–251, 254–256, 262, 267, 268, 270–273, 277, 278, 281, 286–288, 296–300, 302, 304, 311–314, 316, 317, 321–324, 326–329, 332, 334, 335, 339–341, 344, 345, 358, 362, 363, 366, 368, 370, 371, 375, 379, 380, 383, 388, 390, 392, 396, 397, 399, 404, 408, 409, 411, 412

Index

Contracting, 30, 31, 75, 79, 104, 107, 139, 140, 148, 151, 157, 186, 199, 200, 209, 212, 230, 231, 248, 254, 256, 263, 276, 279, 375
Contract management, 126, 130, 133, 136, 137, 141, 157, 186, 197, 209, 217, 248, 262, 273, 276, 292, 296, 306, 309, 312, 313, 374, 378–381, 397
Contractor's Quality Control Plan, 22, 25, 27, 35, 273, 278, 291, 292, 296–298, 358
Coordination drawing, 85
Corrective Action, 63, 67, 83, 84, 94, 102, 117, 119, 120, 132, 134, 181, 273, 298, 303, 312, 313, 320, 323, 330, 335, 336, 338, 341, 346, 347, 354, 357, 379, 390, 393, 395, 404, 405, 411
Cost control, 126, 131, 133, 135–137, 150, 273, 279, 280, 294, 344, 345, 353, 354
Cost loading, 128, 295
Cost reimbursement, 104, 107
Critical Path Method (CPM), 46, 49, 51, 294
Customer needs, 27, 87, 94, 95, 138, 163, 165, 168, 355
Customer privacy, 4
Customer satisfaction, 17, 23, 24, 26, 27, 82, 99, 138, 162, 355

D

Data collection, 80, 89, 95, 97, 105, 112, 114, 119, 157, 158, 162, 165, 171, 174, 179, 182, 185, 202, 208, 209, 217, 235, 236
Day lighting, 4, 10, 148, 163, 168
Defect prevention, 129, 357
Defined Scope, 3, 18, 19, 21, 27, 83, 227, 329, 355
Definite beginning, 14, 175
Delphi techniques, 69, 70, 72, 147
Dependencies, 47, 48, 128
Dependency relationship, 48
Design-Build, 151, 153, 155, 380
Design development, 29, 32–34, 104, 112, 127, 128, 132, 155, 179, 181, 184, 203, 209, 211, 214, 215, 234, 236, 237, 250, 258, 357
Design Professional, 16, 158, 203, 211, 212, 218
Design team, 18, 29, 156–160, 179, 180, 185, 186, 188, 189, 192, 201, 208–210, 357
Detail design, 32–34, 76, 77, 116, 127, 128, 132, 152, 155, 175, 178, 185, 203, 208–210, 212–218, 233–238, 246–251, 254, 255, 282, 340
Detailed Engineering, 185, 203, 293, 298, 303, 315
Detection, 62, 63, 199, 309, 315, 403
Disposal, 76, 124, 137, 376, 386

Documentation QMS, 20, 23, 25
Documents tendering, 104, 114, 118, 130, 153

E

Ecofriendly, 4, 10, 191, 214, 289
Economical, 1–4, 8, 10, 20, 103, 107, 109, 122–124, 131, 192, 296, 341
Economical factors, 1, 3, 12
Economy, 3, 5, 113, 157, 166, 183, 366
Ecosystem, 1, 2, 4, 6
Efficient procurement, 4
Elément, 3, 4, 6, 8, 9, 14, 24, 26, 63, 75, 77, 82, 83, 95, 97, 102–107, 112, 118, 122, 124–137, 139, 146, 148, 149, 155, 159, 165, 167, 169–172, 180, 183, 184, 191, 194, 200, 205, 214, 218, 221, 237, 250, 259, 289, 336, 338, 339, 355
End user, 16, 19, 21, 29, 108, 109, 146, 150, 175, 210, 272, 276, 306, 327, 334, 355, 357, 374, 378, 393, 396, 398–400, 409–412, 414
Energy efficiency, 4, 9, 123, 157, 161, 246
Environment, 1–3, 5–7, 10, 11, 26, 35, 88, 123, 126, 130, 133, 136, 140, 143, 148, 163, 165, 199
Environmental issues, 1, 3, 11, 12, 115, 139, 179, 180, 187
Environmental protection, 4, 5, 133, 143, 163, 182, 236, 308, 316, 376
EPC, 151
Equity and equality, 4
Estimate(d) cost, 128, 145, 152, 186, 194, 200
Estimate time, 45, 49, 145, 152, 293
Ethics, 2, 4
Evaluation bid, 261, 263, 265, 269
Evaluation risk, 377
Executed works, 22, 48, 119, 272, 278, 290, 291, 327, 338, 357, 386, 390, 392, 396
Executing, 36, 88, 98, 125, 131–133, 316, 327, 334, 382

F

Facility high performance, 10, 11
Fair labor practices, 4, 123
Feasibility study, 30, 103, 105, 106, 109, 138, 141, 142, 144–146, 157, 177, 380
Flowcharting, 59, 64, 65
Forecasted, 134, 135, 337
Functional, 9, 19, 21, 26, 29, 33, 89, 91, 92, 107, 108, 113, 115–117, 120, 132, 140, 141, 147, 154, 157, 160, 162, 163, 179, 183, 184, 186, 187, 197, 199, 201, 204, 205, 234, 256, 329, 331, 357, 378, 403, 411, 412, 414

G

Global warming, 1, 4, 122, 123, 143
Goals, 5, 6, 8, 9, 11–13, 16, 24, 30, 33, 55, 89–91, 94, 96, 97, 103, 106, 108, 113, 115, 124–127, 138–142, 144, 146–151, 154, 157, 162, 167, 172, 174, 180, 182, 183, 187, 203, 205, 209, 233, 234, 236, 248, 256, 270, 275, 277, 304, 330, 337, 338
Greenhouse, 4

H

Handover, 8, 22, 29, 35, 36, 104, 118, 127, 128, 137, 177, 272, 275, 278, 295, 298, 328, 329, 334, 338, 358, 393, 395–398, 400, 408, 409
Hazard(ous), 74, 78, 133, 182, 236, 287, 289, 312, 314, 316, 327, 376, 382, 383, 386, 414
HSE, 36, 126, 130, 133, 141, 156–158, 161, 174, 181, 182, 185, 186, 197, 200, 202, 208, 209, 217, 236, 247, 248, 258, 262, 272, 273, 275, 278, 284, 292, 296, 297, 312, 314–316, 364, 382, 386, 397
Humanity, 1, 4
Human resources, 129, 294, 299, 361–363
Human rights, 4, 5, 7, 123

I

Identification of need, 103, 106, 108, 141, 142
Initiating, 125, 126, 141, 157, 186, 209, 248, 262, 273, 314, 334, 377, 397
Inspection, 8, 10, 18, 22, 35, 36, 81, 83–85, 121, 129, 135, 136, 186, 272, 273, 277, 278, 282, 290, 291, 293, 298, 301, 303, 308–310, 312, 313, 315, 320, 328–331, 333, 338, 339, 341, 352, 357, 358, 366, 379, 390–392, 394–396, 398, 400, 403–405, 409, 411
Instrument, 85, 160, 282, 284, 308, 327, 391, 400, 410, 412
Insurance, 11, 104, 110, 118, 131, 136, 159, 266, 275, 283, 286, 307, 332
Integration, 11, 26, 27, 96, 125, 127, 132, 134, 137, 141, 157, 165, 186, 209, 224, 227, 248, 262, 272, 273, 334, 378, 397
Interdisciplinary coordination, 10, 34, 179, 209, 234, 235, 237–245, 248, 256
Investment, 17, 20, 131, 139, 143, 374
ISO 9000, 21, 23–26, 297
ISO 9001, 6, 24, 25, 28
ISO 14000, 5, 6, 26, 28
ISO 26000, 5

Issues environmental, 1, 3, 11, 12, 115, 180, 187
ITP (Inspection and Testing Procedure), 22, 35, 298, 357

K

Kick off meeting, 114, 127, 273–275

L

Legal, 26, 109, 112, 124, 145, 146, 165, 254, 260, 276, 376
Lesson learned, 137, 295, 297
Litigation, 267, 313, 342, 376, 389
Log(s), 127, 130, 132–134, 300, 303, 308, 310, 317, 332, 333, 336, 349–351, 364, 365, 368, 370–373, 382
Lump sum, 104, 107

M

Main contractor, 276, 281, 302, 363, 384, 385, 388, 396, 414
Maintainability, 113, 122, 162
Maintenance, 89
Manage, HSE, 247
Manage, Risk, 133, 185, 208, 247, 262
Manageability, 122
Manpower, 20, 22, 29, 38, 39, 44, 61, 66, 84, 106, 120, 133, 177, 201, 234, 248, 267, 273, 289, 295, 299, 342, 349–352, 357, 361
Manual, Quality, 23–25, 35
Material approval, 50, 52, 132, 273, 304, 318, 320, 321
Matrix for site administration, 299–302, 304, 368
Meeting, coordination, 127, 135, 157, 186, 209, 248, 278, 352
Meeting, progress, 127, 134, 135, 180, 333, 350
Meeting, safety, 135, 314, 353, 383
Method statement, 28, 35, 36, 55, 83, 85, 102, 132, 259, 265, 273, 288, 297, 298, 301, 308, 326–328, 333, 337, 357, 405
Milestone, 49, 93, 98, 114, 128, 145, 177, 198, 200, 259, 294, 336, 347, 351
Mistake proofing, 75, 79, 80, 105, 234, 235
Mitigate, mitigation, 106, 133, 136, 183, 205, 259, 371
Mobilization, 8, 50, 52, 118, 119, 132, 133, 137, 272, 273, 275, 277, 281, 286, 293, 312, 332, 334, 353, 375, 397
Mock-up, 111, 132, 301
Models, 25, 33, 67, 95, 104, 111–113, 152, 157, 162, 172, 174, 179, 184, 189, 190, 193, 199, 200, 203, 204, 207, 209, 216, 231, 237, 348, 349, 409

Index

Monitoring, 4, 10, 24, 46, 124, 277, 312, 314, 315, 330, 348, 349, 354, 374, 377–379, 397
Monitoring and controlling, 2, 8, 29, 115, 119, 125, 134–136, 277, 278, 329, 334–336, 346, 347, 358

N

Natural, 2, 4, 10, 49, 122, 163, 304, 376
Need analysis, 103, 106, 126, 140–144, 154, 380
Need statement, 103, 106, 108, 126, 127, 140–142, 144, 151, 154
Negotiation, 136, 262, 280, 341, 375, 381
Non-conformance, 60, 82, 83, 301, 309, 333, 351, 390, 393, 398
Non-repetitive, 12, 17, 20, 27, 37, 131
Notice to proceed, 104, 118, 132, 133, 272–274, 300, 307

O

Objective, 8, 9, 13, 14, 16, 21, 23, 25, 26, 29, 30, 33, 57, 60, 69, 79, 82, 84, 103, 106, 108–110, 124–127, 138–152, 154, 157, 159, 162, 165, 167, 172, 174, 175, 180, 182–185, 189, 193, 198, 201, 203, 205, 207, 209, 211, 212, 216, 233, 234, 236, 238, 249, 256, 270, 275, 292, 296, 304, 311, 314, 330, 332, 334, 337–339, 342, 346, 355, 357, 363, 374, 379, 380
Objective six sigma, 94, 96, 100
Operational, 4, 7, 22, 33, 75, 76, 97, 109, 256, 338, 409
Optimization, 148, 163, 192, 361
Organization, 26, 27, 32, 35, 56, 59, 75, 80, 88, 90, 91, 97, 98, 100, 107, 123, 138, 143, 159, 160, 179, 181, 185, 188, 193, 201, 216, 230, 233, 266, 271, 275, 282, 283, 298, 306, 315, 330, 332, 352, 355, 364, 374, 379
Organizational breakdown structure (OBS), 127, 294
Owner's requirements, 9, 19, 21, 22, 32, 34, 83, 86, 117, 147, 156, 157, 166, 167, 175, 179, 188, 190, 197, 201, 203, 208, 212, 213, 246, 247, 249, 250, 258, 262, 278, 289, 399

P

Personal protective equipment (PPE), 327, 386
Planning process group, 125–131
Plantation, 10, 12, 87, 161, 165, 187, 191, 196, 214, 218, 253, 289
Pollution, 1, 2, 9, 123, 161, 165, 182, 191, 196, 214–216, 218, 236, 328, 376

Preferred alternative, 103, 107, 108, 139, 140, 148, 149, 151, 154
Prequalification, 130, 263, 282–285
Prevention, 10, 62, 63, 75, 80, 82, 85, 117, 129, 136, 170, 196, 273, 315, 357, 383
Preventive action, 85, 119, 132, 134, 181, 298, 313, 316, 341, 347, 390, 393
Prioritization, 46, 55–57, 95, 103, 106
Probability, 304, 374
Process type projects, 15
Procurement procedure, 318, 320
Progress payment, 128, 129, 132, 135, 273, 292, 308, 313, 333, 336, 382, 386–388
Project charter, 31, 95, 104, 107, 126, 137, 139–141, 147–151, 156, 159, 162, 164, 175, 179–181, 183
Project delivery system, 30, 103, 104, 107, 108, 126, 127, 130, 136, 141, 148, 150, 151, 153, 155, 157, 158, 187, 210, 230, 237, 246, 247, 256, 258, 261, 329, 375, 379–381
Project management consultant (PMC), 22, 151, 181, 275, 300–303, 343, 370, 371
Project manager, 16, 18, 20, 55, 117, 125, 129, 131, 147, 150, 151, 153, 158–160, 188, 189, 211, 237, 248, 249, 258, 272, 276, 278, 284, 299, 305, 317, 324, 330, 340, 350, 351, 359–362, 369–371, 380, 383–386, 396, 411
Project team, 19, 21, 31, 85, 86, 95, 97, 104, 107, 108, 111, 115, 117, 126, 136, 137, 141, 146, 147, 152, 155, 157, 159, 181, 187, 232, 233, 247, 255, 263, 264, 273, 274, 299, 303, 306, 311, 313, 330, 337, 341, 361, 367, 370, 375, 379, 381, 389
Punch list, 137, 338, 397, 398, 400, 405–407, 411, 413
Purchasing contract, 17
Purchasing cycle, 337

Q

QMS, 6, 23, 25–28
QMS (Integrated), 28
Qualitative, 20, 38, 147, 149, 152, 218, 296, 377, 378
Quality assurance, 22, 27, 29–31, 36, 129, 132, 138, 179, 181, 201, 211, 233, 234, 255, 256, 268, 269, 273, 296, 328, 355, 357, 358, 403, 405
Quality auditing, 35, 180, 373, 398
Quality control, 17, 18, 22, 25, 27–29, 31, 34–36, 85, 129, 138, 163, 177, 179–181, 209, 211, 232, 233, 248, 255, 273, 278, 291, 292, 296–298, 300, 308, 328, 333, 335, 353, 355, 357–360, 390, 393, 405

Quality control plan, 22, 25, 27, 28, 35, 177, 179, 180, 232, 273, 278, 291, 292, 296–298, 300, 308, 333, 358, 369
Quality definition, 17
Quality management plan, 27, 29, 129, 211, 275, 348, 355, 364
Quality manager, 159, 160, 188, 189, 211
Quality matrix, 85, 129, 355
Quality planning, 29–32, 83
Quality procedure, 24, 25, 298
Quality trilogy, 19
Quantitative, 38, 68, 86, 147, 149, 377, 378
Questionnaire, 70, 72, 263, 264, 266, 267, 281–285, 363

R

Record books, 398, 399, 408
Reducing waste, 1, 4, 75, 82
Removal, 137, 307, 309, 386
Renewable energy, 4, 148, 161, 192, 224
Request for information (RFI), 65, 117, 263, 288, 300, 308, 333, 336, 339, 343, 351, 382
Request for Proposal (RFP), 104, 118, 344
Request for Qualification (RFQ), 263
Residential, 13–15, 266
Responsibilities matrix, 127, 185, 208, 247, 337
Risk management, 4, 28, 35, 125, 126, 130, 133, 136, 141, 150, 157, 186, 197, 209, 217, 248, 262, 273, 275, 289, 292, 296, 298, 304, 306, 333, 364, 371, 374, 377, 378, 397
Risk register, 130, 133, 235

S

Safety management, 26, 130, 150, 273, 278, 308, 312, 313, 315, 333
Schedule management, 125, 128, 132, 135, 141, 157, 186, 209, 248, 262, 273, 343, 397
Scheduling, 46, 49, 84, 103, 106, 109, 118, 125, 144, 145, 149, 150, 177, 272, 277, 291, 294, 343, 346
Schematic design, 32, 104, 105, 112, 115, 127, 128, 132, 152, 155, 184, 186, 188, 193–196, 201, 202, 205, 209, 212, 213, 340
Scope management, 125, 127, 132, 134, 141, 157, 186, 209, 248, 262, 273, 337, 397
Scope of work, 8, 18, 21, 27, 32, 83, 110, 112, 117, 153, 158, 166, 182, 192, 201, 215, 217, 218, 232, 233, 235, 236, 238, 246, 251, 255, 258, 270, 275, 277, 288, 290, 313, 327, 329, 332, 355, 357, 364, 375, 399, 404
S-Curve, 135, 291, 292, 295, 296, 336, 353, 354, 356

Security practices, 4, 161
Shop drawings preparation, 323, 325, 326
Site facilities, 272, 286
Site safety, 73, 74, 130, 133, 273, 301, 352, 353, 360, 383
Site survey, 251, 286
Site work instructions (SWI), 134, 273, 278, 304, 311, 333, 341, 344, 353, 414
SMART, 9, 30, 106, 141
Social elements, 122
Spare parts, 137, 232, 302, 309, 397, 398, 400, 406–410, 414
Staffing, 14, 127, 277, 280
Stakeholder management, 337
Statutory, 30, 107, 148, 184, 375
Study stage, 29–31, 105–108, 124, 138–141, 144, 150, 304
Subconsultant, 159, 180
Subcontractor, 28, 35, 129, 133, 135, 137, 252, 254, 266, 267, 270, 272, 273, 276, 279, 281, 292, 293, 298–300, 307, 312, 314, 315, 318, 322, 333, 349, 350, 352, 353, 358, 362–364, 374, 375, 380, 385, 386, 388, 391, 396, 398, 399, 410, 412, 413
Submittals, 35, 111, 132, 133, 135, 158, 179, 184, 188, 207, 210, 232, 249, 272, 273, 298, 300, 306, 316, 317, 321, 324, 333, 350, 351, 353, 359, 364, 368, 370, 371, 375, 389
Substantial completion, 48, 137, 278, 302, 309, 393, 395, 398, 408, 409, 411, 412
Sustainability development goals, 5, 6, 11, 12
Sustainability elements, 4, 131, 180
Sustainability requirements, 3, 8, 138, 142–144, 146, 147, 152, 154, 156–159, 161, 166, 177, 179, 183, 185–187, 191, 197, 199, 201, 202, 205, 208, 209, 214, 246–248, 250, 272, 289, 337, 395, 397, 399
Sustainable, 1, 3, 5–9, 11, 12, 88, 96, 111, 122, 163, 192, 259
Systems engineering, 124

T

Technical manual, 137, 393, 399, 408, 414
Temporary facilities, 286, 353, 398
Tender analysis, 118
Tender documents, 104, 114, 118, 130, 247–249, 251, 254, 255, 260–263, 265, 268, 273, 334
Terms of Reference (TOR), 80, 104, 105, 107, 110, 111, 126, 127, 141, 147, 150–155, 157, 166, 181, 235, 238, 340, 357
Testing, commissioning, and handover, 29, 35, 104, 118, 127, 128, 177, 278, 298, 338, 393, 395–398, 408

Index

Traffic, 112, 161, 166, 169, 171, 191, 195, 197, 202, 206, 228, 236
Training, 2, 11, 13, 19, 26, 74, 83, 85, 89–92, 98, 101, 117, 187, 273, 302, 314, 315, 361, 364, 393, 398, 400, 406, 407, 409, 414
Transmittals, 22, 275, 314, 333, 357, 370, 371, 382
Treatment, 11, 15, 123, 289, 328, 377, 378, 401, 403
Trilogy, 18, 19

U

Update cost, 204, 352
Update documents, 65, 132, 134, 389
Update schedule, 247

V

Validate, 97, 98, 181, 204, 291, 338, 340, 388

Value Engineering, 27, 28, 104, 111, 112, 115, 180, 185–188, 190, 192–194, 203, 204, 207, 211, 288, 338, 340, 342
Variation, 64, 65, 134–136, 159, 273, 276, 278, 307, 308, 312, 333, 336, 337, 344, 345, 353, 379
Vendor selection procedure, 298, 313, 318, 319

W

Warranties, 137, 232, 302, 399, 406–408, 414
Waste management, 15, 123, 130, 133, 161, 162, 165, 191, 196, 215, 216, 218, 222, 236, 273, 316
Work breakdown structure (WBS), 124, 125, 128, 212, 294
Workforce, 133, 259, 273, 283, 284, 286, 299, 362, 363, 375, 376, 382, 385
Workshop drawings, 99, 168, 187, 204, 308, 325

Made in the USA
Monee, IL
03 May 2026

49437831R00254